AIR CONDITIONING
and
REFRIGERATION ENGINEERING

AIR CONDITIONING
and
REFRIGERATION
ENGINEERING

Shan K. Wang

Zalman Lavan

Paul Norton

CRC Press
Taylor & Francis Group
Boca Raton London New York

CRC Press is an imprint of the
Taylor & Francis Group, an **informa** business

CRC Press
Taylor & Francis Group
6000 Broken Sound Parkway NW, Suite 300
Boca Raton, FL 33487-2742

First issued in paperback 2019

ISBN-13: 978-0-8493-0057-8 (hbk)
ISBN-13: 978-0-367-39917-7 (pbk)
Library of Congress Card Number 99-050066

Library of Congress Cataloging-in-Publication Data

Air conditioning and refrigeration engineering / editor-in-chief, Frank Kreith.
      p. cm.
   Includes bibliographical references and index.
   ISBN 0-8493-0057-6 (alk. paper)
   1. Air conditioning. 2. Refrigeration and refrigerating machinery. I. Kreith, Frank.
TH7687.A4813 1999
697.9′3—dc21
                                                                       99-050066
                                                                                       CIP

Visit the Taylor & Francis Web site at
http://www.taylorandfrancis.com

and the CRC Press Web site at
http://www.crcpress.com

Preface

The purpose of *Air Conditioning and Refrigeration Engineering* is to provide, in a single volume, a ready reference for the practicing engineer in industry, government, and academia, with relevant background and up-to-date information on the most important topics of modern air conditioning systems. This book supplies the basics of design, from selecting the optimum system and equipment to preparing the drawings and specifications. It discusses the four phases of preparing a project: gathering information, developing alternatives, evaluating alternatives, and selling the best solution. In addition the author breaks down the responsibilities of the engineer, design documents, computer-aided design, and government codes and standards.

Air Conditioning and Refrigeration Engineering provides the reader with an easy reference to all aspects of the topic. The book addresses the most current areas of interest, such as computer-aided design and drafting, desiccant air conditioning, and energy conservation. It is a thorough and convenient guide to air conditioning and refrigeration engineering.

Frank Kreith
Editor-in-Chief

Contributors

Zalman Lavan
Illinois Institute of Technology
Evanston, Illinois

Paul Norton
National Renewable Energy
 Laboratory
Golden, Colorado

Shan K. Wang
Consultant
Alhambra, California

Frank Kreith, P.E.
Engineering Consultant
Boulder, Colorado

Contents

CHAPTER 1   Introduction

Shan K. Wang   1

Air-conditioning • Air-Conditioning Systems • Air-Conditioning Project Development and System Design

CHAPTER 2   Psychrometrics

Shan K. Wang   11

IMoist Air • Humidity and Enthalpy • Moist Volume, Density, Specific Heat, and Dew Point • Thermodynamic Wet Bulb Temperature and Wet Bulb Temperature • Psychometric Charts

CHAPTER 3   Air-Conditioning Processes and Cycles

Shan K. Wang   19

Air-Conditioning Processes • Space Conditioning, Sensible Cooling, and Sensible Heating Processes • Humidifying and Cooling and Dehumidifying Processes • Air-Conditioning Cycles and Operating Modes

CHAPTER 4   Refrigerants and Refrigeration Cycles

Shan K. Wang   35

Refrigeration and Refrigeration Systems • Refrigerants, Cooling Mediums, and Absorbents • Classification of Refrigerants • Required Properties of Refrigerants • Ideal Single-Stage Vapor Compression Cycle • Coefficient of Performance of Refrigeration Cycle • Subcooling and Superheating • Refrigeration Cycle of Two-Stage Compound Systems with a Flash Cooler • Cascade System Characteristics

CHAPTER 5   Outdoor Design Conditions and Indoor Design Criteria

Shan K. Wang   51

Outdoor Design Conditions • Indoor Design Criteria and Thermal Comfort • Indoor Temperature, Relative Humidity, and Air Velocity • Indoor Air Quality and Outdoor Ventilation Air Requirements

CHAPTER 6   Load Calculations

Shan K. Wang   59

ISpace Loads • Moisture Transfer in Building Envelope • Cooling Load Calculation Methodology • Conduction Heat Gains • Internal Heat Gains • Conversion of Heat Gains into Cooling Load by TFM • Heating Load

CHAPTER 7   Air HandlingUnits and Packaged Units

Shan K. Wang   71

Terminals and Air Handling Units • Packaged Units • Coils • Air Filters • Humidifers

CHAPTER 8   Refrigeration Components and Evaporative Coolers

Shan K. Wang   83

Refrigeration Compressors • Refrigeration Condensers • Evaporators and Refrigerant Flow Control Devices • Evaporative Coolers

CHAPTER 9   Water Systems

Shan K. Wang   95

Types of Water Systems • Basics • Water Piping • Plant-Building Loop • Plant-Distribution-Building Loop

CHAPTER 10   Heating Systems

Shan K. Wang   103

Types of Heating Systems

CHAPTER 11   Refrigeration Systems

Shan K. Wang   111

Classifications of Refrigeration Systems

CHAPTER 12   Thermal Storage Systems

Shan K. Wang   123

Thermal Storage Systems and Off-Peak Air-Conditioning Systems • Ice-Storage Systems • Chilled-Water Storage Systems

CHAPTER 13   Air System Basics

Shan K. Wang   129

Fan-Duct Systems • System Effect • Modulation of Air Systems • Fan Combinations in Air-Handling Units and Packaged Units • Fan Energy Use • Year-Round Operation and Economizers • Outdoor Ventilation Air Supply

CHAPTER 14   Absorption Systems

Shan K. Wang   141

Double-Effect Direct-Fired Absorption Chillers • Absorption Cycles, Parallel-, Series-, and Reverse-Parallel Flow

Chapter 15   Air-Conditioning Systems and Selection

Shan K. Wang   147

  Basics in Classification • Individual Systems • Packaged Systems • Central Systems • Air-Conditioning System Selection • Comparison of Various Systems • Subsystems • Energy Conservation Recommendations

Chapter 16   Dessicant Dehumidification and Air-Conditioning

Zalman Lavan   165

  Introduction • Sorbents and Desiccants • Dehumidification • Liquid Spray Tower • Solid Packed Tower • Rotary Desiccant Dehumidifiers • Hybrid Cycles • Solid Desiccant Air-Conditioning • Conclusions

APPENDICES

APPENDIX A   A-1

APPENDIX B   B-35

APPENDIX C   C-38

APPENDIX D   D-74

APPENDIX E   E-75

INDEX   I-1

1

Introduction

Shan K. Wang

Consultant

1   Introduction 1
Air-conditioning • Air-Conditioning Systems • Air-
Conditioning Project Development and System Design

Introduction

Air-Conditioning

Air-conditioning is a process that simultaneously conditions air; distributes it combined with the outdoor air to the conditioned space; and at the same time controls and maintains the required space's temperature, humidity, air movement, air cleanliness, sound level, and pressure differential within predetermined limits for the health and comfort of the occupants, for product processing, or both.

The acronym HVAC&R stands for heating, ventilating, air-conditioning, and refrigerating. The combination of these processes is equivalent to the functions performed by air-conditioning.

Because I-P units are widely used in the HVAC&R industry in the U.S., I-P units are used in this chapter.

Air-Conditioning Systems

An *air-conditioning* or *HVAC&R system* consists of components and equipment arranged in sequential order to heat or cool, humidify or dehumidify, clean and purify, attenuate objectionable equipment noise, transport the conditioned outdoor air and recirculate air to the conditioned space, and control and maintain an indoor or enclosed environment at optimum energy use.

The types of buildings which the air-conditioning system serves can be classified as:

- Institutional buildings, such as hospitals and nursing homes
- Commercial buildings, such as offices, stores, and shopping centers
- Residential buildings, including single-family and multifamily low-rise buildings of three or fewer stories above grade
- Manufacturing buildings, which manufacture and store products

Types of Air-Conditioning Systems

In institutional, commercial, and residential buildings, air-conditioning systems are mainly for the occupants' health and comfort. They are often called *comfort air-conditioning systems*. In manufacturing buildings, air-conditioning systems are provided for product processing, or for the health and comfort of workers as well as processing, and are called *processing air-conditioning systems*.

Based on their size, construction, and operating characteristics, air-conditioning systems can be classified as the following.

Individual Room or Individual Systems. An individual air-conditioning system normally employs either a single, self-contained, packaged room air conditioner (installed in a window or through a wall) or separate indoor and outdoor units to serve an individual room, as shown in Figure 1.1. "Self-contained, packaged" means factory assembled in one package and ready for use.

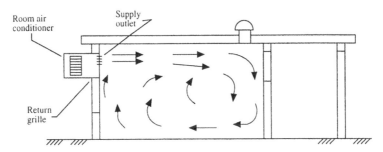

FIGURE 1.1 An individual room air-conditioning system.

Space-Conditioning Systems or Space Systems. These systems have their air-conditioning—cooling, heating, and filtration—performed predominantly in or above the conditioned space, as shown in Figure 1.2. Outdoor air is supplied by a separate outdoor ventilation system.

Unitary Packaged Systems or Packaged Systems. These systems are installed with either a single self-contained, factory-assembled packaged unit (PU) or two split units: an indoor air handler, normally with ductwork, and an outdoor condensing unit with refrigeration compressor(s) and condenser, as shown in Figure 1.3. In a packaged system, air is cooled mainly by direct expansion of refrigerant in coils called DX coils and heated by gas furnace, electric heating, or a heat pump effect, which is the reverse of a refrigeration cycle.

Central Hydronic or Central Systems. A central system uses chilled water or hot water from a central plant to cool and heat the air at the coils in an air handling unit (AHU) as shown in Figure 1.4. For energy transport, the heat capacity of water is about 3400 times greater than that of air. Central systems are built-up systems assembled and installed on the site.

Packaged systems are comprised of only air system, refrigeration, heating, and control systems. Both central and space-conditioning systems consist of the following.

Air Systems. An air system is also called an air handling system or the air side of an air-conditioning or HVAC&R system. Its function is to condition the air, distribute it, and control the indoor environment according to requirements. The primary equipment in an air system is an AHU or air handler; both of these include fan, coils, filters, dampers, humidifiers (optional), supply and return ductwork, supply outlets and return inlets, and controls.

Water Systems. These systems include chilled water, hot water, and condenser water systems. A water system consists of pumps, piping work, and accessories. The water system is sometimes called the water side of a central or space-conditioning system.

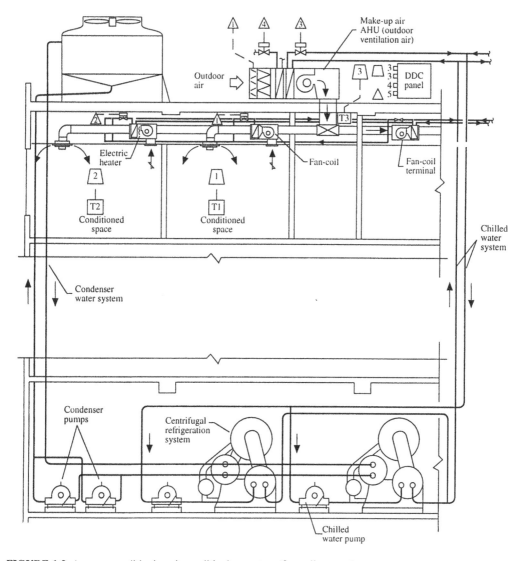

FIGURE 1.2 A space-conditioning air-conditioning system (fan-coil system).

Central Plant Refrigeration and Heating Systems. The refrigeration system in the central plant of a central system is usually in the form of a chiller package with an outdoor condensing unit. The refrigeration system is also called the refrigeration side of a central system. A boiler and accessories make up the heating system in a central plant for a central system, and a direct-fired gas furnace is often the heating system in the air handler of a rooftop packaged system.

Control Systems. Control systems usually consist of sensors, a microprocessor-based direct digital controller (DDC), a control device, control elements, personal computer (PC), and communication network.

Based on Commercial Buildings Characteristics 1992, Energy Information Administration (EIA) of the Department of Energy of United States in 1992, for commercial buildings having a total floor area of 67,876 million ft², of which 57,041 million ft² or 84% is cooled and 61,996 million ft² or 91% is heated, the air-conditioning systems for cooling include:

Individual systems	19,239 million ft²	(25%)
Packaged systems	34,753 million ft²	(49%)
Central systems	14,048 million ft²	(26%)

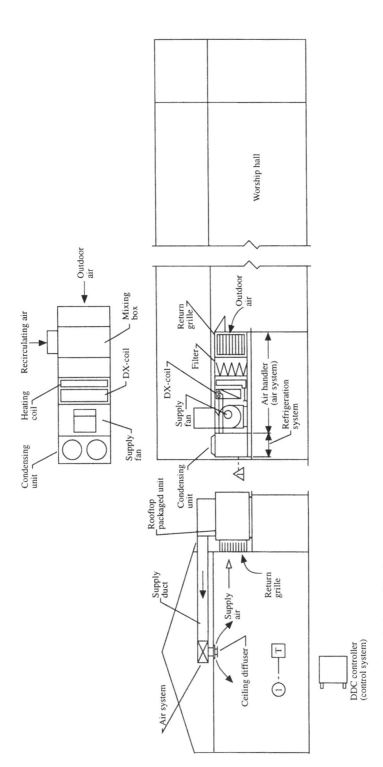

FIGURE 1.3 A packaged air-conditioning system.

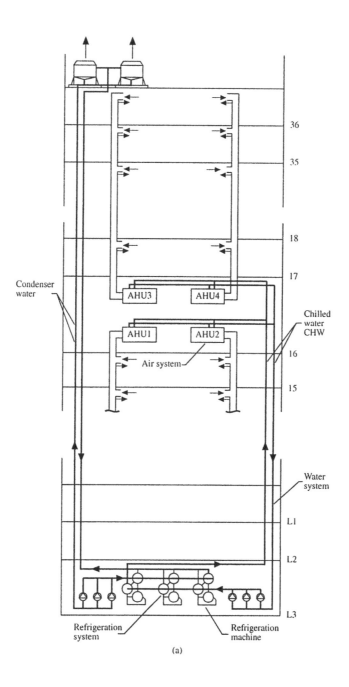

FIGURE 1.4a A central air-conditioning system: schematic diagram.

Space-conditioning systems are included in central systems. Part of the cooled floor area has been counted for both individual and packaged systems. The sum of the floor areas for these three systems therefore exceeds the total cooled area of 57,041 million ft².

Air-Conditioning Project Development and System Design

The goal of an air-conditioning/HVAC&R system is to provide a healthy and comfortable indoor environment with acceptable indoor air quality, while being energy efficient and cost effective.

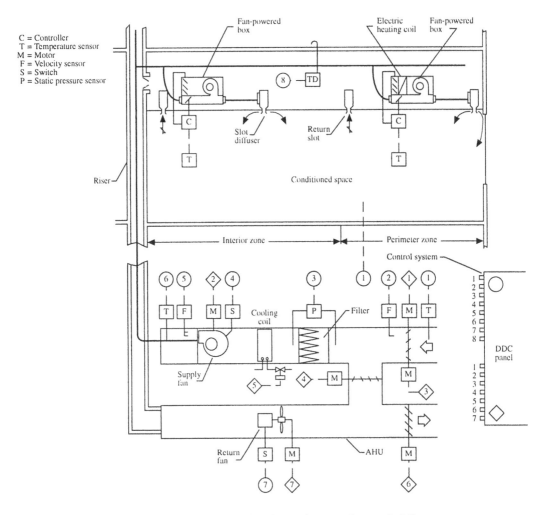

FIGURE 1.4b A central air-conditioning system: air and control systems for a typical floor.

ASHRAE Standard 62-1989 defines *acceptable indoor air quality* as "air in which there are no known contaminants at harmful concentrations as determined by cognizant authorities and with which a substantial majority (80% or more) of the people exposed do not express dissatisfaction."

The basic steps in the development and use of an air-conditioning project are design, installation, commissioning, operation, and maintenance. There are two types of air-conditioning projects: *design-bid* and *design-build*. A design-bid project separates the design (engineering consultant) and installation (contractors) responsibilities. In a design-build project, the design is also done by the installation contractor. A design-build project is usually a small project or a project having insufficient time to go through normal bidding procedures.

In the building construction industry, air-conditioning or HVAC&R is one of the *mechanical services*; these also include plumbing, fire protection, and escalators.

Air-conditioning design is a process of selecting the optimum system, subsystem, equipment, and components from various alternatives and preparing the drawings and specifications. Haines (1994) summarized this process in four phases: gather information, develop alternatives, evaluate alternatives, and sell the best solution. Design determines the basic operating characteristics of a system. After an air-conditioning system is designed and constructed, it is difficult and expensive to change its basic characteristics.

The foundation of a successful project is teamwork and coordination between designer, contractor, and operator and between mechanical engineer, electrical engineer, facility operator, architect, and structural engineer.

Field experience is helpful to the designer. Before beginning the design process it is advisable to visit similar projects that have operated for more than 2 years and talk with the operator to investigate actual performance.

Mechanical Engineer's Responsibilities

The normal procedure in a design-bid construction project and the mechanical engineer's responsibilities are

1. Initiation of a project by owner or developer
2. Organizing a design team
3. Determining the design criteria and indoor environmental parameters
4. Calculation of cooling and heating loads
5. Selection of systems, subsystems, and their components
6. Preparation of schematic layouts; sizing of piping and ductwork
7. Preparation of contract documents: drawings and specifications
8. Competitive biddings by various contractors; evaluation of bids; negotiations and modifications
9. Advice on awarding of contract
10. Monitoring, supervision, and inspection of installation; reviewing shop drawings
11. Supervision of commissioning
12. Modification of drawings to the as-built condition; preparation of the operation and maintenance manual
13. Handing over to the property management for operation

Design Documents

Drawings and *specifications* are legal documents of a construction contract. The designer conveys the owner's or developer's requirements to the contractor through these documents. Drawings and specifications complement each other.

Drawings should clearly and completely show, define, and present the work. Adequate plan and sectional views should be drawn. More often, isometric drawings are used to show the flow diagrams for water or the supply, return, and exhaust air.

Specifications include the legal contract between the owner and the contractor, installer, or vendor and the technical specifications, which describe in detail what kind of material and equipment should be used and how they are to be installed.

Most projects now use a format developed by the Construction Specifications Institute (CSI) called the Masterformat for Specifications. It includes 16 divisions. The 15000 Mechanical division is divided into the following:

Section No.	Title
15050	Basic Mechanical Materials and Methods
15250	Mechanical Insulation
15300	Fire Protection
15400	Plumbing
15500	Heating, Ventilating, and Air-Conditioning
15550	Heat Generation
15650	Refrigeration
15750	Heat Transfer
15850	Air Handling
15880	Air Distribution
15950	Controls
15990	Testing, Adjusting, and Balancing

Each section includes general considerations, equipment and material, and field installation. Design criteria and selected indoor environmental parameters that indicate the performance of the HVAC&R system must be clearly specified in the general consideration of Section 15500.

There are two types of specifications: the performance specification, which depends mainly on the required performance criteria, and the or-equal specification, which specifies the wanted vendor. Specifications should be written in simple, direct, and clear language without repetition.

Computer-Aided Design and Drafting

With the wide acceptance of the PC and the availability of numerous types of engineering software, the use of *computer-aided drafting* (CAD) and *computer-aided design and drafting* (CADD) has increased greatly in recent years. According to the 1994 CADD Application and User Survey of design firms reported in *Engineering Systems* (1994[6]), "15% of the design firms now have a computer on every desk" and "Firms with high productivity reported that they perform 95% on CADD." Word processing software is widely used to prepare specifications.

Drafting software used to reproduce architectural drawings is the foundation of CADD. Automated CAD (AutoCAD) is the leading personal computer-based drafting tool software used in architectural and engineering design firms.

In "Software Review" by Amistadi (1993), duct design was the first HVAC&R application to be integrated with CAD.

- Carrier Corp. DuctLINK and Softdesk HVAC 12.0 are the two most widely used duct design software. Both of them convert the single-line duct layout drawn with CAD to two-dimensional (2D) double-line drawings with fittings, terminals, and diffusers.
- Tags and schedules of HVAC&R equipment, ductwork, and duct fittings can be produced as well.
- DuctLINK and Softdesk can also interface with architectural, electrical, and plumbing drawings through AutoCAD software.

Software for piping system design and analysis can also be integrated with CAD. The software developed at the University of Kentucky, KYCAD/KYPIPE, is intended for the design and diagnosis of large water piping systems, has extensive hydraulic modeling capacities, and is the most widely used. Softdesk AdCADD Piping is relative new software; it is intended for drafting in 2D and 3D, linking to AutoCAD through design information databases.

Currently, software for CADD for air-conditioning and HVAC&R falls into two categories: engineering and product. The engineering category includes CAD (AutoCAD integrated with duct and piping system), load calculations and energy analysis, etc. The most widely used software for load calculations and energy analysis is Department of Energy DOE-2.1D, Trane Company's TRACE 600, and Carrier Corporation's softwares for load calculation, E20-II Loads.

Product categories include selection, configuration, performance, price, and maintenance schedule. Product manufacturers provide software including data and CAD drawings for their specific product.

Codes and Standards

Codes are federal, state, or city laws that require the designer to perform the design without violating people's (including occupants and the public) safety and welfare. Federal and local codes must be followed. The designer should be thoroughly familiar with relevant codes. HVAC&R design codes are definitive concerning structural and electrical safety, fire prevention and protection (particularly for gas- or oil-fired systems), environmental concerns, indoor air quality, and energy conservation.

Conformance with *ASHRAE Standards* is voluntary. However, for design criteria or performance that has not been covered in the codes, whether the ASHRAE Standard is followed or violated is the vital criterion, as was the case in a recent indoor air quality lawsuit against a designer and contractor.

For the purpose of performing an effective, energy-efficient, safe, and cost-effective air-conditioning system design, the following ASHRAE Standards should be referred to from time to time:

- ASHRAE/IES Standard 90.1-1989, Energy Efficient Design of New Buildings Except New Low-Rise Residential Buildings
- ANSI/ASHRAE Standard 62-1989, Ventilation for Acceptable Indoor Air Quality
- ANSI/ASHRAE Standard 55-1992, Thermal Environmental Conditions for Human Occupancy
- ASHRAE Standard 15-1992, Safety Code for Mechanical Refrigeration

2

Psychrometrics

Shan K. Wang

Consultant

2 Psychrometrics 11
   Moist Air • Humidity and Enthalpy • Moist Volume, Density,
   Specific Heat, and Dew Point • Thermodynamic Wet Bulb •
   Temperature • Psychrometric Charts

Psychrometrics

Moist Air

Above the surface of the earth is a layer of air called the *atmosphere*, or *atmospheric air.* The lower atmosphere, or homosphere, is composed of moist air, that is, a mixture of dry air and water vapor.

Psychrometrics is the science of studying the thermodynamic properties of moist air. It is widely used to illustrate and analyze the change in properties and the thermal characteristics of the air-conditioning process and cycles.

The composition of dry air varies slightly at different geographic locations and from time to time. The approximate composition of dry air by volume is nitrogen, 79.08%; oxygen, 20.95%; argon, 0.93%; carbon dioxide, 0.03%; other gases (e.g., neon, sulfur dioxide), 0.01%.

The amount of water vapor contained in the moist air within the temperature range 0 to 100°F changes from 0.05 to 3% by mass. The variation of water vapor has a critical influence on the characteristics of moist air.

The equation of state for an ideal gas that describes the relationship between its thermodynamic properties is

$$pv = RT_R \tag{2.1}$$

or

$$pV = mRT_R \tag{2.2}$$

where p = pressure of the gas, psf (1 psf = 144 psi)
          v = specific volume of the gas, ft^3/lb
          R = gas constant, ftlb$_f$/lb$_m$ °R
          T_R = absolute temperature of the gas, °R
          V = volume of the gas, ft^3
          m = mass of the gas, lb

The most exact calculation of the thermodynamic properties of moist air is based on the formulations recommended by Hyland and Wexler (1983) of the U.S. National Bureau of Standards. The psychrometric charts and tables developed by ASHRAE are calculated and plotted from these formulations. According to Nelson et al. (1986), at a temperature between 0 and 100°F, enthalpy and specific volume calculations using ideal gas Equations (2.1) and (2.2) show a maximum deviation of 0.5% from the results of Hyland and Wexler's exact formulations. Therefore, ideal gas equations are used in the development and calculation of psychrometric formulations in this handbook.

Although air contaminants may seriously affect human health, they have little effect on the thermodynamic properties of moist air. For thermal analysis, moist air may be treated as a binary mixture of dry air and water vapor.

Applying Dalton's law to moist air:

$$p_{at} = p_a + p_w \tag{2.3}$$

where p_{at} = atmospheric pressure of the moist air, psia
          p_a = partial pressure of dry air, psia
          p_w = partial pressure of water vapor, psia

Dalton's law is summarized from the experimental results and is more accurate at low gas pressure. Dalton's law can also be extended, as the Gibbs-Dalton law, to describe the relationship of internal energy, enthalpy, and entropy of the gaseous constituents in a mixture.

Humidity and Enthalpy

The *humidity ratio* of moist air, w, in lb/lb is defined as the ratio of the mass of the water vapor, m_w to the mass of dry air, m_a, or

$$w = m_w/m_a = 0.62198 p_w/\left(p_{at} - p_w\right) \tag{2.4}$$

The *relative humidity* of moist air, φ, or RH, is defined as the ratio of the mole fraction of water vapor, x_w, to the mole fraction of saturated moist air at the same temperature and pressure, x_{ws}. Using the ideal gas equations, this relationship can be expressed as:

$$\varphi = x_w/x_{ws}\big|_{T,p} = p_w/p_{ws}\big|_{T,p} \tag{2.5}$$

and

$$x_w = n_w/\left(n_a + n_w\right); \quad x_{ws} = n_{ws}/\left(n_a + n_{ws}\right)$$

$$x_a + x_w = 1 \tag{2.6}$$

where p_{ws} = pressure of saturated water vapor, psia

T = temperature, °F

n_a, n_w, n_{ws} = number of moles of dry air, water vapor, and saturated water vapor, mol

Degree of saturation μ is defined as the ratio of the humidity ratio of moist air, w, to the humidity ratio of saturated moist air, w_s, at the same temperature and pressure:

$$\mu = w/w_s\big|_{T,p} \tag{2.7}$$

The difference between φ and μ is small, usually less than 2%.

At constant pressure, the difference in specific enthalpy of an ideal gas, in Btu/lb, is $\Delta h = c_p\Delta T$. Here c_p represents the specific heat at constant pressure, in Btu/lb. For simplicity, the following assumptions are made during the calculation of the *enthalpy* of moist air:

1. At 0°F, the enthalpy of dry air is equal to zero.
2. All water vapor is vaporized at 0°F.
3. The enthalpy of saturated water vapor at 0°F is 1061 Btu/lb.
4. The unit of the enthalpy of the moist air is Btu per pound of dry air and the associated water vapor, or Btu/lb.

Then, within the temperature range 0 to 100°F, the enthalpy of the moist air can be calculated as:

$$h = c_{pd}T + w\left(h_{g0} + c_{ps}T\right)$$
$$= 0.240T + w(1061 + 0.444T) \tag{2.8}$$

where c_{pd}, c_{ps}=specific heat of dry air and water vapor at constant pressure, Btu/lb°F. Their mean values can be taken as 0.240 and 0.444 Btu/lb°F, respectively.

h_{g0} =specific enthalpy of saturated water vapor at 0°F.

Moist Volume, Density, Specific Heat, and Dew Point

The specific *moist volume* v, in ft³/lb, is defined as the volume of the mixture of dry air and the associated water vapor when the mass of the dry air is exactly 1 lb:

$$v = V/m_a \tag{2.9}$$

where V = total volume of the moist air, ft³. Since moist air, dry air, and water vapor occupy the same volume,

$$v = R_a T_R / p_{at}(1 + 1.6078w) \tag{2.10}$$

where R_a = gas constant for dry air.

Moist air density, often called *air density* ρ, in lb/ft³, is defined as the ratio of the mass of dry air to the total volume of the mixture, or the reciprocal of the moist volume:

$$\rho = m_a/V = 1/v \tag{2.11}$$

The *sensible heat of moist air* is the thermal energy associated with the change of air temperature between two state points. In Equation (2.8), $(c_{pd} + wc_{ps})T$ indicates the sensible heat of moist air, which depends on its temperature T above the datum 0°F. *Latent heat of moist air*, often represented by wh_{fg0}, is the thermal energy associated with the change of state of water vapor. Both of them are in Btu/lb.

Within the temperature range 0 to 100°F, if the average humidity ratio w is taken as 0.0075 lb/lb, the *specific heat of moist air* c_{pa} can be calculated as:

$$c_{pa} = c_{pd} + wc_{ps} = 0.240 + 0.0075 \times 0.444 = 0.243 \text{ Btu/lb °F} \qquad (2.12)$$

The *dew point temperature* T_{dew}, in °F, is the temperature of saturated moist air of the moist air sample having the same humidity ratio at the same atmospheric pressure. Two moist air samples of similar dew points T_{dew} at the same atmospheric pressure have the same humidity ratio w and the same partial pressure of water vapor p_w.

Thermodynamic Wet Bulb Temperature and Wet Bulb Temperature

The *thermodynamic wet bulb temperature* of moist air, T^*, is equal to the saturated state of a moist air sample at the end of a constant-pressure, ideal adiabatic saturation process:

$$h_1 + \left(w_s^* - w_1\right)h_w^* = h_s^* \qquad (2.13)$$

where h_1, h_s^* =enthalpy of moist air at the initial state and enthalpy of saturated air at the end of the constant-pressure, ideal adiabatic saturation process, Btu/lb

w_1, w_s^* =humidity ratio of moist air at the initial state and humidity ratio of saturated air at the end of the constant-pressure, ideal adiabatic saturation process, lb/lb

h_w^* =enthalpy of water added to the adiabatic saturation process at temperature T^*, Btu/lb

An *ideal adiabatic saturation process* is a hypothetical process in which moist air at initial temperature T_1, humidity ratio w_1, enthalpy h_1, and pressure p flows over a water surface of infinite length through a well-insulated channel. Liquid water is therefore evaporated into water vapor at the expense of the sensible heat of the moist air. The result is an increase of humidity ratio and a drop of temperature until the moist air is saturated at the thermodynamic wet bulb temperature T^* during the end of the ideal adiabatic saturation process.

The thermodynamic wet bulb temperature T^* is a unique fictitious property of moist air that depends only on its initial properties, T_1, w_1, or h_1.

A sling-type *psychrometer*, as shown in Figure 2.1, is an instrument that determines the temperature, relative humidity, and thus the state of the moist air by measuring its dry bulb and wet bulb temperatures. It consists of two mercury-in-glass thermometers. The sensing bulb of one of them is dry and is called the dry bulb. Another sensing bulb is wrapped with a piece of cotton wick, one end of which dips into a water tube. This wetted sensing bulb is called the wet bulb and the temperature measured by it is called the *wet bulb temperature T'*.

When unsaturated moist air flows over the surface of the wetted cotton wick, liquid water evaporates from its surface. As it absorbs sensible heat, mainly from the surrounding air, the wet bulb temperature drops. The difference between the dry and wet bulb temperatures is called *wet bulb depression* $(T - T')$. Turning the handle forces the surrounding air to flow over the dry and wet bulbs at an air velocity between 300 to 600 fpm. Distilled water must be used to wet the cotton wick.

At steady state, if heat conduction along the thermometer stems is neglected and the temperature of the wetted cotton wick is equal to the wet bulb temperature of the moist air, as the sensible heat transfer from the surrounding moist air to the cotton wick exactly equals the latent heat required for evaporation, the heat and mass transfer per unit area of the wet bulb surface can be evaluated as:

$$h_c(T - T') + h_r(T_{ra} - T) = h_d(w_s' - w_1) \qquad (2.14)$$

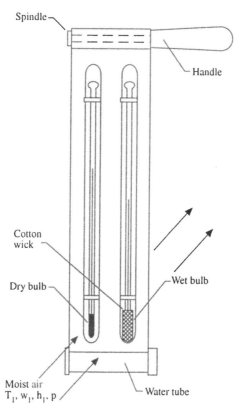

Spindle

Handle

Cotton wick

Dry bulb

Wet bulb

Moist air
T_1, w_1, h_1, p

Water tube

FIGURE 2.1 A sling psychrometer.

where h_c, h_r = mean conductive and radiative heat transfer coefficient, Btu/hr ft²°F

$\quad\quad h_d$ = mean convective mass transfer coefficient, lb/hr ft²

$\quad\quad T$ = temperature of undisturbed moist air at a distance from the wet bulb, °F

$\quad\quad T_{ra}$ = mean radiant temperature (covered later), °F

$\quad w_1$, w_s' = humidity ratio of the moist air and the saturated film at the interface of cotton wick and surrounding air, lb/lb

$\quad\quad h_{fg}'$ = latent heat of vaporization at the wet bulb temperature, Btu/lb

The humidity ratio of the moist air is given by:

$$w_1 = w_s' - K'(T - T')\left(1 + \left\{h_r\left(T_{ra} - T'\right)/\left[h_c(T - T')\right]\right\}\right)$$

$$K' = c_{pa}Le^{0.6667}/h_{fg}' \tag{2.15}$$

where $\quad K'$ = wet bulb constant, which for a sling psychrometer = 0.00218 1/°F

$\quad\quad Le$ = Lewis number

The wet bulb temperature T' depends not only on its initial state but also on the rate of heat and mass transfer at the wet bulb. Therefore, the thermodynamic wet bulb temperature is used in ASHRAE psychrometric charts.

According to Threlkeld (1970), for a sling psychrometer whose wet bulb diameter is 1 in. and for air flowing at a velocity of 400 fpm over the wet bulb, if the dry bulb temperature is 90°F and the measured wet bulb temperature is 70°F, the difference between the measured wet bulb and the thermodynamic wet bulb $(T' - T^*)/(T^* - T')$ is less than 1%.

Psychrometric Charts

A *psychrometric chart* is a graphical presentation of the thermodynamic properties of moist air and various air-conditioning processes and air-conditioning cycles. A psychrometric chart also helps in calculating and analyzing the work and energy transfer of various air-conditioning processes and cycles.

Psychrometric charts currently use two kinds of basic coordinates:

1. *h-w* charts. In *h-w* charts, enthalpy *h*, representing energy, and humidity ratio *w*, representing mass, are the basic coordinates. Psychrometric charts published by ASHRAE and the Charted Institution of Building Services Engineering (CIBSE) are *h-w* charts.
2. *T-w* charts. In *T-w* charts, temperature *T* and humidity ratio *w* are basic coordinates. Psychrometric charts published by Carrier Corporation, the Trane Company, etc. are *T-w* charts.

Figure 2.2 shows an abridged ASHRAE psychrometric chart. In the ASHRAE chart:

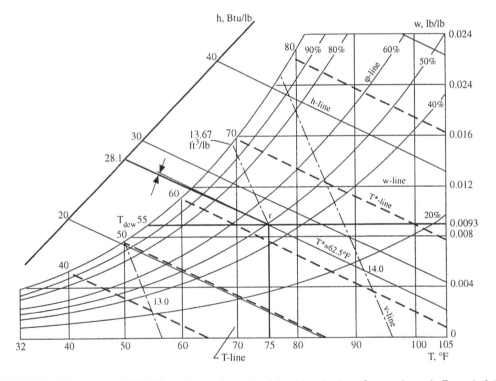

FIGURE 2.2 The abridged ASHRAE psychrometric chart and the determination of properties as in Example 2.1.

- A normal temperature chart has a temperature range of 32 to 120°F, a high-temperature chart 60 to 250°F, and a low-temperature chart –40 to 50°F. Since enthalpy is the basic coordinate, temperature lines are not parallel to each other. Only the 120°F line is truly vertical.
- Thermodynamic properties of moist air are affected by atmospheric pressure. The standard atmospheric pressure is 29.92 in. Hg at sea level. ASHRAE also published charts for high altitudes of 5000 ft, 24.89 in. Hg, and 7500 ft, 22.65 in. Hg. Both of them are in the normal temperature range.
- Enthalpy *h*-lines incline downward to the right-hand side (negative slope) at an angle of 23.5° to the horizontal line and have a range of 12 to 54 Btu/lb.

- Humidity ratio w-lines are horizontal lines. They range from 0 to 0.28 lb/lb.
- Relative humidity φ-lines are curves of relative humidity 10%, 20%, ... 90% and a saturation curve. A saturation curve is a curve of the locus of state points of saturated moist air, that is, φ = 100%. On a saturation curve, temperature T, thermodynamic wet temperature bulb T^*, and dew point temperature T_{dew} have the same value.
- Thermodynamic wet bulb T^*-lines have a negative slope slightly greater than that of the h-lines. A T^*-line meets the T-line of the same magnitude on the saturation curve.
- Moist volume v-lines have a far greater negative slope than h-lines and T^*-lines. The moist volume ranges from 12.5 to 15 ft³/lb.

Moist air has seven independent thermodynamic properties or property groups: h, T, φ, T^*, p_{at}, $\rho - v$, and $w - p_w - T_{dew}$. When p_{at} is given, any additional two of the independent properties determine the state of moist air on the psychrometric chart and the remaining properties.

Software using AutoCAD to construct the psychrometric chart and calculate the thermodynamic properties of moist air is available. It can also be linked to the load calculation and energy programs to analyze the characteristics of air-conditioning cycles.

Refer to Wang's *Handbook of Air Conditioning and Refrigeration* (1993) and *ASHRAE Handbook, Fundamentals* (1993) for details of psychrometric charts and psychrometric tables that list thermodynamic properties of moist air.

Example 2.1

An air-conditioned room at sea level has an indoor design temperature of 75°F and a relative humidity of 50%. Determine the humidity ratio, enthalpy, density, dew point, and thermodynamic wet bulb temperature of the indoor air at design condition.

Solution

1. Since the air-conditioned room is at sea level, a psychrometric chart of standard atmospheric pressure of 14.697 psi should be used to find the required properties.
2. Plot the state point of the room air at design condition r on the psychrometric chart. First, find the room temperature 75°F on the horizontal temperature scale. Draw a line parallel to the 75°F temperature line. This line meets the relative humidity curve of 50% at point r, which denotes the state point of room air as shown in Figure 2.2.
3. Draw a horizontal line toward the humidity ratio scale from point r. This line meets the ordinate and thus determines the room air humidity ratio φ_r = 0.0093 lb/lb.
4. Draw a line from point r parallel to the enthalpy line. This line determines the enthalpy of room air on the enthalpy scale, h_r = 28.1 Btu/lb.
5. Draw a line through point r parallel to the moist volume line. The perpendicular scale of this line indicates v_r = 13.67 ft³/lb.
6. Draw a horizontal line to the left from point r. This line meets the saturation curve and determines the dew point temperature, T_{dew} = 55°F.
7. Draw a line through point r parallel to the thermodynamic wet bulb line. The perpendicular scale to this line indicates that the thermodynamic wet bulb temperature T^* = 62.5°F.

3

Air-Conditioning Processes and Cycles

Shan K. Wang

Individual Consultant

3   Air-Conditioning Processes and Cycles 19
     Air-Conditioning Processes • Space Conditioning, Sensible
     Cooling, and Sensible Heating Processes • Humidifying and
     Cooling and Dehumidifying Processes • Air-Conditioning
     Cycles and Operating Modes

Air-Conditioning Processes and Cycles

Air-Conditioning Processes

An *air-conditioning process* describes the change in thermodynamic properties of moist air between the initial and final stages of conditioning as well as the corresponding energy and mass transfers between the moist air and a medium, such as water, refrigerant, absorbent or adsorbent, or moist air itself. The energy balance and conservation of mass are the two principles used for the analysis and the calculation of the thermodynamic properties of the moist air.

Generally, for a single air-conditioning process, heat transfer or mass transfer is positive. However, for calculations that involve several air-conditioning processes, heat supplied to the moist air is taken as positive and heat rejected is negative.

The *sensible heat ratio* (SHR) of an air-conditioning process is defined as the ratio of the change in absolute value of sensible heat to the change in absolute value of total heat, both in Btu/hr:

$$\mathrm{SHR} = \left| q_{\mathrm{sen}} \right| / \left| q_{\mathrm{total}} \right| = \left| q_{\mathrm{sen}} \right| / \left(\left| q_{\mathrm{sen}} \right| + \left| q_{\mathrm{l}} \right| \right) \tag{3.1}$$

For any air-conditioning process, the sensible heat change

$$q_{\mathrm{sen}} = 60 \, \mathring{V}_{\mathrm{s}} \, \rho_{\mathrm{s}} c_{\mathrm{pa}} \left(T_2 - T_1 \right) = 60 \mathring{m}_{\mathrm{a}} \, c_{\mathrm{pa}} \left(T_2 - T_1 \right) \tag{3.2}$$

where $\mathring{V}_{\mathrm{s}}$ = volume flow rate of supply air, cfm
         ρ_{s} = density of supply air, lb/ft³

T_2, T_1=moist air temperature at final and initial states of an air-conditioning process, °F
and the mass flow rate of supply air

$$\mathring{m}_s = \mathring{V}_s \rho_s \tag{3.3}$$

The latent heat change is

$$q_l \approx 60 \mathring{V}_s \rho_s \left(w_2 - w_1 \right) h_{lg.58} = 1060 \times 60 \mathring{V}_s \rho_s \left(w_2 - w_1 \right) \tag{3.4}$$

where w_2, w_1 = humidity ratio at final and initial states of an air-conditioning process, lb/lb.

In Equation (3.4), $h_{lg58} \approx 1060$ Btu/lb represents the latent heat of vaporization or condensation of water at an estimated temperature of 58°F, where vaporization or condensation occurs in an air-handling unit or packaged unit. Therefore

$$\text{SHR} = \mathring{m}_a c_{pa} \left(T_2 - T_1 \right) \Big/ \left[\mathring{m}_a c_{pa} \left(T_2 - T_1 \right) + \mathring{m}_a \left(w_2 - w_1 \right) h_{lg.58} \right] \tag{3.5}$$

Space Conditioning, Sensible Cooling, and Sensible Heating Processes

In a *space conditioning process*, heat and moisture are absorbed by the supply air at state s and then removed from the conditioned space at the state of space air r during summer, as shown by line sr in Figure 3.1, or heat or moisture is supplied to the space to compensate for the transmission and infiltration losses through the building envelope as shown by line $s'r'$. Both processes are aimed at maintaining a desirable space temperature and relative humidity.

The space cooling load q_{rc}, in Btu/hr, can be calculated as:

$$q_{rc} = 60 \mathring{m}_a \left(h_r - h_s \right) = 60 \mathring{V}_s \rho_s \left(h_r - h_s \right) \tag{3.6}$$

where h_r, h_s = enthalpy of space air and supply air, Btu/lb.

The space sensible cooling load q_{rs}, in Btu/hr, can be calculated from Equation (3.2) and the space latent load q_{rl}, in Btu/hr, from Equation (3.1). In Equation (3.4), T_2 should be replaced by T_r and T_1 by T_s. Also in Equation (3.1), w_2 should be replaced by w_r and w_1 by w_s. The space heating load q_{rh} is always a sensible load, in Btu/hr, and can be calculated as:

$$q_{rh} = 60 \mathring{m}_a c_{pa} \left(T_s - T_r \right) = 60 \mathring{V}_s \rho_s c_{pa} \left(T_s - T_r \right) \tag{3.7}$$

where T_s, T_r = temperature of supply and space air, °F.

A *sensible heating process* adds heat to the moist air in order to increase its temperature; its humidity ratio remains constant, as shown by line 12 in Figure 3.1. A sensible heating process occurs when moist air flows over a heating coil. Heat is transferred from the hot water inside the tubes to the moist air. The rate of heat transfer from the hot water to the colder moist air is often called the heating coil load q_{rh}, in Btu/hr, and is calculated from Equation (3.2).

A *sensible cooling process* removes heat from the moist air, resulting in a drop of its temperature; its humidity ratio remains constant, as shown by line 1'2' in Figure 3.1. The sensible cooling process occurs when moist air flows through a cooling coil containing chilled water at a temperature equal to or greater than the dew point of the entering moist air. The sensible cooling load can also be calculated from Equation (3.2). T_2 is replaced by T_1 and T_1 by T_2'.

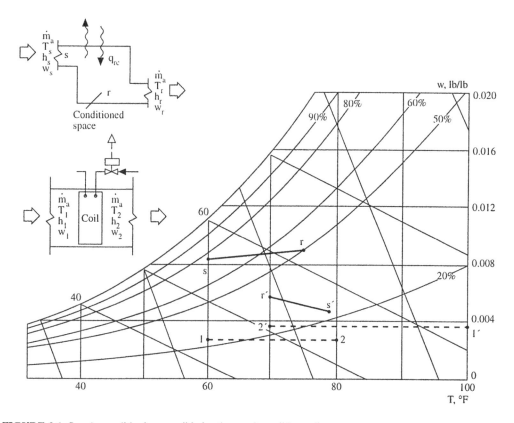

FIGURE 3.1 Supply conditioning, sensible heating, and sensible cooling processes.

Humidifying and Cooling and Dehumidifying Processes

In a *humidifying process*, water vapor is added to moist air and increases the humidity ratio of the moist air entering the humidifier if the moist air is not saturated. Large-scale humidification of moist air is usually performed by steam injection, evaporation from a water spray, atomizing water, a wetted medium, or submerged heating elements. Some details of their construction and characteristics are covered in a later section. Dry steam in a steam injection humidifying process is often supplied from the main steam line to a grid-type humidifier and injected into the moist air directly through small holes at a pressure slightly above atmospheric, as shown by line 12 in Figure 3.2(a) and (b). The humidifying capacity $\mathring{m}_{hu}$, in lb/min, is given by:

$$\mathring{m}_{hu} = \mathring{V}_s \rho_s \left(w_{hl} - w_{he} \right) \tag{3.8}$$

where w_{hl}, w_{he} = humidity ratio of moist air leaving and entering the humidifier, lb/lb. The slight inclination at the top of line 12 is due to the high temperature of the steam. The increase in temperature of the moist air due to steam injection can be calculated as:

$$\left(T_2 - T_1 \right) = w_{sm} c_{ps} T_s \big/ \left(c_{pd} + w_{12} c_{ps} \right) \tag{3.9}$$

where T_2, T_1 = temperature of moist air at initial and final states, °F
  w_{sm} = ratio of mass flow rate of injected steam to moist air, $\mathring{m}_s / \mathring{m}_a$
  T_s = temperature of injected steam, °F
  w_{12} = average humidity ratio of moist air, lb/lb

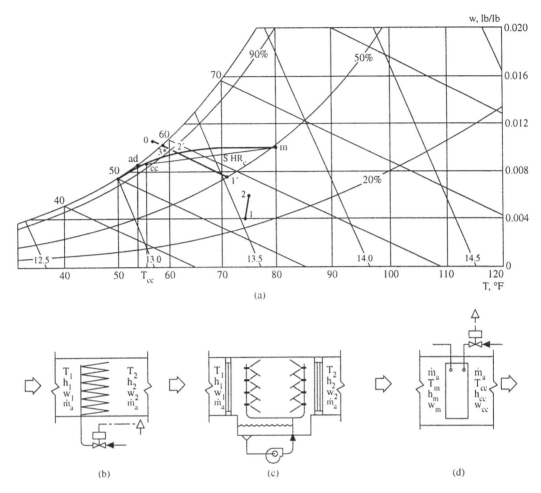

(a)

(b)   (c)   (d)

FIGURE 3.2 Humidifying and cooling and dehumidifying processes: (a) process on psychrometric chart, (b) steam humidifier, (c) air washer, and (d) water cooling or DX coil.

An *air washer* is a device that sprays water into moist air in order to humidify, to cool and dehumidify, and to clean the air, as shown in Figure 3.2(c). When moist air flows through an air washer, the moist air is humidified and approaches saturation. This actual adiabatic saturation process approximately follows the thermodynamic wet bulb line on the psychrometric chart as shown by line 1′2′. The humidity ratio of the moist air is increased while its temperature is reduced. The cooling effect of this adiabatic saturation process is called *evaporative cooling*.

Oversaturation occurs when the amount of water present in the moist air w_{os}, in lb/lb, exceeds the saturated humidity ratio at thermodynamic wet bulb temperature w_s^*, as shown in Figure 3.2(a). When moist air leaves the air washer, atomizing humidifier, or centrifugal humidifier after humidification, it often contains unevaporated water droplets at state point 2′, w_w, in lb/lb. Because of the fan power heat gain, duct heat gain, and other heat gains providing the latent heat of vaporization, some evaporation takes place due to the heat transfer to the water drops, and the humidity ratio increases further. Such evaporation of oversaturated drops is often a process with an increase of both humidity ratio and enthalpy of moist air. Oversaturation can be expressed as:

$$w_{os} = w_o - w_s^* = \left(w_{2'} + w_w\right) - w_s^* \qquad (3.10)$$

where $w_{2'}$ = humidity ratio at state point 2', lb/lb

$\quad\quad\quad w_o$ = sum of $w_{2'}$ and w_w, lb/lb

The magnitude of w_w depends mainly on the construction of the humidifier and water eliminator, if any. For an air washer, w_w may vary from 0.0002 to 0.001 lb/lb. For a pulverizing fan without an eliminator, w_w may be up to 0.00135 lb/lb.

Cooling and Dehumidifying Process

In a cooling and dehumidifying process, both the humidity ratio and temperature of moist air decrease. Some water vapor is condensed in the form of liquid water, called a *condensate*. This process is shown by curve m cc on the psychrometric chart in Figure 3.2(a). Here m represents the entering mixture of outdoor and recirculating air and cc the conditioned air leaving the cooling coil.

Three types of heat exchangers are used in a cooling and dehumidifying process: (1) water cooling coil as shown in Figure 3.2(d); (2) direct expansion DX coil, where refrigerant evaporates directly inside the coil's tubes; and (3) air washer, in which chilled water spraying contacts condition air directly.

The temperature of chilled water entering the cooling coil or air washer T_{wc}, in °F, determines whether it is a sensible cooling or a cooling and dehumidifying process. If T_{wc} is smaller than the dew point of the entering air T''_{ae} in the air washer, or T_{wc} makes the outer surface of the water cooling coil $T_{s.t} < T''_{ae}$, it is a cooling and dehumidifying process. If $T_{wc} \geq T''_{ae}$, or $T_{s.t} \geq T''_{ae}$, sensible cooling occurs. The cooling coil's load or the cooling capacity of the air washer q_{cc}, in Btu/hr, is

$$q_{cc} = 60 \mathring{V}_s \rho_s \left(h_{ae} - h_{cc} \right) - 60 \mathring{m}_w h_w \tag{3.11a}$$

where h_{ae}, h_{cc}=enthalpy of moist air entering and leaving the coil or washer, Btu/lb

$\quad\quad \mathring{m}_w$=mass flow rate of the condensate, lb/min

$\quad\quad h_w$=enthalpy of the condensate, Btu/lb

Since the thermal energy of the condensate is small compared with q_{cc}, in practical calculations the term $60\mathring{m}_w h_w$ is often neglected, and

$$q_{cc} = 60 \mathring{V}_s \rho_s \left(h_{ae} - h_{cc} \right) \tag{3.11b}$$

The sensible heat ratio of the cooling and dehumidifying process SHR_c can be calculated from

$$SHR_c = q_{cs}/q_{cc} \tag{3.12}$$

where q_{cs} = sensible heat removed during the cooling and dehumidifying process, Btu/hr. SHR_c is shown by the slope of the straight line joining points m and cc.

The relative humidity of moist air leaving the water cooling coil or DX coil depends mainly on the outer surface area of the coil including pipe and fins. For coils with ten or more fins per inch, if the entering moist air is around 80°F dry bulb and 68°F wet bulb, the relative humidity of air leaving the coil (off-coil) may be estimated as:

Four-row coil	90 to 95%
Six-row and eight-row coils	96 to 98%

Two-Stream Mixing Process and Bypass Mixing Process

For a *two-stream adiabatic mixing process*, two moist air streams, 1 and 2, are mixed together adiabatically and a mixture m is formed in a mixing chamber as shown by line 1 m1 2 in Figure 3.3. Since the

AHU or PU is well insulated, the heat transfer between the mixing chamber and ambient air is small and is usually neglected. Based on the principle of heat balance and conservation of mass:

$$\mathring{m}_1 h_1 + \mathring{m}_2 h_2 = \mathring{m}_m h_m$$

$$\mathring{m}_1 w_1 + \mathring{m}_2 w_2 = \mathring{m}_m w_m \tag{3.13}$$

$$\mathring{m}_1 T_1 + \mathring{m}_2 T_2 = \mathring{m}_m T_m$$

In Equation (3.13), $\mathring{m}$ represents the mass flow rate of air, lb/min; h the enthalpy, in Btu/lb; w the humidity ratio, in lb/lb; and T the temperature, in °F. Subscripts 1 and 2 indicate air streams 1 and 2 and m the mixture; also,

$$\mathring{m}_1 / \mathring{m}_m = \left(h_2 - h_m\right)/\left(h_2 - h_1\right) = \left(w_2 - w_m\right)/\left(w_2 - w_1\right)$$

$$= (\text{line segment m1 2})/(\text{line segment 12}) \tag{3.14}$$

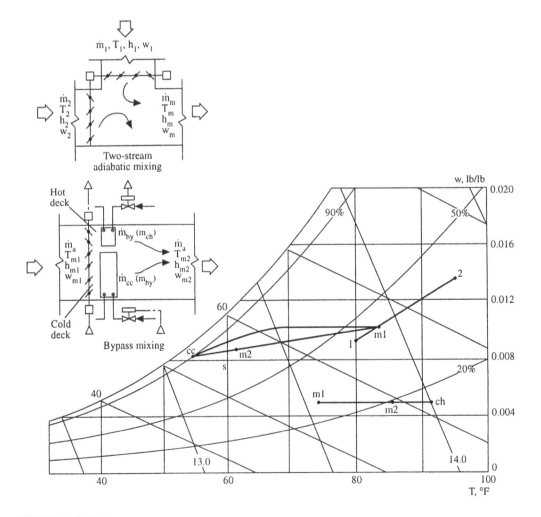

FIGURE 3.3 Mixing processes.

Similarly,

$$\dot{m}_2/\dot{m}_m = (h_m - h_1)/(h_2 - h_1) = (w_m - w_1)/(w_2 - w_1)$$

$$= (\text{line segment } 1\,m1)/(\text{line segment } 12) \tag{3.15}$$

Mixing point m must lie on the line that joins points 1 and 2 as shown in Figure 3.3.

If the differences between the density of air streams 1 and 2 and the density of the mixture are neglected,

$$\dot{V}_1 h_1 + \dot{V}_2 h_2 = \dot{V}_m h_m$$

$$\dot{V}_1 w_1 + \dot{V}_2 w_2 = \dot{V}_m w_m \tag{3.16}$$

$$\dot{V}_1 T_1 + \dot{V}_2 T_2 = \dot{V}_m T_m$$

$$\dot{V}_1 + \dot{V}_2 = \dot{V}_m \tag{3.17}$$

In a *bypass mixing process*, a conditioned air stream is mixed with a bypass air stream that is not conditioned. The cold conditioned air is denoted by subscript cc, the heated air ch, and the bypass air by.

Equations (3.14) and (3.17) can still be used but subscript 1 should be replaced by cc or ch and subscript 2 by "by" (bypass).

Let $K_{cc} = \dot{m}_{cc}/\dot{m}_m$ and $K_{ch} = \dot{m}_{ch}/\dot{m}_m$; then the cooling coil's load q_{cc} and heating coil's load q_{ch}, both in Btu/hr, for a bypass mixing process are

$$q_{cc} = K_{cc} \dot{V}_s \rho_s (h_m - h_{cc})$$

$$q_{ch} = K_{ch} \dot{V}_s \rho_s (h_2 - h_1) \tag{3.18}$$

In Equation (3.18), subscript s denotes the supply air and m the mixture air stream.

Air-Conditioning Cycle and Operating Modes

An *air-conditioning cycle* comprises several air-conditioning processes that are connected in a sequential order. An air-conditioning cycle determines the operating performance of the air system in an air-conditioning system. The *working substance* to condition air may be chilled or hot water, refrigerant, desiccant, etc.

Each type of air system has its own air-conditioning cycle. Psychrometric analysis of an air-conditioning cycle is an important tool in determining its operating characteristics and the state of moist air at various system components, including the volume flow rate of supply air, the coil's load, and the humidifying and dehumidifying capacity.

According to the cycle performance, air-conditioning cycles can be grouped into two categories:

- *Open cycle,* in which the moist air at its end state does not resume its original state. An air-conditioning cycle with all outdoor air is an open cycle.
- *Closed cycle,* in which moist air resumes its original state at its end state. An air-conditioning cycle that conditions the mixture of recirculating and outdoor air, supplies it, recirculates part of the return air, and mixes it again with outdoor air is a closed cycle.

Based on the outdoor weather and indoor operating conditions, the operating modes of air-conditioning cycles can be classified as:

- *Summer mode*: when outdoor and indoor operating parameters are in summer conditions.
- *Winter mode*: when outdoor and indoor operating parameters are in winter conditions.
- *Air economizer mode*: when all outdoor air or an amount of outdoor air that exceeds the minimum amount of outdoor air required for the occupants is taken into the AHU or PU for cooling. The air economizer mode saves energy use for refrigeration.

Continuous modes operate 24 hr a day and 7 days a week. Examples are systems that serve hospital wards and refrigerated warehouses. An *intermittently operated mode* usually shuts down once or several times within a 24-hr operating cycle. Such systems serve offices, class rooms, retail stores, etc. The 24-hr day-and-night cycle of an intermittently operated system can again be divided into:

1. *Cool-down or warm-up period.* When the space is not occupied and the space air temperature is higher or lower than the predetermined value, the space air should be cooled down or warmed up before the space is occupied.
2. *Conditioning period.* The air-conditioning system is operated during the occupied period to maintain the required indoor environment.
3. *Nighttime shut-down period.* The air system or terminal is shut down or only partly operating to maintain a set-back temperature.

Summer, winter, air economizer, and continuously operating modes consist of *full-load* (design load) and part-load operations. *Part load* occurs when the system load is less than the design load. The capacity of the equipment is selected to meet summer and winter system design loads as well as system loads in all operating modes.

Basic Air-Conditioning Cycle — Summer Mode

A *basic air-conditioning system* is a packaged system of supply air at a constant volume flow rate, serving a single zone, equipped with only a single supply/return duct. A *single zone* is a conditioned space for which a single controller is used to maintain a unique indoor operating parameter, probably indoor temperature. A *basic air-conditioning cycle* is the operating cycle of a basic air-conditioning system. Figure 1.3 shows a basic air-conditioning system. Figure 3.4 shows the basic air-conditioning cycle of this system. In summer mode at design load, recirculating air from the conditioned space, a worship hall, enters the packaged unit through the return grill at point ru. It is mixed with the required minimum amount of outdoor air at point o for acceptable indoor air quality and energy saving. The mixture m is then cooled and dehumidified to point cc at the DX coil, and the conditioned air is supplied to the hall through the supply fan, supply duct, and ceiling diffuser. Supply air then absorbs the sensible and latent load from the space, becoming the space air r. Recirculating air enters the packaged unit again and forms a closed cycle. *Return air* is the air returned from the space. Part of the return air is exhausted to balance the outdoor air intake and infiltration. The remaining part is the *recirculating air* that enters the PU or AHU.

The summer mode operating cycle consists of the following processes:

1. Sensible heating process, represented by line r ru, due to the return system gain $q_{r.s}$, in Btu/hr, when recirculating air flows through the return duct, ceiling plenum, and return fan, if any. In this packaged system, the return system heat gain is small and neglected.
2. Adiabatic mixing process of recirculating air at point ru and outdoor air at point o in the mixing box, represented by line ru m o.
3. Cooling and dehumidifying process m cc at the DX coil whose coil load determines the cooling capacity of the system calculated from Equation (3.11).
4. Sensible heating process related to the supply system heat gain $q_{s.s}$, in Btu/hr, represented by line cc sf s. $q_{s.s}$ consists of the fan power heat gain q_{sf}, line cc sf, and duct heat gain q_{sd}, line sf s, that is:

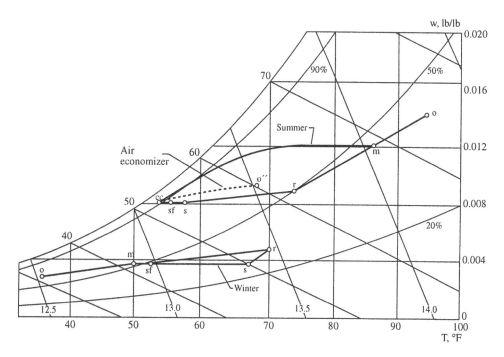

FIGURE 3.4 Basic air-conditioning cycle — summer, winter, and air economizer mode.

$$q_{s.s} = q_{sf} + q_{sd} = \mathring{V}_s \rho_s c_{pa} \Delta T_{s.s} \tag{3.19}$$

It is more convenient to use the temperature rise of the supply system $\Delta T_{s.s}$ in psychrometric analysis.

5. Supply conditioning process line sr.

Design Supply Volume Flow Rate

Design supply volume flow rate and cooling and heating capacities are primary characteristics of an air-conditioning system. Design supply volume flow rate is used to determine the size of fans, grills, outlets, air-handling units, and packaged units. For most comfort systems and many processing air-conditioning systems, *design supply volume flow rate* $\mathring{V}_{s.d}$, in cfm, is calculated on the basis of the capacity to remove the space cooling load at summer design conditions to maintain a required space temperature T_r:

$$\mathring{V}_{s.d} = q_{rc.d} \big/ \left[60\rho_s (h_r - h_s)\right] = q_{rs.d} \big/ \left[60\rho_s c_{pa} (T_r - T_s)\right] \tag{3.20}$$

where $q_{rc.d}$, $q_{rs.d}$ = design space cooling load and design sensible cooling load, Btu/hr. In Equation (3.20), the greater the $q_{rs.d}$, the higher $\mathring{V}_s$ will be. Specific heat c_{pa} is usually considered constant. Air density ρ_s may vary with the various types of air systems used. A greater ρ_s means a smaller $\mathring{V}_{s.d}$ for a given supply mass flow rate. For a given $q_{rs.d}$, the supply temperature differential $\Delta T_s = (T_r - T_s)$ is an important parameter that affects $\mathring{V}_{s.d}$. Conventionally, a 15 to 20°F ΔT_s is used for comfort air-conditioning systems. Recently, a 28 to 34°F ΔT_s has been adopted for cold air distribution in ice-storage central systems. When ΔT_s has a nearly twofold increase, there is a considerable reduction in $\mathring{V}_s$ and fan energy use and saving in investment on ducts, terminals, and outlets.

The summer cooling load is often greater than the winter heating load, and this is why q_{rc} or $q_{rs.d}$ is used to determine $\mathring{V}_{s.d}$, except in locations where the outdoor climate is very cold.

Sometimes the supply volume flow rate may be determined from the following requirements:

- To dilute the concentration of air contaminants in the conditioned space C_i, in mg/m³, the design supply volume flow rate is

$$\mathring{V}_{s.d} = 2118\,\mathring{m}_{par}\big/\left(C_i - C_s\right) \qquad (3.21)$$

where C_s = concentration of air contaminants in supply air, mg/m³

$\mathring{m}_{par}$ = rate of contaminant generation in the space, mg/sec

- To maintain a required space relative humidity φ_r and a humidity ratio w_r at a specific temperature, the design supply volume flow rate is

$$\mathring{V}_{s.d} = q_{rl.d}\big/\left[60\rho_s\left(w_r - w_s\right)h_{fg.58}\right] \qquad (3.22)$$

where $q_{rl.d}$ = design space latent load, Btu/hr.

- To provide a required air velocity v_r, in fpm, within the working area of a clean room, the supply volume flow rate is given by

$$\mathring{V}_{s.d} = A_r v_r \qquad (3.23a)$$

where A_r = cross-sectional area perpendicular to the air flow in the working area, ft².

- To exceed the outdoor air requirement for acceptable air quality for occupants, the supply volume flow rate must be equal to or greater than

$$\mathring{V}_s \geq n\,\mathring{V}_{oc} \qquad (3.23b)$$

where n = number of occupants

$\mathring{V}_{oc}$ = outdoor air requirement per person, cfm/person

- To exceed the sum of the volume flow rate of exhaust air $\mathring{V}_{ex}$ and the exfiltrated or relief air $\mathring{V}_{ef}$, both in cfm,

$$\mathring{V}_s \geq \mathring{V}_{ex} + \mathring{V}_{ef} \qquad (3.24)$$

The design supply volume flow rate should be the largest of any of the foregoing requirements.

Rated Supply Volume Flow Rate

For an air system at atmospheric pressure, since the required mass flow rate of the supply air is a function of air density and remains constant along the air flow,

$$\mathring{m}_a = \mathring{V}_{cc}\,\rho_{cc} = \mathring{V}_s\,\rho_s = \mathring{V}_{sf}\,\rho_{sf}$$

$$\mathring{V}_{sf} = \mathring{V}_s\,\rho_s\big/\rho_{sf} \qquad (3.25)$$

where $\mathring{V}_{sf}$ = volume flow rate at supply fan outlet, cfm

ρ_{sf} = air density at supply fan outlet, lb/ft³

A supply fan is rated at *standard air conditions*, that is, dry air at a temperature of 70°F, an atmospheric pressure of 29.92 in. Hg (14.697 psia), and an air density of 0.075 lb/ft³. However, a fan is a constant-volume machine at a given fan size and speed; that is, $\mathring{V}_{sf} = \mathring{V}_{sf.r}$. Here $\mathring{V}_{sf.r}$ represents the rated volume flow rate of a fan at standard air conditions. Therefore,

$$\mathring{V}_{sf.r} = \mathring{V}_{sf} = q_{rs.d} \Big/ \Big[60 \rho_{sf} c_{pa} (T_r - T_s) \Big] \tag{3.26}$$

- For conditioned air leaving the cooling coil at $T_{cc} = 55°F$ with a relative humidity of 92% and T_{sf} of 57°F, $\rho_{sf.r} = 1/v_{sf} = 1/13.20 = 0.0758$ lb/ft³. From Equation (3.26):

$$\mathring{V}_{sf.r} = q_{rs.d} \Big/ \Big[60 \times 0.0758 \times 0.243 (T_r - T_s) \Big] = q_{rs.d} \Big/ \Big[1.1 (T_r - T_s) \Big] \tag{3.26a}$$

Equation (3.26a) is widely used in calculating the supply volume flow rate of comfort air-conditioning systems.

- For cold air distribution, $T_{cc} = 40°F$ and $\varphi_{cc} = 98\%$, if $T_{sf} = 42°F$, then $v_{sf} = 12.80$ ft³/lb, and the rated supply volume flow rate:

$$\mathring{V}_{sf.r} = 12.80 q_{rs.d} \Big/ \Big[60 \times 0.243 (T_r - T_s) \Big] = q_{rs.d} \Big/ \Big[1.14 (T_r - T_s) \Big] \tag{3.26b}$$

- For a blow-through fan in which the fan is located upstream of the coil, if $T_{sf} = 82°F$ and $\varphi_{sf} = 43\%$, then $v_{sf} = 13.87$ ft³/lb, and the rated supply volume flow rate:

$$\mathring{V}_{sf.r} = 13.87 q_{rs.d} \Big/ \Big[60 \times 0.243 (T_r - T_s) \Big] = q_{rs.d} \Big/ \Big[1.05 (T_r - T_s) \Big] \tag{3.26c}$$

Effect of the Altitude

The higher the altitude, the lower the atmospheric pressure and the air density. In order to provide the required mass flow rate of supply air, a greater $\mathring{V}_{sf.r}$ is needed. For an air temperature of 70°F:

$$\mathring{V}_{x.ft} = \mathring{V}_{sf.r} (p_{sea}/p_{x.ft}) = \mathring{V}_{sf.r} (\rho_{sea}/\rho_{x.ft}) \tag{3.27}$$

where $\mathring{V}_{x.ft}$ = supply volume flow rate at an altitude of x ft, cfm

$p_{sea}, p_{x.ft}$ = atmospheric pressure at sea level and an altitude of x ft, psia

$\rho_{sea}, \rho_{x.ft}$ = air density at sea level and an altitude of x ft, psia

Following are the pressure or air density ratios at various altitudes. At 2000 ft above sea level, the rated supply volume flow rate $\mathring{V}_{r.2000} = \mathring{V}_{sf.r} (p_{sea}/p_{x.ft}) = 1.076 \mathring{V}_{sf.r}$ cfm instead of $\mathring{V}_{sf.r}$ cfm at sea level.

Altitude, ft	p_{at}, psia	ρ, lb/ft³	$(p_{sea}/p_{x.ft})$
0	14.697	0.075	1.000
1000	14.19	0.0722	1.039
2000	13.58	0.0697	1.076
3000	13.20	0.0672	1.116
5000	12.23	0.0625	1.200

Off-Coil and Supply Air Temperature

For a given design indoor air temperature T_r, space sensible cooling load q_{rs}, and supply system heat gain $q_{s.s}$, a lower air off-coil temperature T_{cc} as well as supply temperature T_s means a greater supply temperature differential ΔT_s and a lower space relative humidity φ_r and vice versa. A greater ΔT_s decreases the supply volume flow rate $\dot{V}_s$ and then the fan and terminal sizes, duct sizes, and fan energy use. The result is a lower investment and energy cost.

A lower T_{cc} and a greater ΔT_s require a lower chilled water temperature entering the coil T_{we}, a lower evaporating temperature T_{ev} in the DX coil or refrigerating plant, and therefore a greater power input to the refrigerating compressors. When an air-conditioning system serves a conditioned space of a single zone, optimum T_{cc}, T_s, and T_{we} can be selected. For a conditioned space of multizones, T_{cc}, T_s, and T_{we} should be selected to satisfy the lowest requirement. In practice, T_s and T_{we} are often determined according to previous experience with similar projects.

In general, the temperature rise due to the supply fan power system heat gain q_{sf} can be taken as 1 to 3°F depending on the fan total pressure. The temperature rise due to the supply duct system heat gain at design flow can be estimated as 1°F/100 ft insulated main duct length based on 1-in. thickness of duct insulation.

Outside Surface Condensation

The outside surface temperature of the ducts, terminals, and supply outlets T_{sur} in the ceiling plenum in contact with the return air should not be lower than the dew point of the space air T_r'' in °F. The temperature rise due to the fan power heat gain is about 2°F. According to Dorgan (1988), the temperature difference between the conditioned air inside the terminal and the outside surface of the terminal with insulation wrap is about 3°F. For a space air temperature of 75°F and a relative humidity of 50%, its dew point temperature is 55°F. If the outside surface temperature $T_s = (T_{cc} + 2 + 3) \leq 55°F$, condensation may occur on the outside surface of the terminal. Three methods are often used to prevent condensation:

1. Increase the thickness of the insulation layer on the outside surface.
2. Adopt a supply outlet that induces more space air.
3. Equip with a terminal that mixes the supply air with the space air or air from the ceiling plenum.

During the cool-down period, due to the high dew point temperature of the plenum air when the air system is started, the supply air temperature must be controlled to prevent condensation.

Example 3.1

The worship hall of a church uses a package system with a basic air system. The summer space sensible cooling load is 75,000 Btu/hr with a latent load of 15,000 Btu/hr. Other design data for summer are as follows:

Outdoor summer design temperature: dry bulb 95°F and wet bulb 75°F
Summer indoor temperature: 75°F with a space relative humidity of 50%:
Temperature rise: fan power 2°F
supply duct 2°F
Relative humidity of air leaving cooling coil: 93%
Outdoor air requirement: 1800 cfm

Determine the

1. Temperature of supply air at summer design conditions
2. Rated volume flow rate of the supply fan
3. Cooling coil load
4. Possibility of condensation at the outside surface of the insulated branch duct to the supply outlet

Solution

1. From Equation 3.1 the sensible heat ratio of the space conditioning line is

$$\text{SHR}_s = |q_{rs}| / (|q_{rs}| + |q_{rl}|) = 60,000 / (60,000 + 15,000) = 0.8$$

On the psychrometric chart, from given $T_r = 75°F$ and $\varphi_r = 50\%$, plot space point r. Draw a space conditioning line sr from point r with $SHR_s = 0.8$.

Since $\Delta T_{s.s} = 2 + 2 = 4°F$, move line segment cc s (4°F) up and down until point s lies on line sr and point cc lies on the $\varphi_{cc} = 93\%$ line. The state points s and cc are then determined as shown in Figure 3.4:

$T_s = 57.5°F$, $\varphi_s = 82\%$. and $w_s = 0.0082$ lb/lb
$T_{cc} = 53.5°F$, $\varphi_{cc} = 93\%$, $h_{cc} = 21.8$ Btu/lb, and $w_{cc} = 0.0082$ lb/lb

2. Since $T_{sf} = 53.5 + 2 = 55.5°F$ and $w_{sf} = 0.0082$ lb/lb, $\rho_{sf} = 1/v_{sf} = 1/13.15 = 0.076$ lb/ft³. From Equation 3.2, the required rated supply volume flow rate is

$$\dot{V}_{sf.r} = q_{rs.d} / \left[60 \rho_{sf} c_{pa} (T_r - T_s) \right]$$

$$= 60,000 / \left[60 \times 0.076 \times 0.243 (75 - 57.5) \right] = 3094 \text{ cfm}$$

3. Plot outdoor air state point o on the psychrometric chart from given dry bulb 95°F and wet bulb 75°F. Connect line ro. Neglect the density differences between points r, m, and o; then

$$rm/ro = 1800/3094 = 0.58$$

From the psychrometric chart, the length of line ro is 2.438 in. As shown in Figure 3.4, point m is then determined as:

$$T_m = 86.7°F, \qquad h_m = 35 \text{ Btu/lb}$$

From Equation (3.11), the cooling coil load is

$$q_{cc} = 60 \dot{V}_s \rho_s (h_m - h_{cc}) = 60 \times 3094 \times 0.076 (35 - 21.8) = 186,234 \text{ Btu/lb}$$

4. From the psychrometric chart, since the dew point of the space air $T_r'' = 55°F$ and is equal to that of the plenum air, the outside surface temperature of the branch duct $T_s = 53.5 + 2 + 3 = 58°F$ which is higher than $T_r'' = 55°F$. Condensation will not occur at the outside surface of the branch duct.

Basic Air-Conditioning Cycle — Winter Mode

When the basic air-conditioning systems are operated in winter mode, their air-conditioning cycles can be classified into the following four categories:

Cold Air Supply without Space Humidity Control. In winter, for a fully occupied worship hall, if the heat loss is less than the space sensible cooling load, a cold air supply is required to offset the space sensible cooling load and maintain a desirable indoor environment as shown by the lower cycle in Figure 3.4. Usually, a humidifier is not used.

The winter cycle of a cold air supply without humidity control consists of the following air-conditioning processes:

1. Adiabatic mixing process of outdoor air and recirculating air o m r.
2. Sensible heating process due to supply fan power heat gain m sf. Because of the smaller temperature difference between the air in the ceiling plenum and the supply air inside the supply duct, heat transfer through duct wall in winter can be neglected.
3. Supply conditioning line sr.

For a winter-mode basic air-conditioning cycle with a cold air supply without space humidity control, the space relative humidity depends on the space latent load, the humidity ratio of the outdoor air, and the amount of outdoor air intake. In order to determine the space humidity ratio w_r, in lb/lb, and the space relative humidity φ_r, in %, Equations (3.15) and (3.22) should be used to give the following relationships:

$$(w_r - w_m)/(w_r - w_o) = \mathring{V}_o / \mathring{V}_s$$

$$(w_r - w_s) = q_{rl} / \left(60 \mathring{V}_s \rho_s h_{fg\,58} \right)$$

(3.28)

$$w_s = w_m$$

For a cold air supply, if there is a high space sensible cooling load, the amount of outdoor air must be sufficient, and the mixture must be cold enough to satisfy the following relationships:

$$(T_r - T_s) = q_{rs} / \left(60 \mathring{V}_s \rho_s c_{pa} \right)$$

(3.29)

$$(T_r - T_s)/(T_r - T_o) = \mathring{V}_o / \mathring{V}_s$$

The heating coil load for heating of the outdoor air can be calculated using Equation (3.7).

Example 3.2

For the same packaged air-conditioning system using a basic air system to serve the worship hall in a church as in Example 3.1, the space heating load at winter design condition is 10,000 Btu/hr and the latent load is 12,000 Btu/hr. Other winter design data are as follows:

Winter outdoor design temperature	35°F
Winter outdoor design humidity ratio	0.00035 lb/lb
Winter indoor design temperature	70°F
Temperature rise due to supply fan heat gain	2°F
Outdoor air requirement	1800 cfm

Determine (1) the space relative humidity at winter design temperature and (2) the heating coil load.

Solution

1. Assume that the supply air density $\rho_{sf} = 1/v_{sf} = 1/13.0 = 0.0769$ lb/ft³, and the mass flow rate of the supply air is the same as in summer mode. Then from Equation 3.28 the humidity ratio difference is

$$(w_r - w_s) = q_{rl} / \left(60 \mathring{V}_{sf.r} \rho_{sf} h_{fg\,58} \right) = 12{,}000/(60 \times 3094 \times 0.0769 \times 1060) = 0.00079 \text{ lb/lb}$$

From Equation 3.29, the supply air temperature differential is

$$(T_r - T_s) = q_{rs.d} / \left(60 \mathring{V}_{sf.r} \rho_{sf} c_{pa} \right) = 10{,}000/(60 \times 3094 \times 0.0769 \times 0.243) = 2.88°F$$

Since $\mathring{V}_o / \mathring{V}_s = 1800/3094 = 0.58$ and $w_s = w_m$,

$$\left(w_r - w_s\right)/\left(w_r - w_o\right) = 0.00079/\left(w_r - w_o\right) = \mathring{V}_o/\mathring{V}_s = 0.58$$

$$\left(w_r - w_o\right) = 0.00079/0.58 = 0.00136 \text{ lb/lb}$$

And from given information,

$$w_r = 0.00136 + w_o = 0.00136 + 0.0035 = 0.00486 \text{ lb/lb}$$

From the psychrometric chart, for $T_r = 70°F$ and $w_r = 0.00486$ lb/lb, point r can be plotted, and φ_r is about 32% (see Figure 3.4).

2. Since mr/or = 0.58, point m can be determined, and from the psychrometric chart $T_m = 50.0°F$. As $T_s = 70 - 2.88 = 67.12°F$ and $T_{sf} = T_m + 2 = 50.0 + 2 = 52.0°F$, from Equation 3.7 the heating coil's load is

$$q_{ch} = 60\mathring{V}_s \rho_s c_{pa}\left(T_s - T_{sf}\right) = 60 \times 3094 \times 0.0769 \times 0.243(67.12 - 52.0) = 52{,}451 \text{ Btu/hr}$$

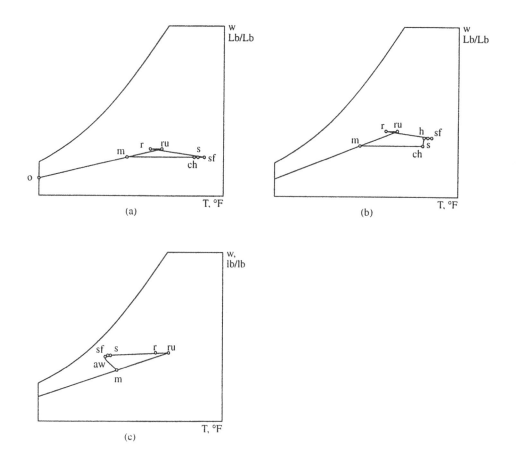

FIGURE 3.5 Basic air-conditioning cycle — winter modes: (a) warm air supply without space humidity control, (b) cold air supply without space humidity control, and (c) cold air supply with space humidity control. ch = air leaving heating coil, h = air leaving humidifer, and aw = air leaving air washer.

Warm Air Supply without Space Humidity Control

When the sum of space heat losses is greater than the sum of the internal heat gains in winter, a warm air supply is needed. For many comfort systems such as those in offices and stores, in locations where winter is not very cold, humidification is usually not necessary. The basic air-conditioning cycle for a warm air supply without space humidity control is shown in Figure 3.5(a). This cycle is similar to the winter mode cycle of a cold air supply without space humidity control shown in Figure 3.4 except that the supply air temperature is higher than space temperature, that is, $T_s > T_r$. To prevent stratification, with the warm supply air staying at a higher level, $(T_s - T_r) > 20°F$ is not recommended.

Warm Air Supply with Space Humidity Control

This operating cycle (see Figure 3.5[b]) is often used for hospitals, nurseries, etc. or in locations where winter is very cold. The state point of supply air must be determined first by drawing a space conditioning line with known SHR_s and then from the calculated supply temperature differential ΔT_s. The difference in humidity ratio $(w_s - w_{ch})$ is the water vapor must be added at the humidifier. Humidifying capacity can be calculated from Equation 3.8.

Cold Air Supply with Space Humidity Control

This operating cycle (shown in Figure 3.5[c]) is widely used in industrial applications such as textile mills where a cold air supply is needed to remove machine load in winter and maintains the space relative humidity required for the manufacturing process. An outdoor air and recirculating air mixture is often used for the required cold air supply. An air washer is adopted for winter humidification.

Air Economizer Mode

In the *air economizer* mode, as shown by the middle dotted line cycle o″-cc-sf-s-r in Figure 3.4, all outdoor air or an outdoor air-recirculating air mixture is used to reduce the refrigeration capacity and improve the indoor air quality during spring, fall, or winter.

When all outdoor air is admitted, it is an open cycle. Outdoor air is cooled and often dehumidified to point cc. After absorbing fan and duct heat gains, it is supplied to the conditioned space. Space air is exhausted entirely through openings, relief dampers, or relief/exhaust fans to the outside. An all-outdoor air-operating mode before the space is occupied is often called an *air purge* operation, used to dilute space air contaminants.

Cool-Down and Warm-Up Modes

In summer, when an air system is shut down during an unoccupied period at night, the space temperature and relative humidity often tend to increase because of infiltration of hot and humid air and heat transfer through the building envelope. The air system is usually started before the space is occupied in cool-down mode to cool the space air until the space temperature falls within predetermined limits.

In winter, the air system is also started before the occupied period to warm up the space air to compensate for the nighttime space temperature setback to 55 to 60°F for energy saving or the drop of space temperature due to heat loss and infiltration.

If dilution of indoor air contaminants is not necessary, only recirculating space air is used during cool-down or warm-up periods in order to save energy.

4

Refrigerants and Refrigeration Cycles

Shan K. Wang

Consultant

4   Refrigerants and Refrigeration Cycles 35
     Refrigeration and Refrigeration Systems • Refrigerants,
     Cooling Mediums, and Absorbents • Classification of
     Refrigerants • Required Properties of Refrigerants • Ideal
     Single-Stage Vapor Compression Cycle • Coefficient of
     Performance of Refrigeration Cycle • Subcooling and
     Superheating • Refrigeration Cycle of Two-Stage Compound
     Systems with a Flash Cooler • Cascade System Characteristics

Refrigerants and Refrigeration Cycles

Refrigeration and Refrigeration Systems

Refrigeration is the cooling effect of the process of extracting heat from a lower temperature heat source, a substance or cooling medium, and transferring it to a higher temperature heat sink, probably atmospheric air and surface water, to maintain the temperature of the heat source below that of the surroundings.

A *refrigeration system* is a combination of components, equipment, and piping, connected in a sequential order to produce the refrigeration effect. Refrigeration systems that provide cooling for air-conditioning are classified mainly into the following categories:

1. *Vapor compression systems.* In these systems, a compressor(s) compresses the refrigerant to a higher pressure and temperature from an evaporated vapor at low pressure and temperature. The compressed refrigerant is condensed into liquid form by releasing the latent heat of condensation to the condenser water. Liquid refrigerant is then throttled to a low-pressure, low-temperature vapor, producing the refrigeration effect during evaporation. Vapor compression is often called *mechanical refrigeration,* that is, refrigeration by mechanical compression.

2. *Absorption systems.* In an absorption system, the refrigeration effect is produced by means of thermal energy input. After liquid refrigerant produces refrigeration during evaporation at very low pressure, the vapor is absorbed by an aqueous absorbent. The solution is heated by a direct-fired gas furnace or waste heat, and the refrigerant is again vaporized and then condensed into

liquid form. The liquid refrigerant is throttled to a very low pressure and is ready to produce the refrigeration effect again.

3. *Gas expansion systems*. In an air or other gas expansion system, air or gas is compressed to a high pressure by compressors. It is then cooled by surface water or atmospheric air and expanded to a low pressure. Because the temperature of air or gas decreases during expansion, a refrigeration effect is produced.

Refrigerants, Cooling Mediums, and Absorbents

A *refrigerant* is a primary working fluid used to produce refrigeration in a refrigeration system. All refrigerants extract heat at low temperature and low pressure during evaporation and reject heat at high temperature and pressure during condensation.

A *cooling medium* is a working fluid cooled by the refrigerant during evaporation to transport refrigeration from a central plant to remote cooling equipment and terminals. In a large, centralized air-conditioning system, it is more economical to pump the cooling medium to the remote locations where cooling is required. Chilled water and brine are cooling media. They are often called secondary refrigerants to distinguish them from the primary refrigerants.

A *liquid absorbent* is a working fluid used to absorb the vaporized refrigerant (water) after evaporation in an absorption refrigeration system. The solution that contains the absorbed vapor is then heated. The refrigerant vaporizes, and the solution is restored to its original concentration to absorb water vapor again.

A numbering system for refrigerants was developed for hydrocarbons and halocarbons. According to ANSI/ASHRAE Standard 34-1992, the first digit is the number of unsaturated carbon–carbon bonds in the compound. This digit is omitted if the number is zero. The second digit is the number of carbon atoms minus one. This is also omitted if the number is zero. The third digit denotes the number of hydrogen atoms plus one. The last digit indicates the number of fluorine atoms. For example, the chemical formula for refrigerant R-123 is $CHCl_2CF_3$. In this compound:

No unsaturated carbon–carbon bonds, first digit is   0
There are two carbon atoms, second digit is $2 - 1 = 1$
There is one hydrogen atom, third digit is $1 + 1 = 2$
There are three fluorine atoms, last digit is   3

To compare the relative ozone depletion of various refrigerants, an index called the *ozone depletion potential* (ODP) has been introduced. ODP is defined as the ratio of the rate of ozone depletion of 1 lb of any halocarbon to that of 1 lb of refrigerant R-11. For R-11, ODP = 1.

Similar to the ODP, halocarbon global warming potential (HGWP) is an index used to compare the global warming effect of a halocarbon refrigerant with the effect of refrigerant R-11.

Classification of Refrigerants

Nontoxic and nonflammable synthetic chemical compounds called *halogenated hydrocarbons*, or simply *halocarbons*, were used almost exclusively in vapor compression refrigeration systems for comfort air-conditioning until 1986. Because chlorofluorocarbons (CFCs) cause ozone depletion and global warming, they must be replaced. A classification of refrigerants based on ozone depletion follows (see Table 4.1):

Hydrofluorocarbons (HFCs)

HFCs contain only hydrogen, fluorine, and carbon atoms and cause no ozone depletion. HFCs group include R-134a, R-32, R-125, and R-245ca.

HFC's Azeotropic Blends or Simply HFC's Azeotropic

An azeotropic is a mixture of multiple components of volatilities (refrigerants) that evaporate and condense as a single substance and do not change in volumetric composition or saturation temperature when they evaporate or condense at constant pressure. HFC's azeotropics are blends of refrigerant with HFCs. ASHRAE assigned numbers between 500 and 599 for azeotropic. HFC's azeotropic R-507, a blend of R-125/R-143, is the commonly used refrigerant for low-temperature vapor compression refrigeration systems.

HFC's Near Azeotropic

A near azeotropic is a mixture of refrigerants whose characteristics are near those of an azeotropic. Because the change in volumetric composition or saturation temperature is rather small for a near azeotropic, such as, 1 to 2°F, it is thus named. ASHRAE assigned numbers between 400 and 499 for zeotropic. R-404A (R-125/R-134a/R-143a) and R-407B (R-32/R-125/R134a) are HFC's near azeotropic. R-32 is flammable; therefore, its composition is usually less than 30% in the mixture. HFC's near azeotropic are widely used for vapor compression refrigeration systems.

Zeotropic or nonazeotropic, including near azeotropic, shows a change in composition due to the difference between liquid and vapor phases, leaks, and the difference between charge and circulation. A shift in composition causes the change in evaporating and condensing temperature/pressure. The difference in dew point and bubble point during evaporation and condensation is called glide, expressed in °F. Near azeotropic has a smaller glide than zeotropic. The midpoint between the dew point and bubble point is often taken as the evaporating and condensing temperature for refrigerant blends.

Hydrochlorofluorocarbons (HCFCs) and Their Zeotropics

HCFCs contain hydrogen, chlorine, fluorine, and carbon atoms and are not fully halogenated. HCFCs have a much shorter lifetime in the atmosphere (in decades) than CFCs and cause far less ozone depletion (ODP 0.02 to 0.1). R-22, R-123, R-124, etc. are HCFCs. HCFCs are the most widely used refrigerants today.

HCFC's near azeotropic and *HCFC's zeotropic* are blends of HCFCs with HFCs. They are transitional or interim refrigerants and are scheduled for a restriction in production starting in 2004.

Inorganic Compounds

These compounds include refrigerants used before 1931, like ammonia R-717, water R-718, and air R-729. They are still in use because they do not deplete the ozone layer. Because ammonia is toxic and flammable, it is used in industrial applications. Inorganic compounds are assigned numbers between 700 and 799 by ASHRAE.

Chlorofluorocarbons, Halons, and Their Azeotropic

CFCs contain only chlorine, fluorine, and carbon atoms. CFCs have an atmospheric lifetime of centuries and cause ozone depletion (ODP from 0.6 to 1). R-11, R-12, R-113, R-114, R-115... are all CFCs.

Halons or BFCs contain bromide, fluorine, and carbon atoms. R-13B1 and R-12B1 are BFCs. They cause very high ozone depletion (ODP for R-13B1 = 10). Until 1995, R-13B1 was used for very low temperature vapor compression refrigeration systems.

Phaseout of CFCs, BFCs, HCFCs, and Their Blends

On September 16, 1987, the European Economic Community and 24 nations, including the United States, signed a document called the Montreal Protocol. It is an agreement to restrict the production and consumption of CFCs and BFCs in the 1990s because of ozone depletion.

The Clean Air Act amendments, signed into law in the United States on November 15, 1990, concern two important issues: the phaseout of CFCs and the prohibition of deliberate venting of CFCs and HCFCs.

In February 1992, President Bush called for an accelerated ban of CFCs in the United States. In late November 1992, representatives of 93 nations meeting in Copenhagen agreed to phase out CFCs beginning January 1, 1996. Restriction on the use of HCFCs will start in 2004, with a complete phaseout by 2030.

In the earlier 1990s, R-11 was widely used for centrifugal chillers, R-12 for small and medium-size vapor compression systems, R-22 for all vapor compression systems, and CFC/HCFC blend R-502 for

TABLE 4.1 Properties of Commonly Used Refrigerants 40°F Evaporating and 100°F Condensing

	Chemical Formula	Molecular Mass	Ozone Depletion Potential (ODP)	Global Warming Potential (HGWP)	Evaporating Pressure, psia	Condensing Pressure, psia	Compression Ratio	Refrigeration Effect, Btu/lb
Hydrofluorocarbons HFCs								
R-32 Difluoromethane	CH_2F_2	52.02	0.0	0.14	135.6	340.2	2.51	37.1
R-125 Pentafluoroethane	CHF_2CF_3	120.03	0.0	0.84	111.9	276.2	2.47	65.2
R-134a Tetrafluoroethane	CF_3CH_2F	102.03	0.0	0.26	49.7	138.8	2.79	
R-143a Trifluoroethane	CH_3CF_3	84.0	0.0					
R-152a Difluoroethane	CH_3CHF_2	66.05	0.0		44.8	124.3	2.77	
R-245ca Pentafluoropropane	$CF_3CF_2CH_3$	134.1	0.0					
HFC's azeotropics								
R-507 R-125/R-143 (45/55)			0.0	0.98				
HFC's near azeotropic								
R-404A R-125/R-143a (44/52/4)			0.0	0.94				
R-407A R-32/R-125/R-134a (20/40/40)			0.0	0.49				
R-407C R-32/R-125/R-134a (23/25/52)			0.0	0.70				
Hydrochlorofluorocarbons HCFCs and their azeotropics								
R-22 Chlorodifluoromethane	$CHClF_2$	86.48	0.05	0.40	82.09	201.5	2.46	69.0
R-123 Dichlorotrifluoroethane	$CHCl_2CF_3$	152.93	0.02	0.02	5.8	20.8	3.59	62.9
R-124 Chlorotetrafluoroethane	$CHFClCF_3$	136.47	0.02		27.9	80.92	2.90	5.21
HCFC's near azeotropics								
R-402A R-22/R-125/R-290 (38/60/2)			0.02	0.63				
HCFC's azeotropics								
R-401A R-22/R-124/R-152a (53/34/13)			0.37	0.22				
R-401B R-22/R-124/R-152a (61/28/11)			0.04	0.24				

TABLE 4.1 Properties of Commonly Used Refrigerants 40°F Evaporating and 100°F Condensing (continued)

	Chemical Formula	Molecular Mass	Ozone Depletion Potential (ODP)	Global Warming Potential (HGWP)	Evaporating Pressure, psia	Condensing Pressure, psia	Compression Ratio	Refrigeration Effect, Btu/lb
Inorganic compounds								
R-717 Ammonia	NH_3	17.03	0	0	71.95	206.81	2.87	467.4
R-718 Water	H_2O	18.02	0					
R-729 Air		28.97	0					
Chlorofluorocarbons CFCs, halons BFCs and their azeotropic								
R-11 Trichlorofluoromethane	CCl_3F	137.38	1.00	1.00	6.92	23.06	3.33	68.5
R-12 Dichlorodifluoromethane	CCl_2F_2	120.93	1.00	3.20	50.98	129.19	2.53	50.5
R-13B1 Bromotrifluoromethane	$CBrF_3$	148.93	10					
R-113 Trichlorotrifluoroethane	CCl_2FCClF_2	187.39	0.80	1.4	2.64	10.21	3.87	54.1
R-114 Dichlorotetrafluoroethane	CCl_2FCF_3	170.94	1.00	3.9	14.88	45.11	3.03	42.5
R-12/R-152a (73.8/26.2)		99.31			59.87	152.77	2.55	60.5
R-22/R-115 (48.8/51.2)		111.63	0.283	4.10				

TABLE 4.1 Properties of Commonly Used Refrigerants 40°F Evaporating and 100°F Condensing (continued)

Replacement of	Trade Name	Specific Volume of Vapor ft³/lb	Compressor Displacement cfm/ton	Power Consumption hp/ton	Critical Temperature °F	Discharge Temperature °F	Flammability	Safety
Hydrofluorocarbons HFCs								
R-32			0.63			173.1		Nonflammable	A1
R-125			0.33			150.9	103	Nonflammable	A1
R134a	R-12		0.95			213.9			A1
R143a									
R-152a			1.64			235.9		Lower flammable	A2
R-245ca									
HFC's azeotropics								
R-507	R-502	Genetron AZ-50							
HFC's near azeotropic								
R-404A	R-22								A1/A1[a]
R-407A	R-22	SUVA HP-62							A1/A1[a]
		KLEA 60							
R-407C	R-22	KLEA 66							A1/A1[a]
Hydrochlorofluorocarbons HCFC's and their azeotropics								
R-22			0.66	1.91	0.696	204.8	127	Nonflammable	A1
R-123	R-11		5.88	18.87	0.663	362.6		Nonflammable	B1
R-124			1.30	5.06	0.698	252.5			A1/A1[a]
HCFC's near azeotropics								
R-402A	R-502	SUVA HP-80							
HCFC's azeotropics								
R-401A	R-12	MP 39							A1/A1[a]
R-401B	R-12	MP 66							A1/A1[a]
Inorganic compounds								
R-717			3.98	1.70	0.653	271.4	207	Lower flammability	B2
R-718								Nonflammable	
R-729								Nonflammable	

TABLE 4.1 Properties of Commonly Used Refrigerants 40°F Evaporating and 100°F Condensing (continued)

	Replacement of	Trade Name	Specific Volume of Vapor ft³/lb	Compressor Displacement cfm/ton	Power Consumption hp/ton	Critical Temperature °F	Discharge Temperature °F	Flammability	Safety
Chlorofluorocarbons CFCs, halons BFCs, and their azeotropics									
R-11			5.43	15.86	0.636	388.4	104	Nonflammable	A1
R-12			5.79	3.08	0.689	233.6	100	Nonflammable	A1
R-13B1			0.21			152.6	103	Nonflammable	A1
R-113			10.71	39.55	0.71	417.4	86	Nonflammable	A1
R-114			2.03	9.57	0.738	294.3	86	Nonflammable	A1
R-500	R-12/R-152a (73.8/26.2)		0.79	3.62	0.692	221.9	105	Nonflammable	A1
R-502	R-22/R-115 (48.8/51.2)						98	Nonflammable	A1

Source: Adapted with permission from ASHRAE Handbooks 1993 Fundamentals. Also from refrigerant manufacturers.
[a] First classification is that safety classification of the formulated composition. The second is the worst case of fractionation.

low-temperature vapor compression systems. Because of the phaseout of CFCs and BFCs before 1996 and HCFCs in the early years of the next century, alternative refrigerants have been developed to replace them:

- R-123 (an HCFC of ODP = 0.02) to replace R-11 is a short-term replacement that causes a slight reduction in capacity and efficiency. R-245ca (ODP = 0) may be the long-term alternative to R-11.
- R-134a (an HFC with ODP = 0) to replace R-12 in broad applications. R-134a is not miscible with mineral oil; therefore, a synthetic lubricant of polyolester is used.
- R-404A (R-125/R-134a/143a) and R-407C (R-32/R-125/R-134a) are both HFCs near azeotropic of ODP = 0. They are long-term alternatives to R-22. For R-407C, the composition of R-32 in the mixture is usually less than 30% so that the blend will not be flammable. R-407C has a drop of only 1 to 2% in capacity compared with R-22.
- R-507 (R-125/R-143a), an HFC's azeotropic with ODP = 0, is a long-term alternative to R-502. Synthetic polyolester lubricant oil will be used for R-507. There is no major performance difference between R-507 and R-502. R-402A (R-22/R-125/R-290), an HCFC's near azeotropic, is a short-term immediate replacement, and drop-in of R-502 requires minimum change of existing equipment except for reset of a higher condensing pressure.

Required Properties of Refrigerants

A refrigerant should not cause ozone depletion. A low global warming potential is required. Additional considerations for refrigerant selection are

1. *Safety*, including toxicity and flammability. ANSI/ASHRAE Standard 34-1992 classifies the *toxicity* of refrigerants as Class A and Class B. Class A refrigerants are of low toxicity. No toxicity was identified when their time-weighted average concentration was less than or equal to 400 ppm, to which workers can be exposed for an 8-hr workday and 40-hr work week without adverse effect. Class B refrigerants are of higher toxicity and produce evidence of toxicity.

ANSI/ASHRAE Standard 34-1982 classifies the *flammability* of refrigerants as Class 1, no flame propagation; Class 2, lower flammability; and Class 3, higher flammability.

The safety classification of refrigerants is based on the combination of toxicity and flammability: A1, A2, A3, B1, B2, and B3. R-134a and R-22 are in the A1 group, lower toxicity and nonflammable; R-123 in the B1 group, higher toxicity and nonflammable; and R-717 (ammonia) in the B2 group, higher toxicity and lower flammability.

2. *Effectiveness of refrigeration cycle*. High effectiveness is a desired property. The power consumed per ton of refrigeration produced, hp/ton or kW/ton, is an index for this assessment. Table 4.1 gives values for an ideal single-stage vapor compression cycle.
3. *Oil miscibility*. Refrigerant should be miscible with mineral lubricant oil because a mixture of refrigerant and oil helps to lubricate pistons and discharge valves, bearings, and other moving parts of a compressor. Oil should also be returned from the condenser and evaporator for continuous lubrication. R-22 is partially miscible. R-134a is hardly miscible with mineral oil; therefore, synthetic lubricant of polyolester will be used.
4. *Compressor displacement*. Compressor displacement per ton of refrigeration produced, in cfm/ton, directly affects the size of the positive displacement compressor and therefore its compactness. Ammonia R-717 requires the lowest compressor displacement (1.70 cfm/ton) and R-22 the second lowest (1.91 cfm/ton).
5. Desired properties:
    - Evaporating pressure p_{ev} should be higher than atmospheric. Then noncondensable gas will not leak into the system.
    - Lower condensing pressure for lighter construction of compressor, condenser, piping, etc.

- A high thermal conductivity and therefore a high heat transfer coefficient in the evaporator and condenser.
- Dielectric constant should be compatible with air when the refrigerant is in direct contact with motor windings in hermetic compressors.
- An inert refrigerant that does not react chemically with material will avoid corrosion, erosion, or damage to system components. Halocarbons are compatible with all containment materials except magnesium alloys. Ammonia, in the presence of moisture, is corrosive to copper and brass.
- Refrigerant leakage can be easily detected. Halide torch, electronic detector, and bubble detection are often used.

Ideal Single-Stage Vapor Compression Cycle

Refrigeration Process

A refrigeration process shows the change of the thermodynamic properties of the refrigerant and the energy and work transfer between the refrigerant and surroundings.

Energy and work transfer is expressed in British thermal units per hour, or Btu/hr. Another unit in wide use is ton of refrigeration, or ton. A ton = 12,000 Btu/hr of heat removed; i.e., 1 ton of ice melting in 24 hr = 12,000 Btu/hr.

Refrigeration Cycles

When a refrigerant undergoes a series of processes like evaporation, compression, condensation, throttling, and expansion, absorbing heat from a low-temperature source and rejecting it to a higher temperature sink, it is said to have undergone a refrigeration cycle. If its final state is equal to its initial state, it is a *closed cycle*; if the final state does not equal the initial state, it is an *open cycle*. Vapor compression refrigeration cycles can be classified as singl e stage, multistage, compound, and cascade cycles.

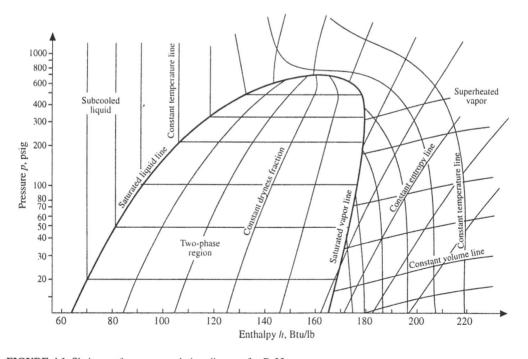

FIGURE 4.1 Skeleton of pressure-enthalpy diagram for R-22.

A *pressure-enthalpy diagram* or *p-h* diagram is often used to calculate the energy transfer and to analyze the performance of a refrigeration cycle, as shown in Figure 4.1.In a *p-h* diagram, pressure *p*, in psia or psig logarithmic scale, is the ordinate, and enthalpy *h*, in Btu/lb, is the abscissa. The saturated liquid and saturated vapor line encloses a two-phase region in which vapor and liquid coexist. The two-phase region separates the subcooling liquid and superheated vapor regions. The constant-temperature line is nearly vertical in the subcooling region, horizontal in the two-phase region, and curved down sharply in the superheated region.

In the two-phase region, a given saturated pressure determines the saturated temperature and vice versa. The constant-entropy line is curved upward to the right-hand side in the superheated region. Each kind of refrigerant has its own *p-h* diagram.

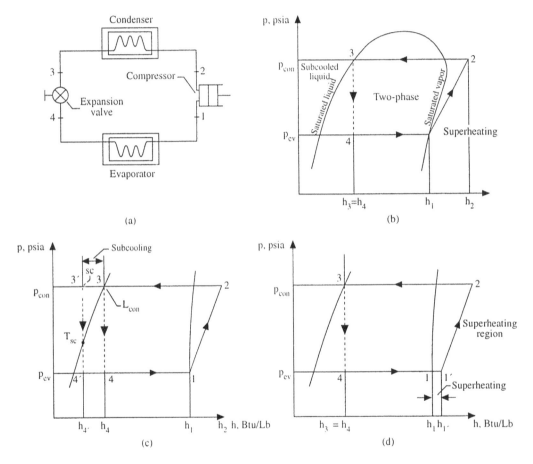

FIGURE 4.2 A single-stage ideal vapor compression refrigeration cycle: (a) schematic diagram, (b) *p-h* diagram, (c) subcooling, and (d) superheating.

Refrigeration Processes in an Ideal Single-Stage Cycle

An *ideal cycle* has isentropic compression, and pressure losses in the pipeline, valves, and other components are neglected. All refrigeration cycles covered in this section are ideal. Single stage means a single stage of compression.

There are four refrigeration processes in an ideal single-stage vapor compression cycle, as shown in Figure 4.2(a) and (b):

1. Isothermal evaporation process 4–1 — The refrigerant evaporates completely in the evaporator and produces refrigeration effect q_{rf}, in Btu/lb:

$$q_{rf} = (h_1 - h_4)$$ (4.1)

where h_1, h_4 = enthalpy of refrigerant at state points 1 and 4, respectively, Btu/lb.

2. Isentropic compression process 1–2 — Vapor refrigerant is extracted by the compressor and compressed isentropically from point 1 to 2. The work input to the compressor W_{in}, in Btu/lb, is

$$W_{in} = (h_2 - h_1)$$ (4.2)

where h_2 = enthalpy of refrigerant at state point 2, Btu/lb.

The greater the difference in temperature/pressure between the condensing pressure p_{con} and evaporating pressure p_{ev}, the higher will be the work input to the compressor.

3. Isothermal condensation process 2–3 — Hot gaseous refrigerant discharged from the compressor is condensed in the condenser into liquid, and the latent heat of condensation is rejected to the condenser water or ambient air. The heat rejection during condensation, q_{2-3}, in Btu/lb, is

$$-q_{2-3} = (h_2 - h_3)$$ (4.3)

where h_3 = enthalpy of refrigerant at state point 3, Btu/lb.

4. Throttling process 3–4 — Liquid refrigerant flows through a throttling device (e.g., an expansion valve, a capillary tube, or orifices) and its pressure is reduced to the evaporating pressure. A portion of the liquid flashes into vapor and enters the evaporator. This is the only irreversible process in the ideal cycle, usually represented by a dotted line. For a throttling process, assuming that the heat gain from the surroundings is negligible:

$$h_3 = h_4$$ (4.4)

The mass flow rate of refrigerant $\dot{m}_r$, in lb/min, is

$$\dot{m}_r = q_{rc}/60q_{rf}$$ (4.5)

where q_{rc} = refrigeration capacity of the system, Btu/hr.

The ideal single-stage vapor compression refrigeration cycle on a p-h diagram is divided into two pressure regions: high pressure (p_{con}) and low pressure (p_{ev}).

Coefficient of Performance of Refrigeration Cycle

The *coefficient of performance* (COP) is a dimensionless index used to indicate the performance of a thermodynamic cycle or thermal system. The magnitude of COP can be greater than 1.

- If a *refrigerator* is used to produce a refrigeration effect, COP_{ref} is

$$COP_{ref} = q_{rf}/W_{in}$$ (4.6)

- If a *heat pump* is used to produce a useful heating effect, its performance denoted by COP_{hp} is

$$COP_{hp} = q_{2-3}/W_{in}$$ (4.7)

- For a heat recovery system when both refrigeration and heating effects are produced, the COP_{hr} is denoted by the ratio of the sum of the absolute values of q_{rf} and q_{2-3} to the work input, or

$$COP_{hr} = (|q_{rf}| + |q_{2-3}|)/W_{in}$$ (4.8)

Subcooling and Superheating

Condensed liquid is often subcooled to a temperature lower than the saturated temperature corresponding to the condensing pressure p_{con}, in psia or psig, as shown in Figure 4.2(c). *Subcooling* increases the refrigeration effect to $q_{rf.sc}$ as shown in Figure 4.2(c):

$$q_{rf.sc} = \left(h_{4'} - h_1\right) > \left(h_4 - h_1\right) \tag{4.9}$$

The enthalpy of subcooled liquid refrigerant h_{sc} approximately equals the enthalpy of the saturated liquid refrigerant at subcooled temperature $h_{s.sc}$, both in Btu/lb:

$$h_{sc} = h_{3'} = h_{4'} = h_{1.con} - c_{pr}\left(T_{con} - T_{sc}\right) \approx h_{s.sc} \tag{4.10}$$

where $h_{3'}$, $h_{4'}$=enthalpy of liquid refrigerant at state points 3' and 4' respectively, Btu/lb
$\qquad h_{1.con}$=enthalpy of saturated liquid at condensing temperature, Btu/lb
$\qquad c_{pr}$=specific heat of liquid refrigerant at constant pressure, Btu/lb °F
$\qquad T_{con}$=condensing temperature or saturated temperature of liquid refrigerant at condensing pressure, °F
$\qquad T_{sc}$=temperature of subcooled liquid refrigerant, °F

The purpose of *superheating* is to prevent liquid refrigerant flooding back into the compressor and causing slugging damage as shown in Figure 4.2(d). The degree of superheating depends mainly on the types of refrigerant feed, construction of the suction line, and type of compressor. The state point of vapor refrigerant after superheating of an ideal system must be at the evaporating pressure with a specific degree of superheat and can be plotted on a *p-h* diagram for various refrigerants.

Refrigeration Cycle of Two-Stage Compound Systems with a Flash Cooler

A *multistage system* employs more than one compression stage. Multistage vapor compression systems are classified as compound systems and cascade systems. A *compound system* consists of two or more compression stages connected in series. It may have one high-stage compressor (higher pressure) and one low-stage compressor (lower pressure), several compressors connected in series, or two or more impellers connected internally in series and driven by the same motor.

The *compression ratio* R_{com} is defined as the ratio of discharge pressure p_{dis} to the suction pressure at the compressor inlet p_{suc}:

$$R_{com} = p_{dis}/p_{suc} \tag{4.11}$$

Compared to a single-stage system, a multistage has a smaller compression ratio and higher compression efficiency for each stage of compression, greater refrigeration effect, lower discharge temperature at the high-stage compressor, and greater flexibility. At the same time, a multistage system has a higher initial cost and more complicated construction.

The pressure between the discharge pressure of the high-stage compressor and the suction pressure of the low-stage compressor of a multistage system is called *interstage pressure* p_i, in psia. Interstage pressure for a two-stage system is usually determined so that the compression ratios are nearly equal between two stages for a higher COP. Then the interstage pressure is

$$p_i = \sqrt{\left(p_{con}p_{ev}\right)} \tag{4.12}$$

where p_{con}, p_{ev} = condensing and evaporating pressures, psia.

For a multistage system of n stages, then, the compression ratio of each stage is

$$R_{com} = \left(p_{con}/p_{suc}\right)^{1/n} \tag{4.13}$$

Figure 4.3(a) shows a schematic diagram and Figure 4.3(b) the refrigeration cycle of a two-stage compound system with a flash cooler. A *flash cooler*, sometimes called an economizer, is used to subcool the liquid refrigerant to the saturated temperature corresponding to the interstage pressure by vaporizing a portion of the liquid refrigerant in the flash cooler.

Based on the principle of heat balance, the fraction of evaporated refrigerant, x, or quality of the mixture in the flash cooler is

$$x = \left(h_{5'} - h_8\right)/\left(h_7 - h_8\right) \tag{4.14}$$

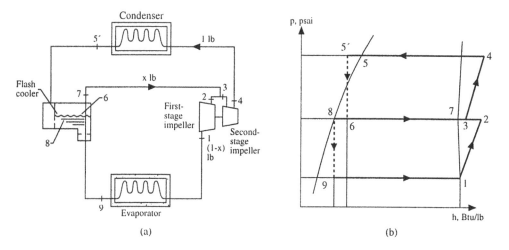

FIGURE 4.3 Two-stage compound system with a flash cooler: (a) schematic diagram and (b) refrigeration cycle.

where $h_{5'}$, h_7, h_8 = enthalpy of the refrigerant at state points 5′, 7, and 8, respectively, Btu/lb. The coefficient of performance of the refrigeration cycle of a two-stage compound system with a flash cooler, COP_{ref}, is given as

$$COP_{ref} = q_{rf}/W_{in} = (1-x)\left(h_1 - h_9\right)/\left[(1-x)\left(h_2 - h_1\right) + \left(h_4 - h_3\right)\right] \tag{4.15}$$

where h_1, h_2, h_3, h_4, h_9 = enthalpy of refrigerant at state points 1, 2, 3, 4, and 9, respectively, Btu/lb. The mass flow rate of refrigerant flowing through the condenser, $\mathring{m}_r$, in lb/min, can be calculated as

$$\mathring{m}_r = q_{rc}/60q_{rf} \tag{4.16}$$

Because a portion of liquid refrigerant is flashed into vapor in the flash cooler and goes directly to the second-stage impeller inlet, less refrigerant is compressed in the first-stage impeller. In addition, the liquid refrigerant in the flash cooler is cooled to the saturated temperature corresponding to the interstage temperature before entering the evaporator, which significantly increases the refrigeration effect of this compound system. Two-stage compound systems with flash coolers are widely used in large central air-conditioning systems.

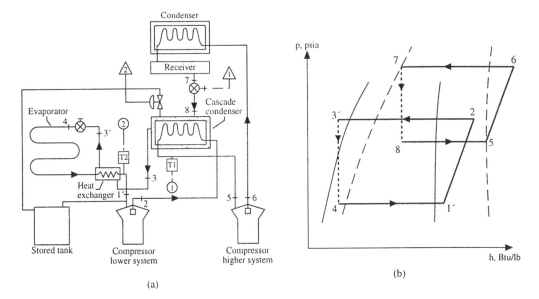

FIGURE 4.4 Cascade system: (a) schematic diagram and (b) refrigeration cycle.

Cascade System Characteristics

A *cascade system* consists of two independently operated single-stage refrigeration systems: a lower system that maintains a lower evaporating temperature and produces a refrigeration effect and a higher system that operates at a higher evaporating temperature as shown in Figure 4.4(a) and (b). These two separate systems are connected by a *cascade condenser* in which the heat released by the condenser in the lower system is extracted by the evaporator in the higher system.

A heat exchanger is often used between the liquid refrigerant from the condenser and the vapor refrigerant leaving the evaporator of the lower system. When the system is shut down in summer, a relief valve connected to a stored tank should be used to relieve the higher pressure of refrigerant at the higher storage temperature.

The main advantages of a cascade system compared with a compound system are that different refrigerants, oils, and equipment can be used for the lower and higher systems. Disadvantages are the overlap of the condensing temperature of the lower system and the evaporating temperature of the higher system because of the heat transfer in the cascade condenser and a more complicated system.

The refrigeration effect q_{rf} of the cascade system is

$$q_{rf} = \left(h_1 - h_4\right) \tag{4.17}$$

where h_1, h_4 = enthalpy of the refrigerant leaving and entering the evaporator of the lower system, Btu/lb. The total work input to the compressors in both higher and lower systems W_{in}, in Btu/lb, can be calculated as

$$W_{in} = \left(h_2 - h_{1'}\right) + \dot{m}_h\left(h_6 - h_5\right)/\dot{m}_1 \tag{4.18}$$

where h_2 = enthalpy of refrigerant discharged from the compressor of the lower system

$h_{1'}$ = enthalpy of the vapor refrigerant leaving the heat exchanger

h_6, h_5 = enthalpy of the refrigerant leaving and entering the high-stage compressor

$\overset{\circ}{m}_h, \overset{\circ}{m}_l$ = mass flow rate of the refrigerant of the higher and lower systems, respectively

The coefficient of performance of a cascade system COP_{ref} is

$$COP_{ref} = q_{rf}/W_{in} = \overset{\circ}{m}_l \left(h_1 - h_4\right) \Big/ \left[\overset{\circ}{m}_l\left(h_2 - h_{1'}\right) + \overset{\circ}{m}_h\left(h_6 - h_5\right)\right] \tag{4.19}$$

5

Outdoor Design Conditions and Indoor Design Criteria

Shan K. Wang

Consultant

5   Outdoor Design Conditions and Indoor
Design Criteria 51
Outdoor Design Conditions • Indoor Design Criteria and
Thermal Comfort • Indoor Temperature, Relative Humidity,
and Air Velocity • Indoor Air Quality and Outdoor Ventilation
Air Requirements

Outdoor Design Conditions and Indoor Design Criteria

Outdoor Design Conditions

In principle, the capacity of air-conditioning equipment should be selected to offset or compensate for the space load so that indoor design criteria can be maintained if the outdoor weather does not exceed the design values. Outdoor and indoor design conditions are used to calculate the design space loads. In energy use calculations, hour-by-hour outdoor climate data of a design day should be adopted instead of summer and winter design values.

ASHRAE Handbook 1993 Fundamentals (Chapter 24 and 27) and *Wang's Handbook of Air Conditioning and Refrigeration* (Chapter 7) both list tables of climate conditions for the U.S. and Canada based on the data from the National Climate Data Center (NCDC), U.S. Air Force, U.S. Navy, and Canadian Atmospheric Environment Service. In these tables:

- *Summer design dry bulb temperature* in a specific location $T_{o.s}$, in °F, is the rounded higher integral number of the statistically determined summer outdoor design dry bulb temperature $T_{o.ss}$ so that the average number of hours of occurrence of outdoor dry bulb temperature T_o higher than $T_{o.ss}$ during June, July, August, and September is less than 1, 2.5, or 5% of the total number of hours in these summer months (2928 hr). The data are an average of 15 years. An occurrence of less than 2.5% of 2928 hr of summer months, that is, $0.025 \times 2928 = 73$ hr, is most widely used.

- *Summer outdoor mean coincident wet bulb temperature* $T'_{o,s}$, in °F, is the mean of all the wet bulb temperatures at the specific summer outdoor design dry bulb temperature $T_{o,s}$ during the summer months.
- *Summer outdoor 2.5% design wet bulb temperature* is the design wet bulb temperature that has an average annual occurrence of $T'_o > T'_{o,s}$ less than 73 hr. This design value is often used for evaporative cooling design.
- *Mean daily range*, in °F, is the difference between the average daily maximum and the average daily minimum temperature during the warmest month.
- In *ASHRAE Handbook 1993 Fundamentals*, *solar heat gain factors* (SHGFs), in Btu/h.ft², are the average solar heat gain per hour during cloudless days through double-strength sheet (DSA) glass. The *maximum SHGFs* are the maximum values of SHGFs on the 21st of each month for a specific latitude.
- *Winter outdoor design dry bulb temperature* $T_{o,w}$, in °F, is the rounded lower integral value of the statically determined winter outdoor design temperature $T_{o,ws}$, so that the annual average number of hours of occurrence of outdoor temperature $T_o > T_{o,ws}$ is equal to or exceeds 99%, or 97.5% of the total number of hours in December, January, and February (2160 hr).

A *degree day* is the difference between a base temperature and the mean daily outdoor air temperature $T_{o,m}$ for any one day, in °F. The total numbers of heating degree days HDD65 and cooling degree days CDD65 referring to a base temperature of 65°F per annum are

$$HDD65 = \sum_{n=1} \left(65 - T_{o,m} \right)$$

$$CDD65 = \sum_{m=1} \left(T_{o,m} - 65 \right)$$

(5.1)

where n = number of days for which $T_{o,m} < 65°F$
        m = number of days for which $T_{o,m} > 65°F$

Indoor Design Criteria and Thermal Comfort

Indoor design criteria, such as space temperature, humidity, and air cleanliness, specify the requirements for the indoor environment as well as the quality of an air-conditioning or HVAC&R project.

The human body requires energy for physical and mental activity. This energy comes from the oxidation of food. The rate of heat release from the oxidation process is called the *metabolic rate*, expressed in met (1 met = 18.46 Btu/h.ft²). The metabolic rate depends mainly on the intensity of the physical activity of the human body. Heat is released from the human body by two means: *sensible heat exchange* and *evaporative heat loss*. Experience and experiments all show that there is thermal comfort only under these conditions:

- Heat transfer from the human body to the surrounding environment causes a steady state of thermal equilibrium; that is, there is no heat storage in the body core and skin surface.
- Evaporative loss or regulatory sweating is maintained at a low level.

The physiological and environmental factors that affect the thermal comfort of the occupants in an air-conditioned space are mainly:

1. Metabolic rate M determines the amount of heat that must be released from the human body.
2. *Indoor air temperature* T_r and *mean radiant temperature* T_{rad}, both in °F. The *operating temperature* T_o is the weighted sum of T_r and T_{rad}. T_{rad} is defined as the temperature of a uniform black

enclosure in which the surrounded occupant would have the same radiative heat exchange as in an actual indoor environment. T_r affects both the sensible heat exchange and evaporative losses, and T_{rad} affects only sensible heat exchange. In many indoor environments, $T_{rad} \approx T_r$.

3. Relative humidity of the indoor air φ_r, in %, which is the primary factor that influences evaporative heat loss.
4. Air velocity of the indoor air v_r, in fpm, which affects the heat transfer coefficients and therefore the sensible heat exchange and evaporative loss.
5. *Clothing insulation* R_{cl}, in clo (1 clo = 0.88 h.ft².°F/Btu), affects the sensible heat loss. Clothing insulation for occupants is typically 0.6 clo in summer and 0.8 to 1.2 clo in winter.

Indoor Temperature, Relative Humidity, and Air Velocity

For comfort air-conditioning systems, according to ANSI/ASHRAE Standard 55-1981 and ASHRAE/IES Standard 90.1-1989, the following indoor design temperatures and air velocities apply for conditioned spaces where the occupant's activity level is 1.2 met, indoor space relative humidity is 50% (in summer only), and $T_r = T_{rad}$:

	Clothing insulation (clo)	Indoor temperature (°F)	Air velocity (fpm)
Winter	0.8–0.9	69–74	<30
Summer	0.5–0.6	75–78	<50

If a suit jacket is the clothing during summer for occupants, the summer indoor design temperature should be dropped to 74 to 75°F.

Regarding the indoor humidity:

1. Many comfort air-conditioning systems are not equipped with humidifiers. Winter indoor relative humidity should not be specified in such circumstances.
2. When comfort air-conditioning systems are installed with humidifiers, ASHRAE/IES Standard 90.1-1989 requires that the humidity control prevent "the use of fossil fuel or electricity to produce humidity in excess of 30% ... or to reduce relative humidity below 60%."
3. Indoor relative humidity should not exceed 75% to avoid increasing bacterial and viral populations.
4. For air-conditioning systems that use flow rate control in the water cooling coil, space indoor relative humidity may be substantially higher in part load than at full load.

Therefore, for comfort air-conditioning systems, the recommended indoor relative humidities, in %, are

	Tolerable range	Preferred value
Summer	30–65	40–50
Winter		
With humidifier		25–30
Without humidifier		Not specified

In surgical rooms or similar health care facilities, the indoor relative humidity is often maintained at 40 to 60% year round.

Indoor Air Quality and Outdoor Ventilation Air Requirements

According to the National Institute for Occcupational Safety and Health (NIOSH), 1989, the causes of indoor air quality complaints in buildings are inadequate outdoor ventilation air, 53%; indoor contaminants, 15%; outdoor contaminants, 10%; microbial contaminants, 5%; construction and furnishings, 4%; unknown and others, 13%. For space served by air-conditioning systems using low- and medium-efficiency air filters, according to the U.S. Environmental Protection Agency (EPA) and Consumer Product Safety Commission (CPSC) publication "A Guide to Indoor Air Quality" (1988) and the field investigations reported by Bayer and Black (1988), *indoor air contaminants* may include some of the following:

1. *Total particulate concentration.* This concentration comprises particles from building materials, combustion products, mineral fibers, and synthetic fibers. In February 1989, the EPA specified the allowable indoor concentration of particles of 10 μm and less in diameter (which penetrate deeply into lungs) as:

   50 μg/m³ (0.000022 grain/ft³), 1 year
   150 μg/m³ (0.000066 grain/ft³), 24 hr

   In these specifications, "1 year" means maximum allowable exposure per day over the course of a year.

2. *Formaldehyde and organic gases.* Formaldehyde is a colorless, pungent-smelling gas. It comes from pressed wood products, building materials, and combustion. Formaldehyde causes eye, nose, and throat irritation as well as coughing, fatigue, and allergic reactions. Formaldehyde may also cause cancer. Other organic gases come from building materials, carpeting, furnishings, cleaning materials, etc.

3. *Radon.* Radon, a colorless and odorless gas, is released by the decay of uranium from the soil and rock beneath buildings, well water, and building materials. Radon and its decay products travel through pores in soil and rock and infiltrate into buildings along the cracks and other openings in the basement slab and walls. Radon at high levels causes lung cancer. The EPA believes that levels in most homes can be reduced to 4 pCi/l (picocuries per liter) of air. The estimated national average is 1.5 pCi/l, and levels as high as 200 pCi/l have been found in houses.

4. *Biologicals.* These include bacteria, fungi, mold and mildew, viruses, and pollen. They may come from wet and moist walls, carpet furnishings, and poorly maintained dirty air-conditioning systems and may be transmitted by people. Some biological contaminants cause allergic reactions, and some transmit infectious diseases.

5. *Combustion products.* These include environmental tobacco smoke, nitrogen dioxide, and carbon monoxide. *Environmental tobacco* smoke from cigarettes is a discomfort factor to other persons who do not smoke, especially children. Nicotine and other tobacco smoke components cause lung cancer, heart disease, and many other diseases. *Nitrogen dioxide* and *carbon monoxide* are both combustion products from unvented kerosene and gas space heaters, wood-stoves, and fireplaces.

   Nitrogen dioxide (NO_2) causes eye, nose, and throat irritation; may impair lung function; and increases respiratory infections. Carbon monoxide (CO) causes fatigue at low concentrations; impaired vision, headache, and confusion at higher concentrations; and is fatal at very high concentrations. Houses without gas heaters and gas stoves may have CO levels varying from 0.5 to 5 parts per million (ppm).

6. *Human bioeffluents.* These include the emissions from breath including carbon dioxide exhaled from the lungs, body odors from sweating, and gases emitted as flatus.

There are three basic means of reducing the concentration of indoor air contaminants and improving indoor air quality: (1) eliminate or reduce the source of air pollution, (2) enhance the efficiency of air filtration, and (3) increase the ventilation (outdoor) air intake. Dilution of the concentrations of indoor contaminants by outdoor ventilation air is often the simple and cheapest way to improve indoor air quality. The amount of outdoor air required for metabolic oxidation is rather small.

Abridged outdoor air requirements listed in ANSI/ASHRAE Standard 62-1989 are as follows:

Applications	cfm/person
Hotels, conference rooms, offices	20
Retail stores	0.2–0.3 cfm/ft²
Classrooms, theaters, auditoriums	15
Hospital patient rooms	25

These requirements are based on the analysis of dilution of CO_2 as the representative human bioeffluent to an allowable indoor concentration of 1000 ppm. Field measurements of daily maximum CO_2 levels in office buildings reported by Persily (1993) show that most of them were within the range 400 to 820 ppm. The quality of outdoor air must meet the EPA's National Primary and Secondary Ambient Air Quality Standards, some of which is listed below:

Pollutants	Long-term concentration			Short-term concentration		
	µg/m³	ppm	Exposure	µg/m³	ppm	Exposure
Particulate	75		1 year	260		24 hr
SO_2	80	0.03	1 year	365	0.14	24 hr
CO				40,000	35	1 hr
				10,000	9	8 hr
NO_2	100	0.055	1 year			
Lead	1.5		3 months			

Here exposure means average period of exposure.

If unusual contaminants or unusually strong sources of contaminants are introduced into the space, or recirculated air is to replace part of the outdoor air supply for occupants, then acceptable indoor air quality is achieved by controlling known and specific contaminants. This is called an indoor air quality procedure. Refer to ANSI/ASHRAE Standard 62-1989 for details.

Clean Rooms

Electronic, pharmaceutical, and aerospace industries and operating rooms in hospitals all need strict control of air cleanliness during manufacturing and operations. According to ASHRAE Handbook 1991 HVAC Applications, clean rooms can be classified as follows based on the particle count per ft³:

Class	Particle size	
	0.5 µm and larger	5 µm and larger
	Particle count per ft³ not to exceed	
1	1	0
10	10	0
100	100	
1000	1000	
10,000	10,000	65
100,000	100,000	700

For clean rooms, space temperature is often maintained at $72 \pm 2°F$ and space humidity at $45 \pm 5\%$. Here, $\pm 2°F$ and $\pm 5\%$ are allowable tolerances. Federal Standard 209B specifies that the ventilation (outdoor air) rate should be 5 to 20% of the supply air.

Space Pressure Differential

Most air-conditioning systems are designed to maintain a slightly higher pressure than the surroundings, a positive pressure, to prevent or reduce infiltration and untreated air entering the space directly. For laboratories, restrooms, or workshops where toxic, hazardous, or objectional gases or contaminants are produced, a slightly lower pressure than the surroundings, a negative pressure, should be maintained to prevent or reduce the diffusion of these contaminants' exfiltrate to the surrounding area.

For comfort air-conditioning systems, the recommended pressure differential between the indoor and outdoor air is 0.02 to 0.05 in in. WG. WG indicates the pressure at the bottom of a top-opened water column of specific inches of height; 1 in. WG = 0.03612 psig.

For clean rooms, Federal Standard No. 209B, Clean Rooms and Work Stations Requirements (1973), specifies that the minimum positive pressure between the clean room and any adjacent area with lower cleanliness requirements should be 0.05 in. WG with all entryways closed. When the entryways are open, an outward flow of air is to be maintained to prevent migration of contaminants into the clean room. In comfort systems, the space pressure differential is usually not specified in the design documents.

Sound Levels

Noise is any unwanted sound. In air-conditioning systems, noise should be attenuated or masked with another less objectionable sound.

Sound power is the capability to radiate power from a sound source exited by an energy input. The intensity of sound power is the output from a sound source expressed in watts (W). Due to the wide variation of sound output at a range of 10^{20} to 1, it is more convenient to use a logarithmic expression to define a *sound power level* L_w, in dB:

$$L_w = 10\log\left(w/10^{-12}\ \text{W}\right)\ \text{re } 1\ \text{pW} \tag{5.2}$$

where w = sound source power output, W.

The human ear and microphones are sound pressure sensitive. Similarly to the sound power level, the *sound pressure level* L_p, in dB, is defined as:

$$L_p = 20\log\left(p/2\times10^{-5}\ \text{Pa}\right)\ \text{re } 20\ \mu\text{Pa} \tag{5.3}$$

where p = sound pressure, Pa.

The sound power level of any sound source is a fixed output. It cannot be measured directly; it can only be calculated from the measured sound pressure level. The sound pressure level at any one point is affected by the distance from the source and the characteristics of the surroundings.

Human ears can hear frequencies from 20 Hz to 20 kHz. For convenience in analysis, sound frequencies are often subdivided into eight octave bands. An *octave* is a frequency band in which the frequency of the upper limit of the octave is double the frequency of the lower limit. An octave band is represented by its center frequency, such as 63, 125, 250, 500, 1000, 2000, 4000, and 8000 Hz. On 1000 Hz the octave band has a higher limit of 1400 Hz and a lower limit of 710 Hz. Human ears do not respond in the same way to low frequencies as to high frequencies.

The object of noise control in an air conditioned space is to provide background sound low enough that it does not interfere with the acoustical requirements of the occupants. The distribution of background sound should be balanced over a broad range of frequencies, that is, without whistle, hum, rumble, and beats.

The most widely used criteria for sound control are the noise criteria NC curves. The shape of NC curves is similar to the equal-loudness contour representing the response of the human ear. NC curves also intend to indicate the permissible sound pressure level of broad-band noise at various octave bands rated by a single NC curve. NC curves are practical and widely used.

Other criteria used are room criteria RC curves and A-weighted sound level, dBA. RC curves are similar to NC curves except that the shape of the RC curves is a close approximation to a balanced, bland-sounding spectrum. The A-weighted sound level is a single value and simulates the response of the human ear to sound at low sound pressure levels.

Type of area	Recommended NC or RC range (dB)
Hotel guest rooms	30–35
Office	
Private	30–35
Conference	25–30
Open	30–35
Computer equipment	40–45
Hospital, private	25–30
Churches	25–30
Movie theaters	30–35

The following are abridged indoor design criteria, NC or RC range, listed in *ASHRAE Handbook 1987 Systems and Applications*:

For industrial factories, if the machine noise in a period of 8 hr exceeds 90 dBA, Occupational Safety and Health Administration Standard Part 1910.95 requires the occupants to use personal protection equipment. If the period is shorter, the dBA level can be slightly higher. Refer to this standard for details.

6

Load Calculations

Shan K. Wang

Consultant

6   Load Calculations 59
     Space Loads • Moisture Transfer in Building Envelope •
     Cooling Load Calculation Methodology • Conduction Heat
     Gains • Internal Heat Gains • Conversion of Heat Gains into
     Cooling Load by TFM • Heating Load

Load Calculations

Space Loads

Space, Room, and Zone

Space indicates a volume or a site without partitions, or a partitioned room or a group of rooms. A *room* is an enclosed or partitioned space that is considered as a single load. An air-conditioned room does not always have an individual zone control system. A *zone* is a space of a single room or group of rooms having similar loads and operating characteristics. An air-conditioned zone is always installed with an individual control system. A typical floor in a building may be treated as a single zone space, or a *multizone space* of perimeter, interior, east, west, south, north, ... zones.

Space and equipment loads can be classified as:

1. *Space heat gain* q_e, in Btu/hr, is the rate of heat transfer entering a conditioned space from an external heat source or heat releases to the conditioned space from an internal source. The rate of sensible heat entering the space is called *sensible heat gain* q_{es}, whereas the rate of latent heat entering the space is called *latent heat gain* q_{el}. In most load calculations, the time interval is often 1 hr, and therefore q_e, q_{es}, and q_{el} are all expressed in Btu/hr.
2. *Space cooling load* or simply *cooling load* q_{rc}, also in Btu/hr, is the rate at which heat must be removed from a conditioned space to maintain a constant space temperature and an acceptable relative humidity. The sensible heat removed is called *sensible cooling load* q_{rs}, and the latent heat removed is called *latent cooling load* q_{rl}, both in Btu/hr.
3. *Space heat extraction rate* q_{ex}, in Btu/hr, is the rate at which heat is removed from the conditioned space. When the space air temperature is constant, $q_{ex} = q_{rc}$.
4. *Space heating load* q_{rh}, in Btu/hr, is the rate at which heat must be added to the conditioned space to maintain a constant temperature.

5. *Coil load q_c*, in Btu/hr, is the rate of heat transfer at the coil. The cooling *coil load q_{cc}* is the rate of heat removal from the conditioned air by the chilled water or refrigerant inside the coil. The *heating coil load q_{ch}* is the rate of heat energy addition to the conditioned air by the hot water, steam, or electric elements inside the coil.

6. *Refrigeration load q_{rl}*, in Btu/hr, is the rate at which heat is extracted by the evaporated refrigerant at the evaporator. For packaged systems using a DX coil, $q_{rl} = q_{cc}$. For central systems:

$$q_{rl} = q_{cc} + q_{pi} + q_{pu} + q_{s.t}$$ (6.1)

where q_{pi} = chilled water piping heat gain, Btu/hr
           q_{pu} = pump power heat gain, Btu/hr
           $q_{s.t}$ = storage tank heat gain, if any, Btu/hr

Heat gains q_{pi} and q_{pu} are usually about 5 to 10% of the cooling coil load q_{cc}.

Convective Heat and Radiative Heat

Heat enters a space and transfer to the space air from either an external source or an internal source is mainly in the form of *convective heat* and *radiative heat* transfer.

Consider radiative heat transfer, such as solar radiation striking the outer surface of a concrete slab as shown in Figure 6.1(a) and (b). Most of the radiative heat is absorbed by the slab. Only a small fraction is reflected. After the heat is absorbed, the outer surface temperature of the slab rises. If the slab and space air are in thermal equilibrium before the absorption of radiative heat, heat is convected from the outer surface of the slab to the space air as well as radiated to other surfaces. At the same time, heat is conducted from the outer surface to the inner part of the slab and stored there when the temperature of the inner part of the slab is lower than that of its outer surface. Heat convected from the outer surface of the concrete slab to the space air within a time interval forms the sensible cooling load.

The sensible heat gain entering the conditioned space does not equal the sensible cooling load during the same time interval because of the stored heat in the building envelope. Only the convective heat becomes cooling load instantaneously. The sum of the convective heats from the outer surfaces, including the outer surfaces of the internal heat gains in a conditioned space, becomes cooling load. This phenomenon results in a smaller cooling load than heat gain, as shown in Figure 6.1(a) and (b). According to *ASHRAE Handbook 1993 Fundamentals*, the percentages of convective and radiative components of the sensible heat gains are as follows:

Load Profile, Peak Load, and Block Load

A *load profile* shows the variation of space, zone, floor, or building load in a certain time period, such as a 24-hr day-and-night cycle. In a load profile, load is always plotted against time. The load profile depends on the outdoor climate as well as the space operating characteristics.

Peak load is the maximum cooling load in a load profile. *Block load* is the sum of the zone loads and floor loads at a specific time. The sum of the zone peak loads in a typical floor does not equal the block load of that floor because the zone peak loads may all not appear at the same time.

Moisture Transfer in Building Envelope

Moisture transfer takes place along two paths:

1. Moisture migrates in the building envelope in both liquid and vapor form. It is mainly liquid if the relative humidity of the ambient air exceeds 50%. Liquid flow is induced by capillary flow and moisture content gradient. Vapor diffusion is induced by vapor pressure gradients. *Moisture content* is defined as the ratio of the mass of moisture contained in a solid to the mass of bone-dry solid. During the migration, the moisture content and the vapor pressure are in equilibrium at a specific temperature and location.

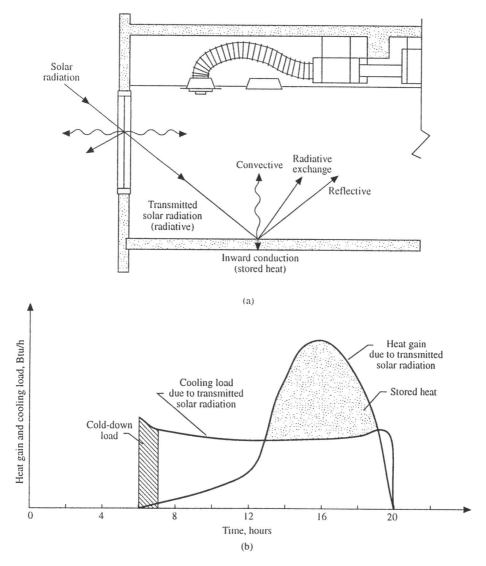

FIGURE 6.1 Solar heat gain from west window and its corresponding space cooling load for a night shutdown air system: (a) convective and radiative heat transfer and (b) heat gain and cooling load curves.

2. Air leakage and its associated water vapor infiltrate or exfiltrate through the cracks, holes, and gaps between joints because of poor construction of the building envelope. The driving potential of this air leakage and associated water vapor is the pressure differential across the building envelope. If the insulating material is of open-cell structure, air leakage and associated water vapor may penetrate the perforated insulating board through cracks and gaps. Condensation, even freezing, will occur inside the perforated insulation board if the temperature of the board is lower than the dew point of the leaked air or the freezing point of the water.

Sensible heat gains	Convective (%)	Radiative (%)
Solar radiation with internal shading	42	58
Fluorescent lights	50	50
Occupants	67	33
External wall, inner surface	40	60

In most comfort air-conditioning systems, usually only the space temperature is controlled within limits. A slight variation of the space relative humidity during the operation of the air system is often acceptable. Therefore, the store effect of moisture is ignored except in conditioned spaces where both temperature and relative humidity need to be controlled or in a hot and humid area where the air system is operated at night shutdown mode. In most cases, latent heat gain is considered equal to latent cooling load instantaneously. For details refer to Wang's *Handbook of Air Conditioning and Refrigeration* (1993), Chapters 6 and 7.

Cooling Load Calculation Methodology

Basic considerations include the following:

- It is assumed that equations of heat transfer for cooling load calculation within a time interval are linear. It is also assumed that the superposition principle holds. When a number of changes occur simultaneously in the conditioned space, they will proceed as if independent of each other. The total change is the sum of the responses caused by the individual changes.
- Space load calculations are often performed by computer-aided design (CAD), with market-available software like DOE-2.1D, TRACE-600, and Carrier E20-II Loads.
- Peak load calculations evaluate the maximum load to size and select the equipment. The energy analysis program compares the total energy use in a certain period with various alternatives in order to determine the optimum one.
- The methodology of various cooling load calculations is mainly due to their differences in the conversion of space radiative heat gains into space cooling loads. Convective heat, latent heat, and sensible heat gains from infiltration are all equal to cooling loads instantaneously.
- Space cooling load is used to calculate the supply volume flow rate and to determine the size of the air system, ducts, terminals, and diffusers. The coil load is used to determine the size of the cooling coil and the refrigeration system. Space cooling load is a component of the cooling coil load.

The Rigorous Approach

The rigorous approach to the calculation of the space cooling load consists of (1) finding the inside surface temperatures of the building structures that enclose the conditioned space due to heat balance at time *t* and (2) calculating the sum of the convective heats transferred from these surfaces as well as from the occupants, lights, appliances, and equipment in the conditioned space at time *t*.

The inside surface temperature of each surface $T_{i,t}$, in °F, can be found from the following simultaneous heat balance equations:

$$q_{i,t} = \left[h_{ci}\left(T_{r,t} - T_{i,t}\right) + \sum_{j=1}^{m} g_{ij}\left(T_{j,t} - T_{i,t}\right) \right] A_i + S_{i,t} + L_{i,t} + P_{i,t} + E_{i,t} \tag{6.2}$$

where h_{ci} =convective heat transfer coefficient, Btu/hr.ft^2.°F
           g_{ij} =radiative heat transfer factor between inside surface *i* and inside surface *j*, Btu/hr.ft^2.°F
           $T_{r,t}$ =space air temperature at time *t*, °F
    $T_{i,t}, T_{j,t}$ =average temperature of inside surfaces *i* and *j* at time *t*, °F
           A_i =area of inside surface *i*, ft^2
$S_{i,t}, L_{i,t}, P_{i,t}, E_{i,t}$=solar radiation transmitted through windows and radiative heat from lights, occupants,
                     and equipment absorbed by inside surface *i* at time *t*, Btu/hr

In Equation (6.2), $q_{i,t}$, in Btu/hr, is the conductive heat that comes to surface *i* at time *t* because of the temperature excitation on the outer opposite surface of *i*. This conductive heat can be found by solving the partial differential equations or by numerical solutions. The number of inside surfaces *i* is usually equal to 6, and surface i is different from j so that radiative exchange can proceed. $q_{i,t}$ could also be expressed in Btu/min or even Btu/sec.

The space sensible cooling load q_{rs}, in Btu/hr, is the sum of the convective heat from the inside surfaces, including the convective heat from the inner window glass due to the absorbed solar radiation and the infiltration:

$$q_{rs} = \left[\sum_{i=1}^{6} h_{ci} \left(T_{i,t} - T_{r,t} \right) \right] A_i + 60 \mathring{V}_{if} \, \rho_o c_{pa} \left(T_{o,t} - T_{r,t} \right) + S_{c,t} + L_{c,t} + P_{c,t} + E_{c,t} \qquad (6.3)$$

where $\mathring{V}_{if}$ =volume flow rate of infiltrated air, cfm
ρ_o =air density of outdoor air, lb/ft³
c_{pa} =specific heat of moist air, Btu/lb.°F
$T_{o,t}$ =temperature of outdoor air at time t, °F
$S_{c,t}, L_{c,t}, P_{c,t}, E_{c,t}$=heat convected from the windows, lights, occupants, and equipment, Btu/hr

Equations (6.2) and (6.3), and partial differential equations to determine conductive heat $q_{i,t}$ must be solved simultaneously. Using a rigorous approach to find the space cooling load requires numerous computer calculations. It is laborious and time consuming. The rigorous approach is impractical and is suitable for research work only.

Transfer Function Method (TFM)

The *transfer function* of a system relates its output in Laplace transform Y to its input in Laplace transform G by a ratio K, that is,

$$K = Y/G = \left(v_0 + v_1 z^{-1} + v_2 z^{-2} + \ldots \right) \Big/ \left(1 + w_1 z^{-1} + w_2 z^{-2} + \ldots \right)$$

$$z = e^{s\Delta} \qquad (6.4)$$

where Δ = time interval.

In Equation (6.4), K, Y, and G are all expressed in z-transforms of the time series function. Coefficients v_n and w_n are called *transfer function coefficients*, or weighting factors. *Weighting factors* are used to weight the importance of the effect of current and previous heat gains as well as the previous space sensible cooling load on the current space sensible cooling load $q_{rs,t}$. Then, the output $q_{rs,t}$ can be related to the input, the space sensible heat gain $q_{es,t}$, through $q_{rs,t} = Kq_{es,t}$.

Mitalas and Stevenson (1967) and others developed a method for determining the transfer function coefficients of a zone of given geometry and details of the calculated space heat gains and the previously known space sensible cooling load through rigorous computation or through tests and experiments. In DOE 2.1A (1981) software for custom weighting factors (tailor made according to a specific parametric zone) is also provided. Sowell (1988) and Spitler et al. (1993) expanded the application of TFM to zones with various parameters: zone geometry, different types of walls, roof, floor, ceilings, building material, and mass locations. Mass of construction is divided into light construction, 30 lb/ft² of floor area; medium construction, 70 lb/ft²; and heavy construction, 130 lb/ft². Data are summarized into groups and listed in tabular form for user's convenience.

Cooling Load Temperature Difference/Solar Cooling Load Factor/Cooling Load Factor (CLTD/SCL/CLF) Method

The *CLTD/SCL/CLF method* is a one-step simplification of the transfer function method. The space cooling load is calculated directly by multiplying the heat gain q_e with CLTD, SCL, or CLF instead of first finding the space heat gains and then converting into space cooling loads through the room transfer function. In the CLTD/SCL/CLF method, the calculation of heat gains is the same as in the transfer function method.

The CLTD/SCL/CLF method was introduced by Rudoy and Duran (1975). McQuiston and Spitler (1992) recommended a new SCL factor. In 1993 they also developed the CLTD and CLF data for different zone geometries and constructions.

Finite Difference Method

Since the development of powerful personal computers, the finite difference or numerical solution method can be used to solve transient simultaneous heat and moisture transfer in space cooling load calculations. Wong and Wang (1990) emphasized the influence of moisture stored in the building structure on the cool-down load during the night shut-down operating mode in locations where the summer outdoor climate is hot and humid. The finite difference method is simple and clear in concept as well as more direct in computation than the transfer function method. Refer to Wang's *Handbook of Air Conditioning and Refrigeration* for details.

Total Equivalent Temperature Differential/Time Averaging (TETD/TA) Method

In this method, the heat gains transmitted through external walls and roofs are calculated from the Fourier series solution of one-dimensional transient heat conduction. The conversion of space heat gains to space cooling loads takes place by (1) averaging the radiative heat gains to the current and successive hours according to the mass of the building structure and experience and (2) adding the instantaneous convective fraction and the allocated radiative fraction in that time period. The TETD/TA method is simpler and more subjective than the TFM.

Conduction Heat Gains

Following are the principles and procedures for the calculation of space heat gains and their conversion to space cooling loads by the TFM. TFM is the method adopted by software DOE-2.1D and is also one of the computing programs adopted in TRACE 600. Refer to *ASHRAE Handbook 1993 Fundamentals*, Chapters 26 and 27, for detail data and tables.

Surface Heat Transfer Coefficients

ASHRAE Handbook 1993 Fundamentals adopts a constant *outdoor heat transfer coefficient* h_o = 3.0 Btu/hr.ft^2.°F and a constant *inside heat transfer coefficient* h_i = 1.46 Btu/hr.ft^2.°F during cooling load calculations. However, many software programs use the following empirical formula (TRACE 600 *Engineering Manual*):

$$h_{o.t} = 2.00 + 0.27 v_{wind.t} \tag{6.5}$$

where $h_{o.t}$ = h_o at time t, Btu/hr.ft^2.°F
$v_{wind.t}$ = wind velocity at time t, ft/sec

Coefficient h_i may be around 2.1 Btu/hr.ft^2.°F when the space air system is operating, and h_i drops to only about 1.4 Btu/hr.ft^2.°F when the air system is shut down. The R *value*, expressed in hr.ft^2.°F/Btu, is defined as the reciprocal of the overall heat transfer coefficient U value, in Btu/hr.ft^2.°F. The R value is different from the thermal resistance R^*, since R^* = $1/UA$, which is expressed in hr.°F/Btu.

Sol-air temperature T_{sol} is a fictitious outdoor air temperature that gives the rate of heat entering the outer surface of walls and roofs due to the combined effect of incident solar radiation, radiative heat exchange with the sky vault and surroundings, and convective heat exchange with the outdoor air. T_{sol}, in °F, is calculated as

$$T_{sol} = T_o + \alpha I_t / h_o - \varepsilon \Delta R / h_o \tag{6.6}$$

where T_o = outdoor air temperature, °F
α = absorptance of incident solar radiation on outer surface

ϵ = hemispherical emittance of outer surface, assumed equal to 1

I_t = total intensity of solar radiation including diffuse radiation, Btu/hr.ft^2

Tabulated sol-air temperatures that have been calculated and listed in *ASHRAE Handbook 1993 Fundamentals* are based on the following: if $\epsilon = 1$ and $h_o = 3$ Btu/hr.ft^2.°F, $\Delta R/h_o$ is -7°F for horizontal surfaces and 0°F for vertical surfaces, assuming that the long-wave radiation from the surroundings compensates for the loss to the sky vault.

External Wall and Roof

The sensible heat gain through a wall or roof at time t, $q_{e,t}$, in Btu/hr, can be calculated by using the sol-air temperature at time t-$n\delta$, $T_{sol,t-n\delta}$, in °F, as the outdoor air temperature and a constant indoor space air temperature T_r, in °F, as

$$q_{e,t} = A\left[\sum_{n=0} b_n T_{sol,t-n\delta} - \sum_{n=1} d_n \left(q_{e,t-n\delta}/A\right)\right] - T_r \sum_{n=0} c_n \tag{6.7}$$

where A = surface area of wall or roof, ft^2

$q_{e,t-n\delta}$ = heat gain through wall or roof at time t-$n\delta$, Btu/hr

n = number of terms in summation

In Equation (6.7), b_n, c_n, and d_n are *conduction transfer function* coefficients.

Ceiling, Floor, and Partition Wall

If the variation of the temperature of the adjacent space T_{aj}, in °F, is small compared with the differential $(T_{aj} - T_r)$, or even when T_{aj} is constant, the heat gain to the conditioned space through ceiling, floor, and partition wall $q_{aj,t}$, in Btu/hr, is

$$q_{aj,t} = UA\left(T_{aj} - T_r\right) \tag{6.8}$$

where U = overall heat transfer coefficient of the ceiling, floor, and partition (Btu/hr.ft^2.°F).

Heat Gain through Window Glass

Shading Devices.   There are two types of shading devices: indoor shading devices and outdoor shading devices. *Indoor shading devices* increase the reflectance of incident radiation. *Venetian blinds* and draperies are the most widely used indoor shading devices. Most horizontal venetian blinds are made of plastic, aluminum, or rigid woven cloth slats spaced 1 to 2 in. apart. Vertical venetian blinds with wider slats are also used. *Draperies* are made of fabrics of cotton, rayon, or synthetic fibers. They are usually loosely hung, wider than the window, often pleated, and can be drawn open and closed as needed. Draperies also increase thermal resistance in winter.

External shading devices include overhangs, side fins, louvers, and pattern grilles. They reduce the sunlit area of the window glass effectively and therefore decrease the solar heat gain. External shading devices are less flexible and are difficult to maintain.

Shading Coefficient (SC).   The shading coefficient is an index indicating the glazing characteristics and the associated indoor shading device to admit solar heat gain. SC of a specific window glass and shading device assembly at a summer design solar intensity and outdoor and indoor temperatures can be calculated as

$$SC = \frac{\text{solar heat gain of specific type of window glass assembly}}{\text{solar heat gain of reference (DSA) glass}} \tag{6.9}$$

Double-strength sheet glass (DSA) has been adopted as the reference glass with a transmittance of τ = 0.86, reflectance ρ = 0.08, and absorptance α = 0.06.

Solar Heat Gain Factors (SHGFs). SHGFs, in Btu/hr.ft^2, are the average solar heat gains during cloudless days through DSA glass. $SHGF_{max}$ is the maximum value of SHGF on the 21st day of each month for a specific latitude as listed in *ASHRAE Handbook 1993 Fundamentals*. At high elevations and on very clear days, the actual SHGF may be 15% higher.

Heat Gain through Window Glass. There are two kinds of heat gain through window glass: heat gain due to the solar radiation transmitted and absorbed by the window glass, $q_{es,t}$, and conduction heat gain due to the outdoor and indoor temperature difference, $q_{ec,t}$, both in Btu/hr:

$$q_{es,t} = \left(A_s \times SHGF \times SC\right) + \left(A_{sh} \times SHGF_{sh} \times SC\right) \tag{6.10}$$

where A_s, A_{sh} = sunlit and shaded areas of the glass (ft^2).

In Equation (6.10), $SHGF_{sh}$ represents the SHGF of the shaded area of the glass having only diffuse radiation. Generally, the SHGF of solar radiation incident on glass facing north without direct radiation can be considered as $SHGF_{sh}$. Because the mass of window glass is small, its heat storage effect is often neglected; then the conduction heat gain at time t is

$$q_{ec,t} = U_g\left(A_s + A_{sh}\right)\left(T_{o,t} - T_r\right) \tag{6.11}$$

where U_g = overall heat transfer coefficient of window glass, Btu/hr.ft^2.°F
         $T_{o,t}$ = outdoor air temperature at time t, °F

The inward heat transfer due to the solar radiation absorbed by the glass and the conduction heat transfer due to the outdoor and indoor temperature difference is actually combined. It is simple and more convenient if they are calculated separately.

Internal Heat Gains

Internal heat gains are heat released from the internal sources.

People

The sensible heat gain and latent heat gain per person, $q_{es,t}$ and $q_{el,t}$, both in Btu/hr, are given as

$$q_{es,t} = N_t q_{os}$$
$$q_{el,t} = N_t q_{ol} \tag{6.12}$$

where N_t = number of occupants in the conditioned space at time t.

Heat gains q_{os} and q_{ol} depend on the occupant's metabolic level, whether the occupant is an adult or a child, male or female, and the space temperature. For a male adult seated and doing very light work, such as system design by computer in an office of space temperature 75°F, q_{os} and q_{ol} are both about 240 Btu/hr.

Lights

Heat gain in the conditioned space because of the electric lights, $q_{e,li}$, in Btu/hr, is calculated as

$$q_{e,li} = \sum_{n=1}^{} \left[3.413 W_{li} N_{li} F_{ul} F_{al}\left(1 - F_{lp}\right)\right] = 3.413 W_{lA} A_{n} \tag{6.13}$$

where W_{li} = watt input to each lamp, W
 W_{lA} = lighting power density, W/ft²
 A_{fl} = floor area of conditioned space, ft²
 N_{li}, n = number of lamps (each type) and number of types of electric lamp
 F_{al}, F_{ul} = use factor of electric lights and allowance factor for ballast loss, usually taken as 1.2
 F_{lp} = heat gain carried away by the return air to plenum or by exhaust device; it varies from 0.2 to 0.5

Machines and Appliances

The sensible and latent heat gains in the conditioned space from machines and appliances, $q_{e.ap}$ and $q_{l.ap}$, both in Btu/hr, can be calculated as

$$q_{e.ap} = 3.413 W_{ap} F_{ua} F_{load} F_{ra} = q_{is} F_{ua} F_{ra}$$

$$q_{l.ap} = q_{il} F_{ua}$$

(6.14)

where W_{ap} = rated power input to the motor of the machines and appliances, W
 F_{ua} = use factor of machine and appliance
 F_{load} = ratio of actual load to the rated power
 F_{ra} = radiation reduction factor because of the front shield of the appliance
 q_{is}, q_{il} = sensible and latent heat input to the appliance, Btu/hr

Infiltration

Infiltration is the uncontrolled inward flow of unconditioned outdoor air through cracks and openings on the building envelope because of the pressure difference across the envelope. The pressure difference is probably caused by wind pressure, stack effect due to outdoor–indoor temperature difference, and the operation of an air system(s).

Today new commercial buildings have their external windows well sealed. If a positive pressure is maintained in the conditioned space when the air system is operating, infiltration is normally considered as zero.

When the air system is shut down, or for hotels, motels, and high-rise residential buildings, ASHRAE/IES Standard 90.1-1989 specifies an infiltration of 0.038cfm/ft² of gross area of the external wall, 0.15 air change per hour (ach) for the perimeter zone.

When exterior windows are not well sealed, the outdoor wind velocity is high at winter design conditions, or there is a door exposed to the outdoors directly, an infiltration rate of 0.15 to 0.4 ach for perimeter zone should be considered.

When the volume flow rate of infiltration is determined, the sensible heat gain due to infiltration $q_{s.if}$ and latent heat gain due to infiltration $q_{l.if}$, in Btu/hr, are

$$q_{s.if} = 60 \overset{\circ}{V}_{if} \rho_o c_{pa} \left(T_o - T_r \right)$$

$$q_{l.if} = 60 \times 1060 \, \overset{\circ}{V}_{if} \rho_o \left(w_o - w_r \right)$$

(6.15)

where $\overset{\circ}{V}_{if}$ = volume flow rate of infiltration, cfm
 ρ_o = air density of outdoor air, lb/ft³
 c_{pa} = specific heat of moist air, 0.243 Btu/lb.°F
 w_o, w_r = humidity ratio of outdoor and space air, lb/lb
 $h_{fg.58}$ = latent heat of vaporization, 1060 Btu/lb

Infiltration enters the space directly and mixes with space air. It becomes space cooling load instantaneously. Ventilation air is often taken at the AHU or PU and becomes sensible and latent coil load components.

Conversion of Heat Gains into Cooling Load by TFM

The space sensible cooling load $q_{rs,t}$, in Btu/hr, is calculated as

$$q_{rs,t} = q_{s-c,t} + q_{s-c,t} \tag{6.16}$$

where $q_{s-c,t}$ = space sensible cooling load converted from heat gains having radiative and convective components, Btu/hr

$q_{s-c,t}$ = space sensible cooling load from convective heat gains, Btu/hr

Based on Equation (6.4), space sensible cooling load $q_{s-c,t}$ can be calculated as

$$q_{s-c,t} = \sum_{i=1}\left(v_o q_{e,t} + v_1 q_{e,t-\delta} + v_2 q_{e,t-2\delta} + \dots \right) - \left(w_1 q_{r,t-\delta} + w_2 q_{r,t-2\delta} + \dots \right) \tag{6.17}$$

where $q_{e,t}$, $q_{e,t-\delta}$, $q_{e,t-2\delta}$=space sensible heat gains having both radiative and convective heats at time t, t-δ, and t-2δ, Btu/hr

$q_{r,t-\delta}$, $q_{r,t-2\delta}$ = space sensible cooling load at time t-δ, and t-2δ, Btu/hr

In Equation (6.17), v_n and w_n are called *room transfer function coefficients* (RTFs). RTF is affected by parameters like zone geometry; wall, roof, and window construction; internal shades; zone location; types of building envelope; and air supply density. Refer to RTF tables in *ASHRAE Handbook 1993 Fundamentals*, Chapter 26, for details.

Space sensible cooling load from convective heat gains can be calculated as

$$q_{s-c,t} = \sum_{k=1} q_{ec,t} \tag{6.18}$$

where $q_{ec,t}$ = each of k space sensible heat gains that have convective heat gains only (Btu/hr). Space latent cooling load at time t, $q_{rl,t}$, in Btu/hr, can be calculated as

$$q_{rl,t} = \sum_{m=1} q_{el,t} \tag{6.19}$$

where $q_{el,t}$ = each of m space latent heat gains (Btu/hr).

Space Air Temperature and Heat Extraction Rate

At equilibrium, the space sensible heat extraction rate at time t, $q_{xs,t}$, is approximately equal to the space sensible cooling load, $q_{rs,t}$, when zero offset proportional plus integral or proportional-integral-derivative control mode is used. During the cool-down period, the sensible heat extraction rate of the cooling coil or DX coil at time t, $q_{xs,t}$, or the sensible cooling coil load, in Btu/hr, is greater than the space sensible cooling load at time t and the space temperature T_r, in °F then drops gradually. According to *ASHRAE Handbook 1993 Fundamentals*, the relationship between the space temperature T_r and the sensible heat extraction rate $q_{xs,t}$ can be expressed as

$$\sum_{i=0}^{1} p_i \left(q_{xs,t} - q_{rs,t-i\delta} \right) = \sum_{i=0}^{2} g_i \left(T_r - T_{r,t-i\delta} \right) \tag{6.20}$$

where $q_{rs,t-i\delta}$ = space sensible cooling load calculated on the basis of constant space air temperature T_r, Btu/hr

$T_{r,t-i\delta}$ = space temperature at time t-$i\delta$

In Equation (6.20), p_i, g_i are called *space air transfer function coefficients*. Space air temperature T_r can be considered an average reference temperature within a time interval.

Cooling Coil Load

Cooling coil load q_{cc}, in Btu/hr, can be calculated from Equation (3.11). The sensible and latent cooling coil loads can then be calculated as

$$q_{cs} = 60\mathring{V}_s \rho_s c_{pa}\left(T_m - T_{cc}\right)$$

$$q_{cl} = 60\mathring{V}_s \rho_s \left(w_m - w_{cc}\right)$$

(6.21)

where q_{cs}, q_{cl} = sensible and latent cooling coil load, Btu/hr
 T_m, T_{cc} = air temperature of the mixture and leaving the cooling coil, °F
 w_m, w_{cc} = humidity ratio of the air mixture and air leaving the cooling coil, lb/lb

Heating Load

The *space heating load* or simply *heating load* is always the possible maximum heat energy that must be added to the conditioned space at winter design conditions to maintain the indoor design temperature. It is used to size and select the heating equipment. In heating load calculations, solar heat gain, internal heat gains, and the heat storage effect of the building envelope are usually neglected for reliability and simplicity.

Normally, space heating load q_{rh}, in Btu/hr, can be calculated as

$$q_{rl} = q_{trans} + q_{if.s} + q_{ma} + q_{hu}$$

$$= \left[\left(\sum_{n=1} AU\right) + 60\mathring{V}_{if}\rho_o c_{pa}\mathring{m}_m c_m\right]\left(T_r - T_o\right) + \mathring{m}_w h_{fg.58}$$

(6.22)

where A = area of the external walls, roofs, glasses, and floors, ft²
 U = overall heat transfer coefficient of the walls, roofs, glasses, and floors, Btu/hr.ft².°F
 c_m, $\mathring{m}_m$ = specific heat and mass flow rate of the cold product entering the space per hour, Btu/lb.°F and lb/hr
 $\mathring{m}_w$ = mass flow rate of water evaporated for humidification, lb/hr
 $h_{fg.58}$ = latent heat of vaporization at 58°F, 1060 Btu/lb

In Equation 6.22, q_{trans} indicates the *transmission loss* through walls, roofs, glasses, and floors, $q_{if.s}$ the *sensible infiltration heat loss*, q_{ma} the heat required to heat the cold product that enters the conditioned space, and q_{hu} the heat required to raise the space air temperature when water droplets from a space humidifier are evaporated in the conditioned space. For details, refer to *ASHRAE Handbook 1993 Fundamentals*.

7

Air Handling Units and Packaged Units

Shan K. Wang

Consultant

7   Air Handling Units and Packaged Units 71
     Terminals and Air Handling Units • Packaged Units • Coils •
     Air Filters • Humidifiers

Air Handling Units and Packaged Units

Terminals and Air Handling Units

A *terminal unit*, or *terminal*, is a device or equipment installed directly in or above the conditioned space to cool, heat, filter, and mix outdoor air with recirculating air. Fan-coil units, VAV boxes, fan-powered VAV boxes, etc. are all terminals.

An *air handling unit* (AHU) handles and conditions the air, controls it to a required state, and provides motive force to transport it. An AHU is the primary equipment of the air system in a central air-conditioning system. The basic components of an AHU include a supply fan with a fan motor, a water cooling coil, filters, a mixing box except in a makeup AHU unit, dampers, controls, and an outer casing. A return or relief fan, heating coil(s), and humidifier are optional depending on requirements. The supply volume flow rate of AHUs varies from 2000 to about 60,000 cfm.

AHUs are classified into the followings groups according to their structure and location.

Horizontal or Vertical Units

Horizontal AHUs have their fan, coils, and filters installed at the same level as shown in Figure 7.1(a). They need more space and are usually for large units. In *vertical units*, as shown in Figure 7.1(b), the supply fan is installed at a level higher than coils and filters. They are often comparatively smaller than horizontal units.

Draw-Through or Blow-Through Units

In a *draw-through unit*, as shown in Figure 7.1(a), the supply fan is located downstream of the coils. Air is evenly distributed over the coil section, and the fan discharge can easily be connected to a supply duct of nearly the same air velocity. In a *blow-through unit*, as shown in Figure 7.1(c), the supply fan

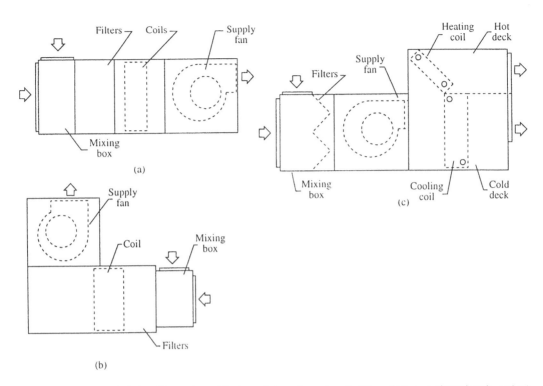

FIGURE 7.1 Type of air handling units: (a) horizontal draw-through unit, (b) vertical draw-through unit, and (c) multizone blow-through unit.

is located upstream of the coils. It usually has hot and cold decks with discharge dampers connected to warm and cold ducts, respectively.

Factory-Fabricated and Field Built-Up Units

Factory-fabricated units are standard in construction and layout, low in cost, of higher quality, and fast in installation. *Field built-up units* or *custom-built units* are more flexible in construction, layout, and dimensions than factory-built standardized units.

Rooftop and Indoor Units

A *rooftop AHU*, sometimes called a penthouse unit, is installed on the roof and will be completely weatherproof. An *indoor AHU* is usually located in a fan room or ceiling and hung like small AHU units.

Make-Up Air and Recirculating Units

A *make-up AHU*, also called a primary-air unit, is used to condition outdoor air entirely. It is a once-through unit. There is no return air and mixing box. *Recirculating units* can have 100% outdoor air intake or mixing of outdoor air and recirculating air.

Packaged Units

A *packaged unit* (PU) is a self-contained air conditioner. It conditions the air and provides it with motive force and is equipped with its own heating and cooling sources. The packaged unit is the primary equipment in a packaged air-conditioning system and is always equipped with a DX coil for cooling, unlike an AHU. R-22, R-134a, and others are used as refrigerants in packaged units. The portion that handles air in a packaged unit is called an *air handler* to distinguish it from an AHU. Like an AHU, an indoor air handler has an indoor fan, a DX coil (indoor coil), filters, dampers, and controls. Packaged units can be classified according to their place of installation: rooftop, indoor, and split packaged units.

Rooftop Packaged Units

A *rooftop packaged unit* is mounted on the roof of the conditioned space as shown in Figure 7.2. From the types of heating/cooling sources provided, rooftop units can be subdivided into:

- Gas/electric rooftop packaged unit, in which heating is provided by gas furnace and cooling by electric power-driven compressors.
- Electric/electric rooftop packaged unit, in which electric heating and electric power-driven compressors provide heating and cooling.
- Rooftop packaged heat pump, in which both heating and cooling are provided by the same refrigeration system using a four-way reversing valve (heat pump) in which the refrigeration flow changes when cooling mode is changed to heating mode and vice versa. Auxiliary electric heating is provided if necessary.

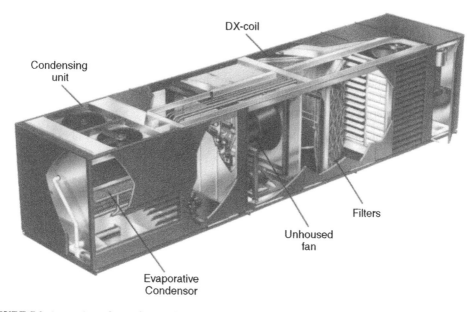

FIGURE 7.2 A cut view of a rooftop package unit. (Source: Mammoth, Inc. Reprinted by permission.)

Rooftop packaged units are single packaged units. Their cooling capacity may vary from 3 to 220 tons with a corresponding volume flow rate of 1200 to 80,000 cfm. Rooftop packaged units are the most widely used packaged units.

Indoor Packaged Units

An *indoor packaged unit* is also a single packaged and factory-fabricated unit. It is usually installed in a fan room or a machinery room. A small or medium-sized indoor packaged unit could be floor mounted directly inside the conditioned space with or without ductwork. The cooling capacity of an indoor packaged unit may vary from 3 to 100 tons and volume flow rate from 1200 to 40,000 cfm.

Indoor packaged units are also subdivided into:

- Indoor packaged cooling units
- Indoor packaged cooling/heating units, in which heating may be provided from a hot water heating coil, a steam heating coil, and electric heating
- Indoor packaged heat pumps

Indoor packaged units have either an air-cooled condenser on the rooftop or a shell-and-tube or double-tube water-cooled condenser inside the unit.

Split Packaged Units

A *split packaged unit* consists of two separate pieces of equipment: an indoor air handler and an outdoor condensing unit. The indoor air handler is often installed in the fan room. Small air handlers can be ceiling hung. The condensing unit is usually located outdoors, on a rooftop or podium or on the ground.

A split packaged unit has its compressors and condenser in its outdoor condensing unit, whereas an indoor packaged unit usually has its compressors indoors. The cooling capacity of split packaged units varies from 3 to 75 tons and the volume flow rate from 1200 to 30,000 cfm.

Rating Conditions and Minimum Performance

Air Conditioning and Refrigeration Institute (ARI) Standards and ASHRAE/IES Standard 90.1-1989 specified the following rating indices:

- Energy efficiency ratio (EER) is the ratio of equipment cooling capacity, in Btu/hr, to the electric input, in W, under rating conditions.
- SEER is the seasonal EER, or EER during the normal annual usage period.
- IPLV is the integrated part-load value. It is the summarized single index of part-load efficiency of PUs based on weighted operations at several load conditions.
- HSPF is the heating seasonal performance factor. It is the total heating output of a heat pump during its annual usage period for heating, in Btu, divided by the total electric energy input to the heat pump during the same period, in watt-hours.

According to ARI standards, the minimum performance for air-cooled, electrically operated single packaged units is

	q_{rc} (Btu/hr)	T_o (°F)	EER	T_o (°F)	IPLV
Air-cooled	<65,000	95	9.5		
	$65,000 \le q_{rc} < 135,000$	95	8.9	80	8.3
	$135,000 \le q_{rc} < 760,000$		8.5		7.5

For water- and evaporatively cooled packaged units including heat pumps, refer to ASHRAE/IES Standard 90.1-1989 and also ARI Standards.

Coils

Coils, Fins, and Water Circuits

Coils are indirect contact heat exchangers. Heat transfer or heat and mass transfer takes place between conditioned air flowing over the coil and water, refrigerant, steam, or brine inside the coil for cooling, heating, dehumidifying, or cooling/dehumidifying. Chilled water, brine, and refrigerants that are used to cool and dehumidify the air are called *coolants*. Coils consist of tubes and external fins arranged in rows along the air flow to increase the contact surface area. Tubes are usually made of copper; in steam coils they are sometimes made of steel or even stainless steel. Copper tubes are staggered in 2, 3, 4, 6, 8, or up to 10 rows.

Fins are extended surfaces often called *secondary surfaces* to distinguish them from the *primary surfaces*, which are the outer surfaces of the tubes. Fins are often made from aluminum, with a thickness $F_t = 0.005$ to 0.008 in., typically 0.006 in. Copper, steel, or sometimes stainless steel fins are also used. Fins are often in the form of continuous plate fins, corrugated plate fins to increase heat transfer, crimped spiral or smooth spiral fins that may be extruded from the aluminum tubes, and spine pipes, which are shaved from the parent aluminum tubes. Corrugated plate fins are most widely used.

Fin spacing S_f is the distance between two fins. *Fin density* is often expressed in fins per inch and usually varies from 8 to 18 fins/in.

In a water cooling coil, *water circuits* or *tube feeds* determine the number of water flow passages. The greater the finned width, the higher the number of water circuits and water flow passages.

Direct Expansion (DX) Coil

In a *direct expansion coil*, the refrigerant, R-22, R-134a, or others, is evaporated and expanded directly inside the tubes to cool and dehumidify the air as shown in Figure 7.3(a). Refrigerant is fed to a distributor and is then evenly distributed to various copper tube circuits typically 0.375 in. in diameter. Fin density is usually 12 to 18 fins/in. and a four-row DX coil is often used. On the inner surface of the copper tubes, microfins, typically at 60 fins/in. and a height of 0.008 in., are widely used to enhance the boiling heat transfer.

Air and refrigerant flow is often arranged in a combination of counterflow and cross flow and the discharge header is often located on the air-entering side. Refrigerant distribution and loading in various circuits are critical to the coil's performance. Vaporized vapor refrigerant is superheated 10 to 20°F in order to prevent any liquid refrigerant from flooding back to the reciprocating compressors and damaging them. Finally, the vapor refrigerant is discharged to the suction line through the header.

For comfort air-conditioning systems, the evaporating temperature of refrigerant T_{ev} inside the tubes of a DX coil is usually between 37 and 50°F. At such a temperature, the surface temperature of the coil is often lower than the dew point of the entering air. Condensation occurs at the coil's outside surface, and the coil becomes a wet coil. A condensate *drain pan* is necessary for each vertically banked DX coil, and a trap should be installed to overcome the negative pressure difference between the air in the coil section and the ambient air.

Face velocity of the DX coil v_a, in fpm, is closely related to the blow-off of the water droplets of the condensate, the heat transfer coefficients, the air-side pressure drop, and the size of the air system. For corrugated fins, the upper limit is 600 fpm, with an air-side pressure drop of 0.20 to 0.30 in. WG/row. A large DX coil is often divided into two refrigerant sections, each with its own expansion valve, distributor, and discharge header.

For a packaged unit of a specific model, size, face velocity and condition of entering air and outdoor air, the DX coil's cooling capacities in nominal tons, number of rows, and fin density are all fixed values.

Water Cooling Coils — Dry–Wet Coils

In a water cooling coil, chilled water at a temperature of 40 to 50°F, brine, or glycol-water at a temperature of 34 to 40°F during cold air distribution enters the coil. The temperature of chilled water, brine, or glycol-water is usually raised 12 to 24°F before it leaves the water cooling coil.

The water tubes are usually copper tubes of 1/2 to 5/8 in. diameter with a tube wall thickness of 0.01 to 0.02 in. They are spaced at a center-to-center distance of 0.75 to 1.25 in. longitudinally and 1 to 1.5 in. transversely. These tubes may be staggered in 2, 3, 4, 6, 8, or 10 rows. Chilled water coils are often operated at a pressure of 175 to 300 psig.

As in a DX coil, the air flow and water flow are in a combination of counterflow and cross flow. The outer surface of a chilled water cooling coil at the air entering side T_{se} is often greater than the dew point of the entering air T''_{ae}, or $T_{se} > T''_{ae}$. The outer surface temperature of coil at the air leaving side T_{sl} may be smaller than T''_{ae}, or $T_{sl} < T''_{ae}$. Then the water cooling coil becomes a dry–wet coil with part of the dry surface on the air entering side and part of the wet surface on the air leaving side. A *dry–wet boundary* divides the dry and wet surfaces. At the boundary, the tube outer surface temperature $T_{sb} = T''_{ae}$ as shown in Figure 7.3(b). A condensate drain pan is necessary for a dry–wet coil.

A water cooling coil is selected from the manufacturer's selection program or from its catalog at (1) a dry and wet bulb of entering air, such as 80°F dry bulb and 67°F wet bulb; (2) an entering water temperature, such as 44 or 45°F; (3) a water temperature rise between 10 and 24°F; and (4) a coil face velocity between 400 and 600 fpm. The number of rows and fins per inch is varied to meet the required sensible and cooling coil load, in Btu/hr.

Water Cooling Coil–Dry Coil

When the temperature of chilled water entering the water cooling coil $T_{we} \geq T''_{ae}$, condensation will not occur on the outer surface of the coil. This coil becomes a sensible cooling–dry coil, and the humidity ratio of the conditioned air w_a remains constant during the sensible cooling process.

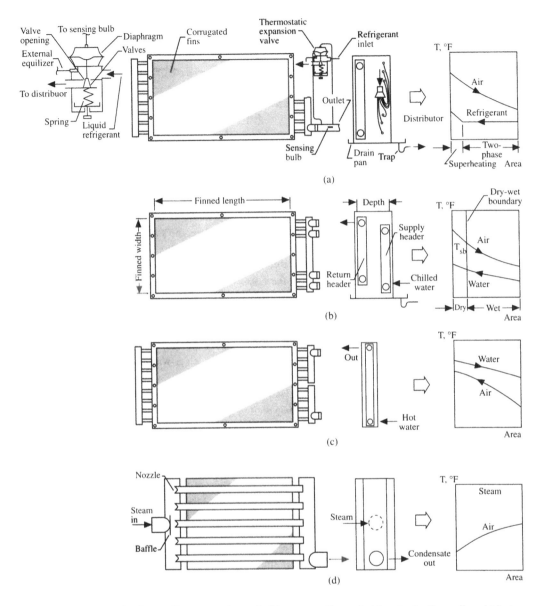

FIGURE 7.3 Types of coils: (a) direct expansion coil, (b) water cooling coil, (c) water heating coil, and (d) steam heating coil.

The construction of a sensible cooling–dry coil, such as material, tube diameter, number of rows, fin density, and fin thickness, is similar to that of a dry–wet coil except that a dry coil always has a poorer surface heat transfer coefficient than a wet coil, and therefore a greater coil surface area is needed; the maximum face velocity of a dry coil can be raised to $v_a \leq 800$ fpm; and the coil's outer surface is less polluted. The effectiveness of a dry coil ϵ_{dry} is usually 0.55 to 0.7.

Water Heating Coil

The construction of a water heating coil is similar to that of a water cooling coil except that in water heating coils hot water is supplied instead of chilled water and there are usually fewer rows, only 2, 3, and 4 rows, than in water cooling coils. Hot water pressure in water heating coils is often rated at 175 to 300 psig at a temperature up to 250°F. Figure 7.3(c) shows a water heating coil.

Steam Heating Coil

In a steam heating coil, latent heat of condensation is released when steam is condensed into liquid to heat the air flowing over the coil, as shown in Figure 7.3(d). Steam enters at one end of the coil, and the condensate comes out from the opposite end. For more even distribution, a baffle plate is often installed after the steam inlet. Steam heating coils are usually made of copper, steel, or sometimes stainless steel.

For a steam coil, the coil core inside the casing should expand or contract freely. The coil core is also pitched toward the outlet to facilitate condensate drainage. Steam heating coils are generally rated at 100 to 200 psig at 400°F.

Coil Accessories and Servicing

Coil accessories include air vents, drain valves, isolation valves, pressure relief valves, flow metering valves, balancing valves, thermometers, pressure gauge taps, condensate drain taps, and even distribution baffles. They are employed depending on the size of the system and operating and serving requirements.

Coil cleanliness is important for proper operation. If a medium-efficiency air filter is installed upstream of the coil, dirt accumulation is often not a problem. If a low-efficiency filter is employed, dirt accumulation may block the air passage and significantly increase the pressure drop across the coil. Coils should normally be inspected and cleaned every 3 months in urban areas when low-efficiency filters are used. Drain pans should be cleaned every month to prevent buildup of bacteria and microorganisms.

Coil Freeze-Up Protection

Improper mixing of outdoor air and recirculating air in the mixing box of an AHU or PU may cause coil freeze-up when the outdoor air temperature is below 32°F. Outdoor air should be guided by a baffle plate and flow in an opposite direction to the recirculating air stream so that they can be thoroughly mixed without stratification.

Run the chilled water pump for the idle coil with a water velocity of 2.5 ft/sec, so that the cooling coil will not freeze when the air temperature drops to 32°F. A better method is to drain the water completely. For a hot water coil, it is better to reset the hot water temperature at part-load operation instead of running the system intermittently. A steam heating coil with inner distributor tubes and outer finned heating tubes provides better protection against freeze-up.

Air Filters

Air Cleaning and Filtration

Air cleaning is the process of removing airborne particles from the air. Air cleaning can be classified into air filtration and industrial air cleaning. Industrial air cleaning involves the removal of dust and gaseous contaminants from manufacturing processes as well as from the space air, exhaust air, and flue gas for air pollution control. In this section, only air filtration is covered.

Air filtration involves the removal of airborne particles presented in the conditioned air. Most of the airborne particles removed by air filtration are smaller than 1 μm, and the concentration of these particles in the airstream seldom exceeds 2 mg/m^3. The purpose of air filtration is to benefit the health and comfort of the occupants as well as meet the cleanliness requirements of the working area in industrial buildings.

An *air filter* is a kind of air cleaner that is installed in AHUs, PUs, and other equipment to filter the conditioned air by inertial impaction or interception and to diffuse and settle fine dust particles on the fibrous medium. The filter medium is the fabricated material that performs air filtration.

Operating performance of air filters is indicated by their:

- *Efficiency* or effectiveness of dust removal
- *Dust holding capacity* m_{dust}, which is the amount of dust held in the air filter, in grains/ft^2
- *Initial pressure drop* when the filter is clean Δp_{fi} and *final pressure drop* Δp_{ff} when the filter's m_{dust} is maximum, both in in. WG
- *Service life*, which is the operating period between Δp_{fi} and Δp_{ff}

Air filters in AHUs and PUs can be classified into low-, medium-, and high-efficiency filters and carbon activated filters.

Test Methods

The performance of air filters is usually tested in a test unit that consists of a fan, a test duct, the tested filter, two samplers, a vacuum pump, and other instruments. Three test methods with their own test dusts and procedures are used for the testing of low-, medium-, and high-efficiency air filters.

The *weight arrestance test* is used for low-efficiency air filters to assess their ability to remove coarse dusts. Standard synthetic dusts that are considerably coarser than atmospheric dust are fed to the test unit. By measuring the weight of dust fed and the weight gain due to the dust collected on the membrane of the sampler after the tested filter, the arrestance can be calculated.

The *atmospheric dust spot efficiency test* is used for medium-efficiency air filters to assess their ability to remove atmospheric dusts. *Atmospheric dusts* are dusts contained in the outdoor air, the outdoor atmosphere. Approximately 99% of atmospheric dusts are dust particles <0.3 μm that make up 10% of the total weight; 0.1% of atmospheric dusts is particles >1 μm that make up 70% of the total weight.

Untreated atmospheric dusts are fed to the test unit. Air samples taken before and after the tested filter are drawn through from identical fiber filter-paper targets. By measuring the light transmission of these discolored white filter papers, the efficiency of the filter can be calculated. Similar atmospheric dust spot test procedures have been specified by American Filter Institute (AFI), ASHRAE Standard 52.1, and former National Bureau of Standards (NBS).

The *DOP penetration and efficiency test* or simply *DOP test* is used to assess high-efficiency filters removing dusts particles of 0.18 μm. According to U.S. Military Standard MIL-STD-282 (1956), a smoke cloud of uniform dioctyl phthalate (DOP) droplets 0.18 μm in diameter, generated from the condensation of the DOP vapor, is fed to the test unit. By measuring the concentration of these particles in the air stream upstream and downstream of the tested filter using an electronic particle counter or laser spectrometer, the penetration and efficiency of the air filter can be calculated.

Low-Efficiency Air Filters

ASHRAE weight arrestance for low-efficiency filters is between 60 and 95%, and ASHRAE dust spot efficiency for low-efficiency filters is less than 20%. These filters are usually in panels as shown in Figure 7.4(a). Their framework is typically 20 × 20 in. or 24 × 24 in. Their thickness varies from 1 to 4 in.

For low-efficiency filters, the filter media are often made of materials such as

- Corrugated wire mesh and screen strips coated with oil, which act as adhesives to enhance dust removal. Detergents may be used to wash off dusts so that the filter media can be cleaned and reused — they are therefore called *viscous and reusable*.

- Synthetic fibers (nylon, terylene) and polyurethane foam can be washed, cleaned, and reused if required — *dry and reusable*.

- Glass fiber mats with fiber diameter greater than 10 μm. The filter medium is discarded when its final pressure drop is reached — *dry and disposable*. The face velocity of the panel filter is usually between 300 and 600 fpm. The initial pressure drop varies from 0.05 to 0.25 in. WG and the final pressure drop from 0.2 to 0.5 in. WG.

Medium-Efficiency Air Filters

These air filters have an ASHRAE dust spot efficiency usually between 20 and 95%. Filter media of medium-efficiency filters are usually made of glass fiber mat with a fiber diameter of 10 to 1 μm using nylon fibers to join them together. They are usually dry and disposable. In addition:

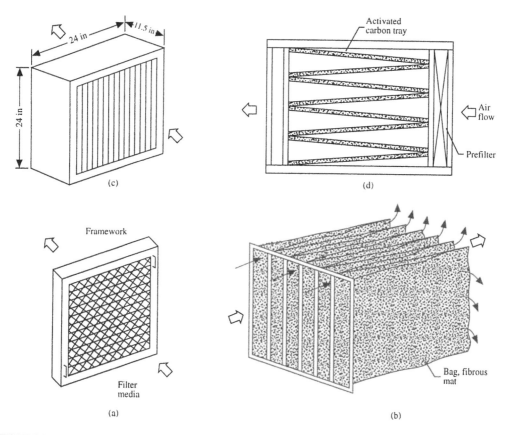

FIGURE 7.4 Various types of air filters: (a) low efficiency, (b) medium efficiency, (c) HEPA and ULPA filters, and (d) activated carbon filter.

- As the dust spot efficiency increases, the diameter of glass fibers is reduced, and they are placed closer together.
- Extended surfaces, such as pleated mats or bags, are used to increase the surface area of the medium as shown in Figure 7.4(b). Air velocity through the medium is 6 to 90 fpm. Face velocity of the air filter is about 500 fpm to match the face velocity of the coil in AHUs and PUs.
- Initial pressure drop varies from 0.20 to 0.60 in. WG and final pressure drop from 0.50 to 1.20 in. WG.

High-Efficiency Particulate Air (HEPA) Filters and Ultra-Low-Penetration Air (ULPA) Filters

HEPA filters have a DOP test efficiency of 99.97% for dust particles ≥0.3 μm in diameter. *ULPA filters* have a DOP test efficiency of 99.999% for dust particles ≥0.12 μm in diameter.

A typical HEPA filter, shown in Figure 7.4(d), has dimensions of 24 × 24 × 11.5 in. Its filter media are made of glass fibers of submicrometer diameter in the form of pleated paper mats. The medium is dry and disposable. The surface area of the HEPA filter may be 50 times its face area, and its rated face velocity varies from 190 to 390 fpm, normally at a pressure drop of 0.50 to 1.35 in. WG for clean filters. The final pressure drop is 0.8 to 2 in. WG. Sealing of the filter pack within its frame and sealing between the frame and the gaskets are critical factors that affect the penetration and efficiency of the HEPA filter.

An ULPA filter is similar to a HEPA filter in construction and filter media. Both its sealing and filter media are more efficient than those of a HEPA filter.

To extend the service life of HEPA filters and ULPA filters, both should be protected by a medium-efficiency filter, or a low-efficiency and a medium-efficiency filter in the sequence low–medium just before the HEPA or ULPA filters. HEPA and ULPA filters are widely used in clean rooms and clean spaces.

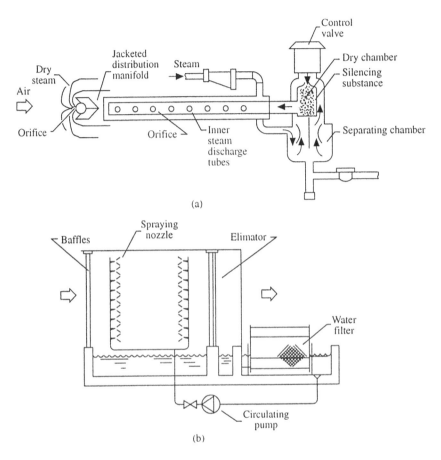

FIGURE 7.5 Steam grid humidifier (a) and air washer (b).

Activated Carbon Filters

These filters are widely used to remove objectional odors and irritating gaseous airborne particulates, typically 0.003 to 0.006 μm in size, from the air stream by adsorption. *Adsorption* is physical condensation of gas or vapor on the surface of an activated substance like activated carbon. Activated substances are extremely porous. One pound of activated carbon contains 5,000,000 ft^2 of internal surface.

Activated carbon in the form of granules or pellets is made of coal, coconut shells, or petroleum residues and is placed in trays to form activated carbon beds as shown in Figure 7.4(d). A typical carbon tray is 23 × 23 × 5/8 in. thick. Low-efficiency prefilters are used for protection. When air flows through the carbon beds at a face velocity of 375 to 500 fpm, the corresponding pressure drop is 0.2 to 0.3 in. WG.

Humidifiers

A *humidifier* adds moisture to the air. Air is humidified by: (1) heating the liquid to evaporate it; (2) atomizing the liquid water into minute droplets by mechanical means, compressed air, or ultrasonic vibration to create a larger area for evaporation; (3) forcing air to flow through a wetted element in which water evaporates; and (4) injecting steam into air directly before it is supplied to the conditioned space.

For comfort air-conditioning systems, a steam humidifier with a separator as shown in Figure 7.5(a) is widely used. Steam is supplied to a jacketed distribution manifold. It enters a separating chamber with its condensate. Steam then flows through a control valve, throttles to a pressure slightly above atmospheric, and enters a dry chamber. Due to the high temperature in the surrounding separating chamber, the steam is superheated. Dry steam is then discharged into the ambient air stream through the orifices on the inner steam discharge tubes.

For an air system of cold air supply with humidity control during winter mode operation, an air washer is economical for large-capacity humidification in many industrial applications.

An air washer is a humidifier, a cooler, a dehumidifier, and an air cleaner. An air washer usually has an outer casing, two banks of spraying nozzles, one bank of guide baffles at the entrance, one bank of eliminators at the exit, a water tank, a circulating pump, a water filter, and other accessories as shown in Figure 7.5(b). Outer casing, baffles, and eliminators are often made of plastics or sometimes stainless steel. Spraying nozzles are usually made of brass or nylon, with an orifice diameter of 1/16 to 3/16 in., a smaller orifice for humidification, and a larger orifice for cooling and dehumidification. An eccentric inlet connected to the discharge chamber of the spraying nozzle gives centrifugal force to the water stream and atomizes the spraying water. Water is supplied to the spraying nozzle at a pressure of 15 to 30 psig. The distance between two spraying banks is 3 to 4.5 ft, and the total length of the air water from 4 to 7 ft. The air velocity inside an air washer is usually 500 to 800 fpm.

Selection of AHUs and PUs

- The size of an AHU is usually selected so that the face velocity of its coil is 600 fpm or less in order to prevent entrained condensate droplets. The cooling and heating capacities of an AHU can be varied by using coils of different numbers of rows and fin densities. The size of a PU is determined by its cooling capacity. Normally, the volume flow rate per ton of cooling capacity in PUs is 350 to 400 cfm. In most packaged units whose supply fans have belt drives, the fan speed can be selected so that the volume flow rate is varied and external pressure is met.

- ASHRAE/IES Standard 90.1-1989 specifies that the selected equipment capacity may exceed the design load only when it is the smallest size needed to meet the load. Selected equipment in a size larger always means a waste of energy and investment.

- To improve the indoor air quality, save energy, and prevent smudging and discoloring building interiors, a medium-efficiency filter of dust spot efficiency ≥50% and an air economizer are preferable for large AHUs and PUs.

8

Refrigeration Components and Evaporative Coolers

Consultant

8   Refrigeration Components and
    Evaporative Coolers 83
    Refrigeration Compressors • Refrigeration Condensers •
    Evaporators and Refrigerant Flow Control Devices •
    Evaporative Coolers

8   Refrigeration Components and Evaporative Coolers

Refrigeration Compressors

A *refrigeration compressor* is the heart of a vapor compression system. It raises the pressure of refrigerant so that it can be condensed into liquid, throttled, and evaporated into vapor to produce the refrigeration effect. It also provides the motive force to circulate the refrigerant through condenser, expansion valve, and evaporator.

According to the compression process, refrigeration compressors can be divided into *positive displacement* and *nonpositive displacement* compressors. A positive displacement compressor increases the pressure of the refrigerant by reducing the internal volume of the compression chamber. Reciprocating, scroll, rotary, and screw compressors are all positive displacement compressors. The centrifugal compressor is the only type of nonpositive displacement refrigeration compressor widely used in refrigeration systems today.

Based on the sealing of the refrigerant, refrigeration compressors can be classified as

- *Hermetic compressors*, in which the motor and the compressor are sealed or welded in the same housing to minimize leakage of refrigerant and to cool the motor windings by using suction vapor
- *Semihermetic compressors*, in which motor and compressor are enclosed in the same housing but are accessible from the cylinder head for repair and maintenance
- *Open compressors*, in which compressor and motor are enclosed in two separate housings

Refrigeration compressors are often driven by motor directly or by gear train.

Performance Indices

Volumetric efficiency η_v of a refrigeration compressor is defined as

$$\eta_v = \dot{V}_{a.v}/\dot{V}_p \tag{8.1}$$

where $\dot{V}_{a.v}$ =actual induced volume of the suction vapor at suction pressure, cfm
   $\dot{V}_p$ =calculated displacement of the compressor, cfm

Isentropic efficiency η_{isen}, *compression efficiency* η_{cp}, compressor efficiency η_{com}, and *mechanical efficiency* η_{mec} are defined as

$$\eta_{isen} = \left(h_2 - h_1\right)/\left(h_2' - h_1\right) = \eta_{cp}\eta_{mec} = \eta_{com}$$

$$\eta_{cp} = W_{sen}/W_v \tag{8.2}$$

$$\eta_{mec} = W_v/W_{com}$$

where h_1, h_2, h_2'=enthalpy of the suction vapor, ideal discharged hot gas, and actual discharged hot gas, respectively, Btu/lb
W_{isen}, W_v, W_{com}= isentropic work = $(h_2 - h_1)$, work delivered to the vapor refrigerant, and work delivered to the compressor shaft, Btu/lb

The actual power input to the compressor P_{com}, in hp, can be calculated as

$$P_{com} = \dot{m}_r\left(h_2 - h_1\right)/\left(42.41\eta_{isen}\eta_{mo}\right)$$

$$\dot{m}_r = \dot{V}_p\,\eta_v\rho_{suc} \tag{8.3}$$

$$\eta_{mo} = P_{com}/P_{mo}$$

where $\dot{m}_r$ = mass flow rate of refrigerant, lb/min
   ρ_{suc} = density of suction vapor, lb/ft³
   P_{mo} = power input to the compressor motor, hp

Power consumption, kW/ton refrigeration, is an energy index used in the HVAC&R industry in addition to EER and COP.

Currently used refrigeration compressors are reciprocating, scroll, screw, rotary, and centrifugal compressors.

Reciprocating Compressors

In a reciprocating compressor, as shown in Figure 8.1(a), a crankshaft connected to the motor shaft drives 2, 3, 4, or 6 single-acting pistons moving reciprocally in the cylinders via a connecting rod.

The refrigeration capacity of a reciprocating compressor is a fraction of a ton to about 200 tons. Refrigerants R-22 and R-134a are widely used in comfort and processing systems and sometimes R-717 in industrial applications. The maximum compression ratio R_{com} for a single-stage reciprocating compressor is about 7. Volumetric efficiency η_v drops from 0.92 to 0.65 when R_{com} is raised from 1 to 6. Capacity control of reciprocating compressor including: on-off and cylinder unloader in which discharge gas is in short cut and return to the suction chamber.

Although reciprocating compressors are still widely used today in small and medium-sized refrigeration systems, they have little room for significant improvement and will be gradually replaced by scroll and screw compressors.

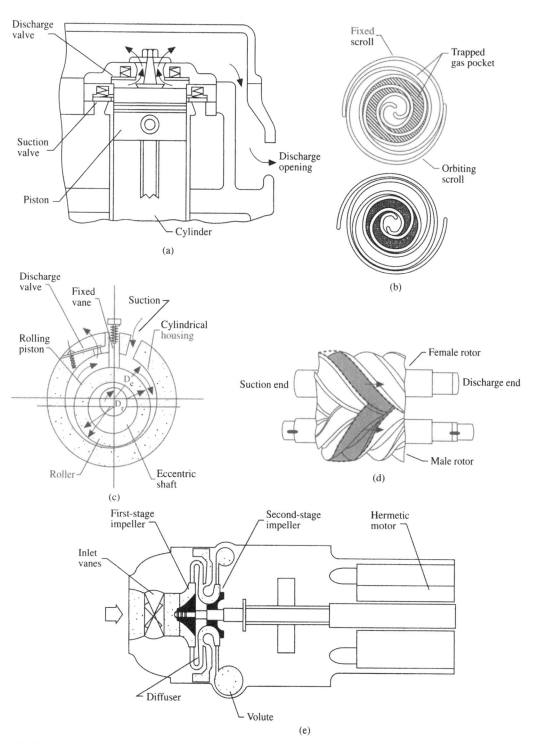

FIGURE 8.1 Various types of refrigeration compressors: (a) reciprocating, (b) scroll, (c) rotary, (d) twin-screw, and (e) centrifugal.

Scroll Compressors

A scroll compressor consists of two identical spiral scrolls assembled opposite to each other, as shown in Figure 9.8.1(b). One of the scrolls is fixed, and the other moves in an orbit around the motor shaft whose amplitude equals the radius of the orbit. The two scrolls are in contact at several points and therefore form a series of pockets.

Vapor refrigerant enters the space between two scrolls through lateral openings. The lateral openings are then sealed and the formation of the two trapped vapor pockets indicates the end of the suction process. The vapor is compressed and the discharge process begins when the trapped gaseous pockets open to the discharge port. Compressed hot gas is then discharged through this opening to the discharge line. In a scroll compressor, the scrolls touch each other with sufficient force to form a seal but not enough to cause wear.

The upper limit of the refrigeration capacity of currently manufactured scroll compressors is 60 tons. A scroll compressor has $\eta_v > 95\%$ at $R_{com} = 4$ and $\eta_{isen} = 80\%$. A scroll compressor also has only about half as many parts as a reciprocating compressor at the same refrigeration capacity. Few components result in higher reliability and efficiency. Power input to the scroll compressor is about 5 to 10% less than to the reciprocating compressor. A scroll compressor also operates more smoothly and is quieter.

Rotary Compressors

Small rotary compressors for room air conditioners and refrigerators have a capacity up to 4 tons. There are two types of rotary compressors: rolling piston and rotating vane. A typical rolling piston rotary compressor is shown in Figure 8.1(c). A rolling piston mounted on an eccentric shaft is kept in contact with a fixed vane that slides in a slot. Vapor refrigerant enters the compression chamber and is compressed by the eccentric motion of the roller. When the rolling piston contacts the top housing, hot gas is squeezed out from the discharge valve.

Screw Compressors

These are also called *helical rotary compressors.* Screw compressors can be classified into single-screw compressors, in which there is a single helical rotor and two star wheels, and twin-screw compressors. Twin-screw compressors are widely used.

A typical twin-screw compressor, as shown in Figure 8.1(d) consists of a four-lobe male rotor and a six-lobe female rotor, a housing with suction and discharge ports, and a sliding valve to adjust the capacity during part load. Normally, the male rotor is the driver. Twin-screw compressors are often direct driven and of hermetic type.

Vapor refrigerant is extracted into the interlobe space when the lobes are separated at the suction port. During the successive rotations of the rotor, the volume of the trapped vapor is compressed. When the interlobe space is in contact with the discharge port, the compressed hot gas discharges through the outlet. Oil injection effectively cools the rotors and results in a lower discharge temperature. Oil also provides a sealing effect and lubrication. A small clearance of 0.0005 in. as well as the oil sealing minimizes leakage of the refrigerant.

The refrigeration capacity of twin-screw compressors is 50 to 1500 tons. The compression ratio of a twin-screw compressor can be up to 20:1. R-22 and R-134a are the most widely used refrigerants in comfort systems. In a typical twin-screw compressor, η_v decreases from 0.92 to 0.87 and η_{isen} drops from 0.82 to 0.67 when R_{com} increases from 2 to 10. Continuous and stepless capacity control is provided by moving a sliding valve toward the discharge port, which opens a shortcut recirculating passage to the suction port.

Twin-screw compressors are more efficient than reciprocating compressors. The low noise and vibration of the twin-screw compressor together with its positive displacement compression results in more applications today.

Centrifugal Compressors

A *centrifugal compressor* is a turbomachine and is similar to a centrifugal fan. A hermetic centrifugal compressor has an outer casing with one, two, or even three impellers internally connected in series and is driven by a motor directly or by a gear train. At the entrance to the first-stage impeller are inlet guide vanes positioned at a specific opening to adjust refrigerant flow and therefore the capacity of the centrifugal compressor.

Figure 8.1(e) shows a two-stage hermetic centrifugal compressor. The total pressure rise in a centrifugal compressor, often called head lift, in psi, is due to the conversion of the velocity pressure into static pressure. Although the compression ratio R_{com} of a single-stage centrifugal compressor using R-123 and R-22 seldom exceeds 4, two or three impellers connected in series satisfy most of the requirements in comfort systems.

Because of the high head lift to raise the evaporating pressure to condensing pressure, the discharge velocity at the exit of the second-stage impeller approaches the acoustic velocity of saturated vapor v_{ac} of R-123, 420 ft/sec at atmospheric pressure and a temperature of 80°F. Centrifugal compressors need high peripheral velocity and rotating speeds (up to 50,000 rpm) to produce such a discharge velocity. It is not economical to manufacture small centrifugal compressors. The available refrigeration capacity for centrifugal compressors ranges from 100 to 10,000 tons. Centrifugal compressors have higher volume flow per unit refrigeration capacity output than positive displacement compressors. Centrifugal compressors are efficient and reliable. Their volumetric efficiency almost equals 1. At design conditions, their η_{isen} may reach 0.83, and it drops to 0.6 during part-load operation. They are the most widely used refrigeration compressors in large air-conditioning systems.

Refrigeration Condensers

A *refrigeration condenser* or simply a *condenser* is a heat exchanger in which hot gaseous refrigerant is condensed into liquid and the latent heat of condensation is rejected to the atmospheric air, surface water, or well water. In a condenser, hot gas is first desuperheated, then condensed into liquid, and finally subcooled.

The capacity of a condenser is rated by its *total heat rejection* Q_{rej}, in Btu/hr, which is defined as the total heat removed from the condenser during desuperheating, condensation, and subcooling. For a refrigeration system using a hermetic compressor, Q_{rej} can be calculated as

$$Q_{rej} = U_{con} A_{con} \Delta T_m = 60 \dot{m}_r (h_2 - h_3') = q_{rl} + (2545 P_{com})/\eta_{mo} \qquad (8.4)$$

where
U_{con} = overall heat transfer coefficient across the tube wall in the condenser, Btu/hr.ft².°F
A_{con} = condensing area in the condenser, ft²
ΔT_m = logarithmic temperature difference, °F
$\dot{m}_r$ = mass flow rate of refrigerant, lb/min
h_2, h_3' = enthalpy of suction vapor refrigerant and hot gas, Btu/lb
q_{rl} = refrigeration load at the evaporator, Btu/hr

A factor that relates Q_{rej} and q_{rl} is the *heat rejection factor* F_{rej}, which is defined as the ratio of total heat rejection to the refrigeration load, or

$$F_{rej} = Q_{rej}/q_{rl} = 1 + (2545 P_{com})/(q_{rl} \eta_{mo}) \qquad (8.5)$$

Fouling factor R_f, in hr.ft².°F/Btu, is defined as the additional resistance caused by a dirty film of scale, rust, or other deposits on the surface of the tube. ARI Standard 550-88 specifies the following for evaporators and condensers:

Field fouling allowance   0.00025 hr.ft².°F/Btu
New evaporators and condensers   0

According to the cooling process used during condensation, refrigeration condensers can be classified as air-cooled, water-cooled, and evaporative-cooled condensers.

Air-Cooled Condensers

In an *air-cooled condenser*, air is used to absorb the latent heat of condensation released during desuperheating, condensation, and subcooling.

An air-cooled condenser consists of a condenser coil, a subcooling coil, condenser fans, dampers, and controls as shown in Figure 8.2(a). There are refrigeration circuits in the condensing coil. Condensing coils are usually made of copper tubes and aluminum fins. The diameter of the tubes is 1/4 to 3/4 in., typically 3/8 in., and the fin density is 8 to 20 fins/in. On the inner surface of the copper tubes, microfins, typically 60 fins/in. with a height of 0.008 in., are used. A condensing coil usually has only two to three rows due to the low pressure drop of the propeller-type condenser fans. A subcooling coil is located at a lower level and is connected to the condensing coil.

Hot gas from the compressor enters the condensing coil from the top. When the condensate increases, part of the condensing area can be used as a subcooling area. A receiver is necessary only when the liquid refrigerant cannot all be stored in the condensing and subcooling coils during the shut-down period in winter.

Cooling air is drawn through the coils by a condenser fan(s) for even distribution. Condenser fans are often propeller fans for their low pressure and large volume flow rate. A damper(s) may be installed to adjust the volume flow of cooling air.

In air-cooled condensers, the volume flow of cooling air per unit of total heat rejection $\dot{V}_{ca}/Q_{u.rej}$ is 600 to 1200 cfm/ton of refrigeration capacity at the evaporator, and the optimum value is about 900 cfm/ton. The corresponding cooling air temperature difference — cooling air leaving temperature minus outdoor temperature $(T_{ca.l} - T_o)$ — is around 13°F.

The condenser temperature difference (CTD) for an air-cooled condenser is defined as the difference between the saturated condensing temperature corresponding to the pressure at the inlet and the air intake temperature, or $(T_{con.i} - T_o)$. Air-cooled condensers are rated at a specific CTD, depending on the evaporating temperature of the refrigeration system T_{ev} in which the air-cooled condenser is installed. For a refrigeration system having a lower T_{ev}, it is more economical to equip a larger condenser with a smaller CTD. For a comfort air-conditioning system having a T_{ev} of 45°F, CTD = 20 to 30°F.

A higher condensing temperature T_{con}, a higher condensing pressure p_{con}, and a higher compressor power input may be due to an undersized air-cooled condenser, lack of cooling air or low $\dot{V}_{ca}/Q_{u.rej}$ value, a high entering cooling air temperature at the roof, a dirty condensing coil, warm air circulation because of insufficient clearance between the condenser and the wall, or a combination of these. The clearance should not be less than the width of the condensing coil.

If p_{con} drops below a certain value because of a lower outdoor temperature, the expansion valve in a reciprocating vapor compression system may not operate properly. At a low ambient temperature T_o, the following controls are often used:

- Duty cycling, turning the condenser fans on and off until all of them are shut down, to reduce cooling air volume flow
- Modulating the air dampers to reduce the volume flow
- Reducing the fan speed

Some manufacturers' catalogs start low ambient control at $T_o = 65°F$ and some specify a minimum operating temperature at $T_o = 0°F$.

Water-Cooled Condensers

In a *water-cooled condenser*, latent heat of condensation released from the refrigerant during condensation is extracted by water. This cooling water, often called condenser water, is taken directly from river, lake, sea, underground well water or a cooling tower.

Two types of water-cooled condensers are widely used for air-conditioning and refrigeration: double-tube condensers and horizontal shell-and-tube condensers.

A *double-tube condenser* consists of two tubes, one inside the other. Condenser water is pumped

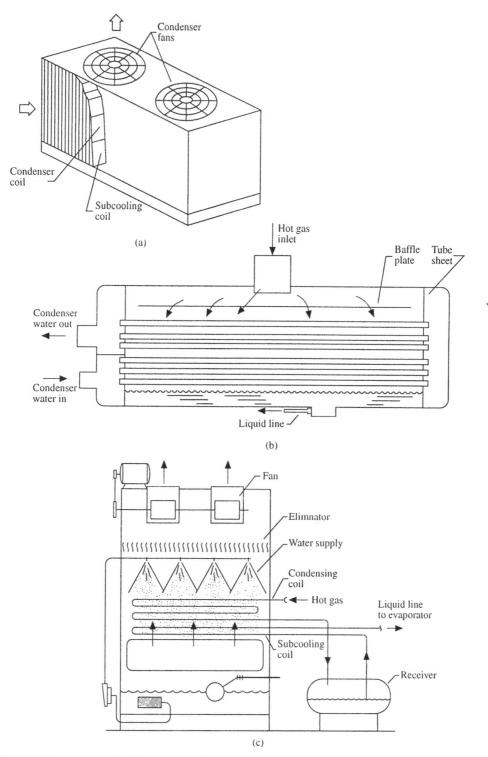

FIGURE 8.2 Various types of refrigeration condensers: (a) air-cooled, (b) shell-and-tube water-cooled, and (c) evaporative cooled.

through the inner tube and refrigerant flows within the space between the inner and outer tubes in a counterflow arrangement. Because of its limited condensing area, the double-tube condenser is used only in small refrigeration systems.

A horizontal *shell-and-tube water-cooled condenser* using halocarbon refrigerant usually has an outer shell in which copper tubes typically 5/8 to 3/4 in. in diameter are fixed in position by tube sheets as shown in Figure 8.2(b). Integral external fins of 19 to 35 fins/in. and a height of 0.006 in. and spiral internal grooves are used for copper tubes to increase both the external and the inner surface area and their heat transfer coefficients.

Hot gas from the compressor enters the top inlet and is distributed along the baffle to fill the shell. Hot gas is then desuperheated, condensed, subcooled into liquid, and discharged into the liquid line at the bottom outlet. Usually one sixth of the volume is filled with subcooled liquid refrigerant. Subcooling depends on the entering temperature of condenser water T_{ce}, in °F, and usually varies between 2 and 8°F.

Condenser water enters the condenser from the bottom for effective subcooling. After extracting heat from the gaseous refrigerant, condenser water is discharged at a higher level. Two-pass or three-pass water flow arrangements are usually used in shell-and-tube water-cooled condensers. The two-pass arrangement means that water flows from one end to the opposite end and returns to the original end. Two-pass is the standard setup. In a shell-and-tube water-cooled condenser, the condensing temperature T_{con} depends mainly on the entering temperature of condenser water T_{ce}, the condenser area, the fouling factor, and the configuration of the copper tube.

Evaporative Condenser

An *evaporative condenser* uses the evaporation of water spray on the outer surface of the condensing tubes to remove the latent heat of condensation of refrigerant during condensation.

An evaporative condenser consists of a condensing coil, a subcooling coil, a water spray, an induced draft or sometimes forced draft fan, a circulating water pump, a water eliminator, a water basin, an outer casing, and controls as shown in Figure 8.2(c). The condensing coil is usually made of bare copper, steel, or sometimes stainless steel tubing.

Water is sprayed over the outside surface of the tubing. The evaporation of a fraction of condenser water from the saturated air film removes the sensible and latent heat rejected by the refrigerant. The wetted outer surface heat transfer coefficient h_{wet} is about four or five times greater than the dry surface heat transfer coefficient h_o, in Btu/hr.ft2.°F. The rest of the spray falls and is collected by the basin. Air enters from the inlet just above the basin. It flows through the condensing coil at a face velocity of 400 to 700 fpm, the water spray bank, and the eliminator. After air absorbs the evaporated water vapor, it is extracted by the fan and discharged at the top outlet. The water circulation rate is about 1.6 to 2 gpm/ton, which is far less than that of the cooling tower.

An evaporative condenser is actually a combination of a water-cooled condenser and a cooling tower. It is usually located on the rooftop and should be as near the compressor as possible. Clean tube surface and good maintenance are critical factors for evaporative condensers. An evaporative condenser also needs low ambient control similar as in an air-cooled condenser.

Comparison of Air-Cooled, Water-Cooled, and Evaporative Condensers

An air-cooled condenser has the highest condensing temperature Tcon and therefore the highest compressor power input. For an outdoor dry bulb temperature of 90°F and a wet bulb temperature of 78°F, a typical air-cooled condenser has Tcon = 110°F. An evaporative condenser has the lowest Tcon and is most energy efficient. At the same outdoor dry and wet bulb temperatures, its Tcon may be equal to 95°F, even lower than that of a water-cooled condenser incorporating with a cooling tower, whose Tcon may be equal to 100°F. An evaporative condenser also consumes less water and pump power. The drawback of evaporative condensers is that the rejected heat from the interior zone is difficult to recover and use as winter heating for perimeter zones and more maintenance is required.

Evaporators and Refrigerant Flow Control Devices

An *evaporator* is a heat exchanger in which the liquid refrigerant is vaporized and extracts heat from the surrounding air, chilled water, brine, or other substance to produce a refrigeration effect.

Evaporators used in air-conditioning can be classified according to the combination of the medium to be cooled and the type of refrigerant feed, as the following.

Direct expansion DX coils are air coolers, and the refrigerant is fed according to its degree of superheat after vaporization. DX coils were covered earlier.

Direct expansion ice makers or *liquid overfeed ice makers* are such that liquid refrigerant is forced through the copper tubes or the hollow inner part of a plate heat exchanger and vaporized. The refrigeration effect freezes the water in the glycol-water that flows over the outside surface of the tubes or the plate heat exchanger. In direct expansion ice makers, liquid refrigerant completely vaporizes inside the copper tubes, and the superheated vapor is extracted by the compressor. In liquid overfeed ice makers, liquid refrigerant floods and wets the inner surface of the copper tubes or the hollow plate heat exchanger. Only part of the liquid refrigerant is vaporized. The rest is returned to a receiver and pumped to the copper tubes or plate heat exchanger again at a circulation rate two to several times greater than the evaporation rate.

Flooded shell-and-tube liquid coolers, or simply *flooded liquid coolers*, are such that refrigerant floods and wets all the boiling surfaces and results in high heat transfer coefficients. A flooded shell-and-tube liquid cooler is similar in construction to a shell-and-tube water-cooled condenser, except that its liquid refrigeration inlet is at the bottom and the vapor outlet is at the top. Water velocity inside the copper tubes is usually between 4 and 12 ft/sec and the water-side pressure normally drops below 10 psi. Flooded liquid coolers can provide larger evaporating surface area and need minimal space. They are widely used in large central air-conditioning systems.

Currently used refrigerant flow control devices include thermostatic expansion valves, float valves, multiple orifices, and capillary tubes.

A *thermostatic expansion valve* throttles the refrigerant pressure from condensing to evaporating pressure and at the same time regulates the rate of refrigerant feed according to the degree of superheat of the vapor at the evaporator's exit. A thermostatic expansion valve is usually installed just prior to the refrigerant distributor in DX coils and direct-expansion ice makers.

A thermostatic expansion valve consists of a valve body, a valve pin, a spring, a diaphragm, and a sensing bulb near the outlet of the DX coil, as shown in Figure 7.3(a). The sensing bulb is connected to the upper part of the diaphragm by a connecting tube.

When the liquid refrigerant passes through the opening of the thermostatic expansion valve, its pressure is reduced to the evaporating pressure. Liquid and a small fraction of vaporized refrigerant then flow through the distributor and enter various refrigerant circuits. If the refrigeration load of the DX coil increases, more liquid refrigerant evaporizes. This increases the degree of superheat of the leaving vapor at the outlet and the temperature of the sensing bulb. A higher bulb temperature exerts a higher saturated pressure on the top of the diaphragm. The valve pin then moves downward and widens the opening. More liquid refrigerant is allowed to enter the DX coil to match the increase of refrigeration load. If the refrigeration load drops, the degree of superheat at the outlet and the temperature of the sensing bulb both drop, and the valve opening is narrower. The refrigeration feed decreases accordingly. The degree of superheat is usually 10 to 20°F. Its value can also be adjusted manually by varying the spring tension.

A *float valve* is a valve in which a float is used to regulate the valve opening to maintain a specific liquid refrigerant level. A lower liquid level causes a lower valve pin and therefore a wider opening and vice versa.

In a centrifugal refrigeration system, two or more orifice plates, *multiple orifices*, are sometimes installed in the liquid line between the condenser and the flash cooler and between the flash cooler and the flooded liquid cooler to throttle their pressure as well as to regulate the refrigerant feed.

A *capillary tube*, sometimes called a *restrictor tube*, is a fixed length of small-diameter tubing installed between the condenser and the evaporator to throttle the refrigerant pressure from p_{con} to p_{ev} and to meter the refrigerant flow to the evaporator. Capillary tubes are usually made of copper. The inside diameter D_{cap} is 0.05 to 0.06 in. and the length L_{cap} from an inch to several feet. There is a trend to use short capillary tubes of L_{cap}/D_{cap} between 3 and 20. Capillary tubes are especially suitable for a heat pump system in which the refrigerant flow may be reversed.

Evaporative Coolers

An evaporative cooling system is an air-conditioning system in which air is cooled evaporatively. It consists of evaporative coolers, fan(s), filters, dampers, controls, and others. A mixing box is optional. An evaporative cooler could be a stand-alone cooler or installed in an air system as a component. There are three types of evaporative coolers: (1) direct evaporative coolers, (2) indirect evaporative coolers, and (3) indirect–direct evaporative coolers.

Direct Evaporative Cooler

In a *direct evaporative cooler*, the air stream to be cooled directly contacts the water spray or wetted medium as shown in Figure 8.3(a). Evaporative pads made of wooden fibers with necessary treatment at a thickness of 2 in., rigid and corrugated plastics, impregnated cellulose, or fiber glass all dripping with water are wetted mediums. The direct evaporation process 12 takes place along the thermodynamic wet bulb line on the psychrometric chart. Saturation effectiveness ε_{sat} is an index that assesses the performance of a direct evaporative cooler:

$$\varepsilon_{sat} = \left(T_{ae} - T_{al}\right)/\left(T_{ae} - T_{ae}^{*}\right) \tag{8.6}$$

where T, T^{*} = temperature and thermodynamic wet bulb temperature of air stream, °F. Subscript ae indicates the entering air and al the leaving air. ε_{sat} usually varies between 0.75 and 0.95 at a water–air ratio of 0.1 to 0.4.

Indirect Evaporative Coolers

In an *indirect evaporative cooler*, the cooled-air stream to be cooled is separated from a wetted surface by a flat plate or tube wall as shown in Figure 8.3(b). A wet-air stream flows over the wetted surface so that liquid water is evaporated and extracts heat from the cooled-air stream through the flat plate or tube wall. The cooled-air stream is in contact with the wetted surface indirectly.

The core part of an indirect evaporative cooler is a plate heat exchanger. It is made of thin polyvinyl chloride plates 0.01 in. thick and spaced from 0.08 to 0.12 in. apart to form horizontal passages for cooled air and vertical passages for wet air and water. As in a direct evaporative cooler, there are also fan(s), water sprays, circulating pump, air intake, dampers, controls, etc.

An indirect evaporative cooling process is represented by a horizontal line on a psychrometric chart, which shows that humidity ratio remains constant. If the space air is extracted and used as the wet air intake, the wet air will be exhausted at point x at nearly saturated state.

The performance of an indirect evaporative cooler can be assessed by its performance factor e_{in}, which is calculated as:

$$e_{in} = \left(T_{ca,e} - T_{ca,l}\right)/\left(T_{ca,e} - T_{s,a}\right) \tag{8.7}$$

where $T_{ca,e}$, $T_{ca,l}$ = temperature of cooled air entering and leaving the indirect evaporative cooler, °F, and $T_{s,a}$ = temperature of the saturated air film on the wet air side and is about 3°F higher than the wet bulb temperature of the entering air, °F.

An indirect evaporative cooler could be so energy efficient as to provide evaporative cooling with an EER up to 50 instead of 9 to 12 for a reciprocating compression refrigeration system.

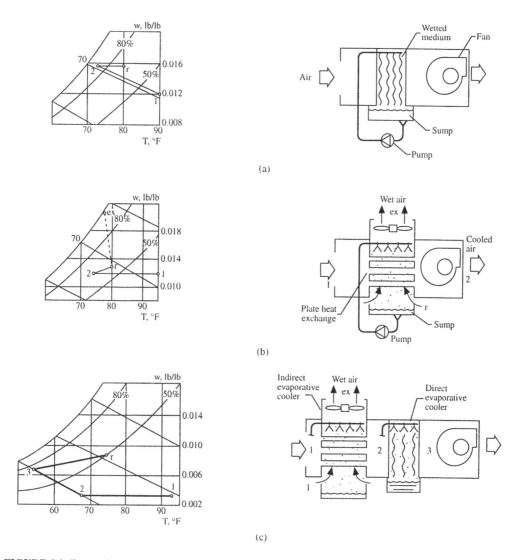

FIGURE 8.3 Types of evaporative coolers: (a) direct, (b) indirect, and (c) indirect–direct.

Direct–Indirect Evaporative Cooler. A direct–indirect evaporative cooler is a two-stage evaporating cooler, in which the first-stage indirect evaporative cooler is connected in series with a second-stage direct evaporative cooler for the purpose of increasing the evaporating effect.

Operating Characteristics. The saturation effectiveness ε_{sat} and performance factor e_{in} are both closely related to the air velocity flowing through the air passages. For a direct evaporative cooler, face velocity is usually less than 600 fpm to reduce drift carryover. For an indirect evaporative cooler, face velocity v_s is usually between 400 to 1000 fpm. A higher v_s results at a greater air-side pressure drop.

Scofield et al. (1984) reported the performance of an indirect–direct evaporative cooler in Denver, Colorado. Outdoor air enters the indirect cooler at a dry bulb of 93°F and a wet bulb of 67.5° and was evaporatively cooled to 67.5°F dry bulb and 49.8°F wet bulb with an $e_{in} = 0.76$ as shown in Figure 8.3(c). In the direct cooler, conditioned air was further cooled to a dry bulb of 53.5°F and the wet bulb remained at 49.8°F at a saturation effectiveness $\varepsilon_{sat} = 0.8$.

In locations where outdoor wet bulb $T_o' \leq 60°F$, a direct evaporative can often provide an indoor environment of 78°F and a relative humidity of 60%. In locations $T_o' \leq 68°F$, an indirect–direct evaporative cooler can maintain a comfortable indoor environment. In locations $T'_o \geq 72°F$, an evaporative cooler with a supplementary vapor compression refrigeration may be cost effective. Because the installation cost of an indirect–direct cooler is higher than that of refrigeration, cost analysis is required to select the right choice. Evaporative coolers are not suitable for dehumidification except in locations where $T'_o \leq 60°F$.

9

Water Systems

Shan K. Wang

Consultant

9   Water Systems..95
     Types of Water Systems • Basics • Water Piping • Plant-
     Building Loop • Plant-Distribution-Building Loop

Water Systems

Types of Water Systems

In central and space conditioning air-conditioning systems, water that links the central plant and the air handling units or terminals, that extracts condensing heat, or that provides evaporative cooling may be classified as

- *Chilled water system*, in which chilled water is first cooled in the centrifugal, screw, and reciprocating chillers in a central plant. Chilled water is then used as a cooling medium to cool the air in the cooling coils in AHUs and terminals.
- *Evaporative-cooled water system*, used to cool air directly or indirectly in evaporative coolers.
- *Hot water system*, in which hot water is heated in the boiler and then used to heat the air through heating coils in AHUs, terminals, or space finned-tube heaters.
- *Dual-temperature water system*, in which chilled water and hot water are supplied to and returned from the coils in AHUs and terminals through separate or common main and branch pipes. Using common main and branch pipes requires a lengthy changeover from chilled water to hot water or vice versa for a period of several hours.
- *Condenser water system*, which is a kind of cooling water system used to extract the latent heat of condensation from the condensing refrigerant in a water-cooled condenser and heat of absorption from the absorber.

Water systems can also be classified according to their operating characteristics.

Closed System

In a closed system, water forms a closed loop for water conservation and energy saving when it flows through the coils, chillers, boilers, heaters, or other heat exchangers and water is not exposed to the atmosphere.

Open System

In an open system, water is exposed to the atmosphere.

Once-Through System

In a once-through system, water flows through a heat exchanger(s) only once without recirculation.

Basics

Volume Flow and Temperature Difference

The rate of heat transfer between water and air or water and refrigerant when water flows through a heat exchanger q_w, in Btu/hr, can be calculated as

$$q_w = 500 \, \mathring{V}_{gal} \left(T_{wl} - T_{we} \right) = 500 \, \mathring{V}_{gal} \, \Delta T_w \tag{9.1}$$

where $\mathring{V}_{gal}$ =volume flow rate of water, gpm
$\quad T_{wl}, T_{we}$ =temperature of water leaving and entering the heat exchanger, °F
$\quad\quad \Delta T_w$ =temperature rise or drop of water when it flows through a heat exchanger, °F

The temperature of chilled water leaving the water chiller T_{el} should not be lower than 38°F in order to prevent freezing in the evaporator. Otherwise, brine or glycol-water should be used. The T_{el} of chilled water entering the coil T_{we} and the temperature difference of chilled water leaving and entering the coil ΔT_w directly affect the temperature of air leaving the cooling coil T_{ce}. The lower T_{we}, the higher will be the compressor power input. The smaller ΔT_w, the greater will be the water volume flow rate, the pipe size, and the pump power. For chilled water in conventional comfort systems, T_{we} is usually 40 to 45°F and ΔT_w 12 to 24°F. Only in cold air distribution, T_{we} may drop to 34°F. For a cooling capacity of 1 ton refrigeration, a ΔT_w of 12°F requires a $\mathring{V}_{gal} = 2$ gpm.

For hot water heating systems in buildings, hot water often leaves the boiler and enters the heating coil or heaters at a temperature T_{we} of 190 to 200°F. It returns at 150 to 180°F. For dual-temperature systems, hot water is usually supplied at 100 to 150°F and returns at a ΔT_w of 20 to 40°F.

Pressure Drop

Usually the pressure drop of water in pipes due to friction for HVAC&R systems, H_f, is in the range 0.75 ft/100 ft length of pipe to 4 ft/100 ft. A pressure loss of 2.5 ft/100 ft is most often used. ASHRAE/IES Standard 90.1-1989 specifies that water piping systems should be designed at a pressure loss of no more than 4.0 ft/100 ft. Figure 9.1(a), (b), and (c) shows the friction charts for steel, copper, and plastic pipes for closed water systems.

Water Piping

The piping materials of various water systems for HVAC&R are as follows:

Chilled water	Black and galvanized steel
Hot water	Black steel, hard copper
Condenser water	Black steel, galvanized ductile iron, polyvinyl chloride (PVC)

The pipe thickness varies from Schedule 10, a light wall pipe, to Schedule 160, a very heavy wall pipe. Schedule 40 is the standard thickness for a pipe of up to 10 in. diameter. For copper tubing, type K is the heaviest, and type L is generally used as the standard for pressure copper tubes.

Steel pipes of small diameter are often joined by threaded cast-iron fittings. Steel pipes of diameter 2 in. and over, welded joints, and bolted flanges are often used.

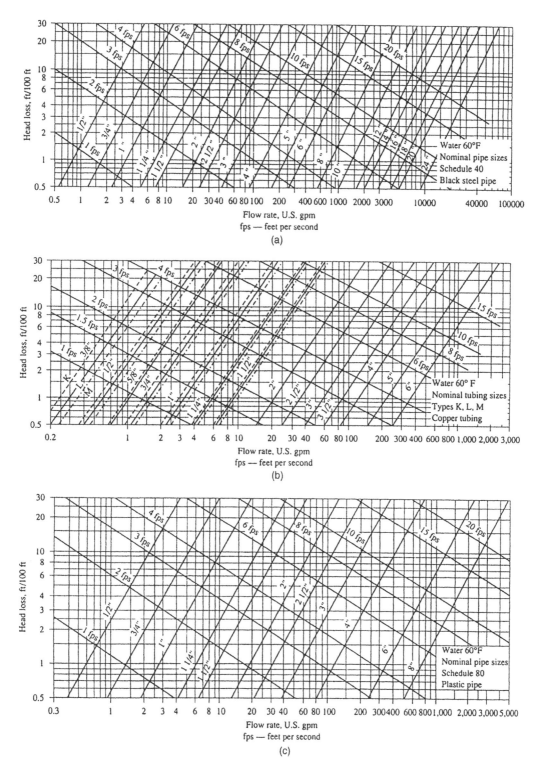

FIGURE 9.1 Friction chart for water in pipes: (a) steel pipe (schedule 40), (b) copper tubing, and (c) plastic pipe (schedule 80). (Souce: *ASHRAE Handbook 1993 Fundamentals.* Reprinted with permission.)

In a water system, the maximum allowable working pressure for steel and copper pipes at 250°F varies from 125 to 400 psig, depending on the pipe wall thickness. Not only pipes, but also their joints and fittings should be considered.

During temperature changes, pipes expand and contract. Both operating and shut-down periods should be taken into consideration. Bends like U-, Z-, and L-bends, loops, and sometimes packed expansion joints, bellows, or flexible metal hose mechanical joints are used.

ASHRAE/IES Standard 90.1-1989 specifies that for chilled water between 40 to 55°F, the minimum thickness of the external pipe insulation varies from 0.5 to 1 in. depending on the pipe diameter. For chilled water temperature below 40°F, the minimum thickness varies from 1 to 1.5 in. For hot water temperatures not exceeding 200°F, the minimum thickness varies from 0.5 to 1.5 in. depending on the pipe diameter and the hot water temperature.

Corrosion, Impurities, and Water Treatments

Corrosion is a destructive process caused by a chemical or electrochemical reaction on metal or alloy. In water systems, dissolved *impurities* cause corrosion and scale and the growth of microbiologicals like algae, bacteria, and fungi. *Scale* is the deposit formed on a metal surface by precipitation of the insoluble constituents. In addition to the dissolved solids, unpurified water may contain suspended solids.

Currently used chemicals include crystal modifiers to change the crystal formation of scale and sequestering chemicals. Growth of bacteria, algae, and fungi is usually treated by biocides to prevent the formation of an insulating layer resulting in lower heat transfer as well as restricted water flow. Chlorine and its compounds are effective and widely used. Blow-down is an effective process in water treatment and should be considered as important as chemical treatments.

Piping Arrangements

MAIN AND BRANCH PIPES. IN A PIPING CIRCUIT AS SHOWN IN FIGURE 9.2(A), chilled water from a chiller or hot water from a boiler is often supplied to a *main pipe* and then distributed to *branch pipes* that connect to coils and heat exchangers. Chilled or hot water from the coils and heat exchangers is accumulated by the return main pipe through return branch pipes and then returned to the chiller or boiler.

Constant Flow and Variable Flow. In a constant-flow water system, the volume flow rate at any cross-sectional plane of the supply and return mains remains constant during the entire operating period. In a variable-flow water system, the volume flow rate varies when the system load changes during the operating period.

Direct Return and Reverse Return. In a *direct return* system, the water supplies to and returns from various coils through various piping circuits. ABCHJKA, ... ABCDEFGHJKA are not equal in length, as shown in Figure 9.2(a). Water flow must be adjusted and balanced by using balance valves to provide required design flow rates at design conditions. In a *reverse-return* system, as shown in Figure 9.2(b), the piping lengths for various piping circuits including the branch and coil are almost equal. Water flow rates to various coils are easier to balance.

Two-Pipe or Four-Pipe. In a dual-temperature water system, the piping from the chiller or boiler to the coils can be either a *two-pipe* system with a supply main and return main as shown in Figure 9.2(a) or (b) or a *four-pipe* system with a chilled water supply main, a hot water supply main, a chilled water return main, and a hot water return main as shown in Figure 9.2(c). The two-pipe system needs a changeover from chilled to hot water and vice versa. A four-pipe system is more expensive to install.

Plant-Building Loop

System Description

The chillers/boilers in a central plant are often located in the basement, machinery room, or on the rooftop of the building. Generally, a fairly constant-volume water flow is required in the evaporator to

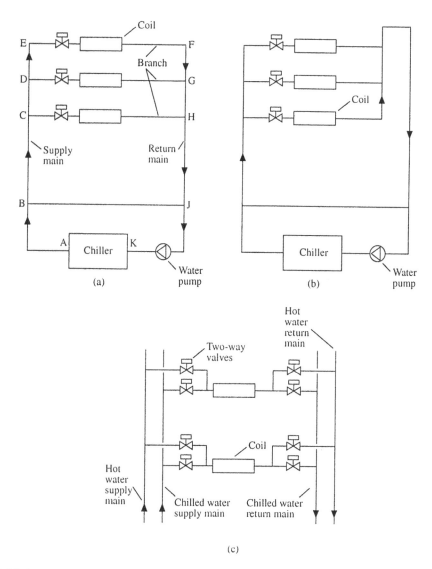

FIGURE 9.2 Piping arrangements: (a) two-pipe direct return system, (b) two-pipe reverse system, and (c) four-pipe system.

protect it from freezing at part load. ASHRAE/IES Standard 90.1-1989 specifies that water systems using control valves to modulate the coil load must be designed for variable flow for energy savings.

Current chilled water systems or dual-temperature water systems often adopt a *plant-building loop* that consists of two piping loops as shown in Figure 9.3(a) for reliable and energy-efficient operation:

Plant Loop. A plant loop ABJKA is comprised of chiller(s)/boiler(s); plant chilled/hot water pump(s) usually of split-case, double suction, or end suction type; a diaphragm expansion tank to allow water expansion and contraction during water temperature changes; an air separator to purge dissolved air from water; corresponding piping and fittings; and control systems. The plant loop is operated at constant flow.

Building Loop. A building loop BCDEFGHJB contains coils; building circulating water pumps, often variable-speed pumps, one being a standby pump; corresponding pipes and fittings; and control systems. A differential pressure transmitter is often installed at the farthest end from the building pump between the supply and return mains. The building loop is operated at variable flow.

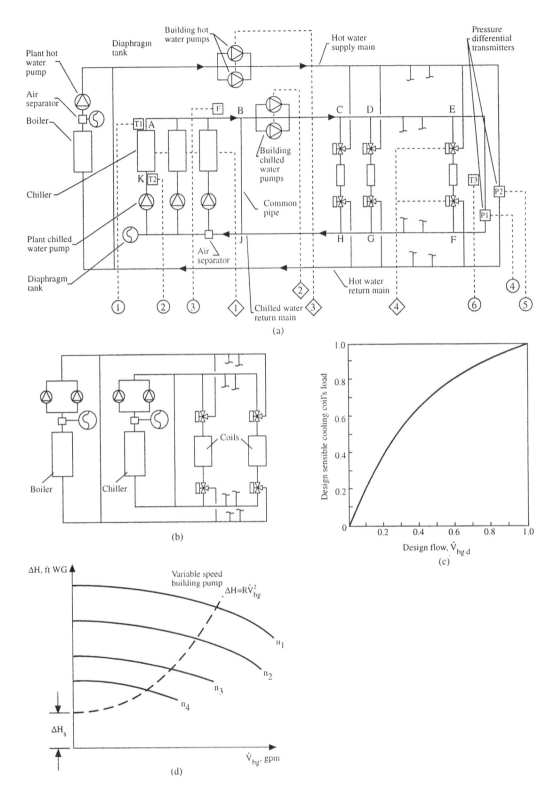

FIGURE 9.3 A dual-temperature water system: (a) plant-building loop, (b) single-loop, (c) design sensible load and design volume flow relationship, and (d) variable-speed building pump curves.

A short common pipe connects these two loops and combines them into a plant-building loop combination. It is also called a *primary-secondary loop,* with the plant loop the primary loop and the building loop the secondary loop. The location of the pump in a water system should be arranged so that the pressure at any point in the system is greater than atmospheric pressure to prevent air leaking into the system.

Operating Characteristics

In a dual-temperature water system, when the coil load q_{cc}, in Btu/hr, drops, two-way valves close to a smaller opening. The pressure drop across the coils then increases. As soon as the increase of the pressure differential is sensed by the transmitter, a DDC controller reduces the flow rate of the variable-speed pump to match the reduction of the coil load during part load. The pressure differential transmitter functions similarly to a duct static pressure sensor in a VAV system.

Since the plant loop is at constant flow, excess chilled water bypasses the building loop and flows back to the chiller(s) through the common pipe. When the reduction of coil loads in the building loop is equal to or greater than the refrigeration capacity of a single chiller, the DDC controller will shut down a chiller and its associated chilled water pump. The operating characteristics of the hot water in a dual-temperature water system are similar to those in a chilled water system.

Due to the coil's heat transfer characteristics, the reduction of a certain fraction of the design sensible cooling coil load $q_{cs.d}$ does not equal the reduction of the same fraction of design water volume flow $\mathring{V}_{w.d}$ (Figure 9.3[c]). Roughly, a 0.7 $q_{cs.d}$ needs only a 0.45 $\mathring{V}_{w.d}$, and the temperature difference $\Delta T_w = (T_{wl} - T_{we})$ will be about 150% of that at the design condition. When an equal-percentage contour two-way control valve is used, the valve stem displacement of the cooling coil's two-way valve is approximately linearly proportional to the sensible cooling coil's load change.

Comparison of a Single-Loop and a Plant-Building Loop

Compared with a single-loop water system as shown in Figure 9.3(b) that has only a single-stage water pump(s), the plant-building loop:

- Provides variable flow at the building loop and thus saves pump power at reduced flow during part load.
- Separates plant and building loops and makes design, operation, maintenance, and control simpler and more stable as stated in Carlson (1968). The pressure drop across the two-way control valve is considerably reduced as the control valve is only a component of the building loop.

Plant-Distribution-Building Loop

In universities, medical centers, and airports, buildings are often scattered from the central plant. A campus-type chilled water system using a plant-distribution-building loop is often reliable in operation, requires less maintenance, has minimal environmental impact, and sometimes provides energy cost savings.

In a plant-distribution-building loop, plant loop water is combined with many building loops through a distribution loop as shown in Figure 9.4(a). The water flow in the plant loop is constant, whereas in both distribution and building loops it is variable. Chilled water often leaves the chiller at a temperature of 40 to 42°F. It is extracted by the distribution pumps and forced into the distribution supply main. At the entrance of each building, chilled water is again extracted by the building pumps and supplied to the cooling coils in AHUs and fan-coil units. From the coils, chilled water is returned through the building return main and the distribution return main and enters the chillers at a temperature around 60°F at design volume flow. System performances of the plant and building loops in a plant-distribution-building loop are similar to those in a plant-building loop.

A campus-type central plant may transport several thousand gallons of chilled water to many buildings, and the farthest building may be several thousand feet away from the central plant. A smaller pressure gradient and end pressure differential between distribution supply and return mains, $\Delta H_{s.d}$, save energy.

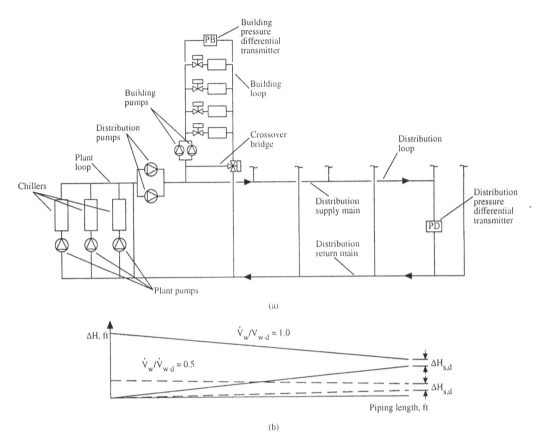

FIGURE 9.4 A chilled water system using a plant-distribution-building loop: (a) schematic diagram and (b) pressure gradient for distribution loop.

Meanwhile, a larger diameter of pipes is needed. For a distribution loop, a pressure drop of 0.5 to 1 ft head loss per 100 ft of piping length may be used. A low pressure differential may be offset at the control valves without affecting the normal operation of the coils. Usually, direct return is used for distribution loops.

Using a two-way distribution from the central plant with two separate supply and return loops often reduces the main pipe length and diameter.

A differential pressure transmitter may be located at the farthest end of the distribution mains. As the pressure loss in the building loop is taken care of by the variable-speed building pump(s), a set point of 5 ft WG pressure differential across the supply and return mains seems appropriate.

Chilled water and hot water distribution pipes are often mounted in underground accessible tunnels or trenches. They are well insulated except when chilled water return temperature $T_{ret} > 55°F$. Under such circumstances, the return main may not be insulated, depending on local conditions and cost analysis.

Many retrofits of campus-type chilled water systems required an existing refrigeration plant(s) and a new developed plant(s) connected to the same plant-distribution-building loop. A stand-alone DDC panel is often used to optimize the turning on and off of the chillers in the existing and developed plants according to the system loads and available sources (chillers).

10

Heating Systems

Shan K. Wang

Consultant

10   Heating Systems... 103
      Types of Heating Systems

Heating Systems

Types of Heating Systems

According to EIA Commercial Buildings Characteristics 1992, for the 57.8 billion ft² of heated commercial buildings in the United States in 1992, the following types of heating systems were used:

Warm air heating systems using warm air furnace	27%
Hot water heating systems using boilers	33%
Heat pumps	13%
District heating	8%
Individual space heaters and others heaters	19%

Modera (1989) reported that nearly 50% of U.S. residential houses are using warm air heating systems with direct-fired warm air furnaces.

Warm Air Furnaces

A *warm air furnace* is a device in which gaseous or liquid fuel is directly fired or electric resistance heaters are used to heat the warm supply air. Natural gas, liquefied petroleum gas (LPG), oil, electric energy, or occasionally wood may be used as the fuel or energy input. Among these, natural gas is most widely used. In a warm air furnace, the warm air flow could be *upflow*, in which the warm air is discharged at the top, as shown in Figure 10.1(a) and (b); downflow, with the warm air discharged at the bottom; or *horizontal flow*, with the warm air discharged horizontally.

Natural Vent Combustion Systems. There are two types of combustion systems in a natural gas-fired warm air furnace: natural vent or power vent combustion systems. In a *natural vent* or *atmospheric vent* combustion system, the buoyancy of the combustion products carries the flue gas flowing through the heat exchanger and draft hood, discharging from the chimney or vent pipe. The gas burner is an *atmospheric burner*. In an atmospheric burner, air is exracted for combustion by the suction effect of the high-velocity discharged gas and the buoyancy effect of the combustion air. An atmospheric burner

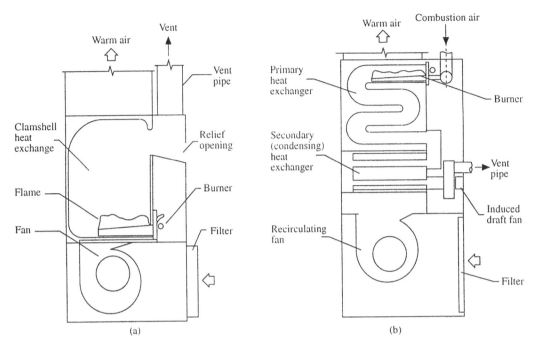

FIGURE 10.1 Upflow warm air gas furnace: (a) a natural-vent gas furnace and (b) a power-vent high-efficiency gas furnace.

can be either an in-shot or an up-shot burner or multiple ports. Atmospheric burners are simple, require only a minimal draft of air, and need sufficient gas pressure for normal functioning.

Two types of ignition have been used in burners: standing pilot and spark ignition. In *standing pilot ignition*, the small pilot flame is monitored by a sensor and the gas supply is shut off if the flame is extinguished. *Spark ignition* fires intermittently only when ignition is required. It saves gas fuel if the furnace is not operating.

In a natural vent combustion system, the heat exchanger is often made from cold-rolled steel or aluminized steel in the shape of a clamshell or S. A fan or blower is always used to force the recirculating air flowing over the heat exchanger and distribute the heated air to the conditioned space. A low-efficiency disposable air filter is often located upstream of the fan to remove dust from the recirculating air. A draft hood is also installed to connect the flue gas exit at the top of the heat exchanger to a vent pipe or chimney. A relief air opening is employed to guarantee that the pressure at the flue gas exit is atmospheric and operates safely even if the chimney is blocked. The outer casing of the furnace is generally made of heavy-gauge steel with access panels.

Power Vent Combustion Systems. In a *power vent* combustion system, either a forced draft fan is used to supply the combustion air or an induced draft fan is used to induce the flue gas to the vent pipe or chimney. A power vent is often used for a large gas furnace or a high-efficiency gas furnace with condensing heat exchangers.

Gas burners in a power vent system are called *power burners*. The gas supply to the power burner is controlled by a pressure regulator and a gas valve to control the firing rate. Intermittent spark ignition and *hot surface ignition* that ignites the main burners directly are often used.

Usually, there are two heat exchangers in a power vent combustion system: a primary heat exchanger and a secondary or condensing heat exchanger. The primary heat exchanger constitutes the heating surface of the combustion chamber. When the water vapor in the flue gas is condensed by indirect contact with the recirculating air, part of the latent heat of condensation released is absorbed by the air. Thus the furnace efficiency is increased in the *secondary* or *condensing heat exchanger*. Both primary and secondary heat exchangers are made from corrosion-resistant steel. A fan is also used to force the recirculating air to flow over the heat exchangers and to distribute the heated air to the conditioned space.

Most natural gas furnaces can use LPG. LPG needs a pressure of 10 in. WG at the manifold, compared with 3 to 4 in. for natural gas. It also needs more primary air for gas burners. Oil furnaces are usually of forced draft and installed with pressure-atomizing burners. The oil pressure and the orifice size of the injection nozzle control the firing rate.

Furnace Performance Indices. The performance of a gas-fired furnace is usually assessed by the following indices:

- Thermal efficiency E_t, in percent, is the ratio of the energy output of heated air or water to the fuel energy input during specific test periods using the same units:

$$E_t = 100(\text{fuel energy output})/(\text{fuel energy input}) \qquad (10.1)$$

- Annual fuel utilization efficiency (AFUE), in percent, is the ratio of the annual output energy from heated air or water to the annual input energy using the same units:

$$\text{AFUE} = (100 \text{ annual output energy})/(\text{annual input energy}) \qquad (10.2)$$

- Steady-state efficiency (SSE) is the efficiency of a given furnace according to an ANSI test procedure, in percent:

$$\text{SSE} = 100(\text{fuel input} - \text{fuel loss})/(\text{fuel input}) \qquad (10.3)$$

Jakob et al. (1986) and Locklin et al. (1987), in a report on ASHRAE Special Project SP43, gave the following performance indices based on a nighttime setback period of 8 hr with a setback temperature of 10°F:

Description	AFUE (%)	SSE (%)
Natural vent		
Pilot ignition	64.5	77
Intermittent ignition	69	77
Intermittent ignition plus vent damper	78	77
Power vent		
Noncondensing	81.5	82.5
Condensing	92.5	93

ASHRAE/IES Standard 90.1-1989 specifies a minimum AFUE of 78% for both gas-fired and oil-fired furnaces of heating capacity <225,000 Btu/hr. For a gas-fired warm air furnace ≥225,000 Btu/hr at maximum rating capacity (steady state) the minimum E_t is 80%, and at minimum rating capacity, E_t is 78%.

Operation and Control. Gas is usually brought from the main to the pressure regulator, where its pressure is reduced to 3.5 in. WG. Gas then flows through a valve and mixes with the necessary amount of outside primary air. The mixture mixes again with the ambient secondary air and is burned. The combustion products flow through the heat exchanger(s) due either to natural draft or power vent by a fan. The flue gas is then vented to the outside atmosphere through a vent pipe or chimney.

Recirculating air is pulled from the conditioned space through the return duct and is mixed with the outdoor air. The mixture is forced through the heat exchanger(s) by a fan and is then heated and distributed to the conditioned space. The fan is often started 1 min after the burner is fired in order to prevent a cold air supply.

At part-load operation, the reduction of the heating capacity of the warm air furnace is usually controlled by the gas valve and the ignition device. For small furnaces, the gas valve is often operated at on–off control. For large furnaces, a two-stage gas valve operates the furnace at 100, 50, and 0% heating capacity as required.

Low NO$_x$ Emissions.   NO$_x$ means nitrogen oxides NO and NO$_2$. They are combustion products in the flue gas and become air pollutants with other emissions like SO$_2$ and CO when they are discharged to the atmosphere. NO$_x$ cause ozone depletion as well as smog.

Southern California regulations required that NO$_x$ emissions should be 30 ppm or less for gas-fired warm air furnaces and boilers. Many gas burner manufactures use induced flue gas recirculation to cool the burner's flame, a cyclonic-type burner to create a swirling high-velocity flame, and other technologies to reduce NO$_x$ and other emissions while retaining high furnace and boiler efficiencies.

Hot Water Boilers

Types of Hot Water Boilers.   A *hot water boiler* is an enclosed pressure vessel used as a heat source for space heating in which water is heated to a required temperature and pressure without evaporation. Hot water boilers are fabricated according to American Society of Mechanical Engineers (ASME) codes for boilers and pressure vessels. Boilers are generally rated on the basis of their gross output delivered at the boiler's outlet. Hot water boilers are available in standard sizes from 50 to 50,000 MBtu/hr (1 MBtu/hr = 1000 Btu/hr).

According to EIA Characteristics of Commercial Buildings (1991), the percentages of floor area served by different kinds of fuel used in hot water and steam boilers in 1989 in the United States are gas-fired, 69%; oil-fired, 19%; electric, 7%; others, 5%.

Hot water boilers can be classified as *low-pressure boilers*, whose working pressure does not exceed 160 psig and working temperature is 250°F or less, and *medium-* and *high-pressure boilers*, whose working pressure is above 160 psig and working temperature above 250°F. Most of the hot water boilers are low-pressure boilers except those in campus-type or district water heating systems.

Based on their construction and material, hot water boilers can be classified as fire tube boilers, water tube boilers, cast iron sectional boilers, and electric boilers. Water tube boilers are used mainly to generate steam. Cast iron sectional boilers consist of many vertical inverted U-shaped cast iron hollow sections. They are lower in efficiency and used mainly for residential and small commercial buildings. Electric boilers are limited in applications because of their higher energy cost in many locations in the United States.

Scotch Marine Boiler. In a *fire tube* hot water boiler, the combustion chamber and flue gas passages are in tubes. These tubes are enclosed by an outer shell filled with water. A recently developed fire tube model is the modified Scotch Marine packaged boiler, which is a compact, efficient, and popular hot water boiler used today.

A *Scotch Marine* boiler, as shown in Figure 10.2, consists of a gas, oil, or gas/oil burner, a mechanical forced-draft fan, a combustion chamber, fire tubes, flue gas vent, outer shell, and corresponding control system. A packaged boiler is a one-piece factory-assembled boiler.

Gas or oil and air are measured, mixed, and injected into the combustion chamber in which they are initially ignited by an ignition device. The mixture burns and the combustion process is sustained when there is a high enough temperature in the combustion chamber. Because the number of fire tubes decreases continuously in the second, third, and fourth passes due to the volume contraction at lower flue gas temperatures, the gas velocity and heat transfer coefficients are maintained at reasonably high values. Return water enters the side of the boiler. It sinks to the bottom, rises again after heating, and then discharges at the top outlet..

The dew point of the flue gas is often 130°F. If the temperature of the return water in a *condensing boiler* is below 125°F, it can be used as a condensing cooling medium to condense the water vapor contained in the flue gas. Latent heat of condensation of water vapor can then be recovered. Corrosion

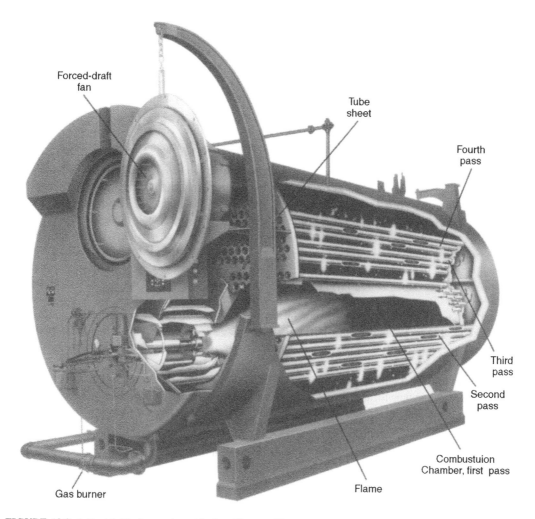

FIGURE 10.2 A Scotch Marine packaged boiler. (Source: Cleaver Brooks. Reprinted by permission.)

in the condensing heat exchanger and the flue gas passage should be avoided.

The *chimney*, or *stack*, is the vertical pipe or structure used to discharge flue gas, which usually has a temperature rise of 300 to 400°F above the ambient temperature. The chimney or stack should be extended to a certain height above adjacent buildings according to local codes.

Operation and Safety Controls. During part-load operation, reduction of heating capacity is achieved by sensing the temperature of return water and controlling the firing rate of the gas burners in on–off, high–low–off, or modulating modes.

For gas burners, two pressure sensors are often provided to maintain the gas pressure within a narrow range. For modulating controls, the ratio of maximum to minimum input is called the *turn-down* ratio. The minimum input is usually 5 to 25% of the maximum input, that is, a turn-down ratio of 20:1 to 4:1. The boiler should be shut off if the input is less than the minimum.

Pressure and temperature relief valves should be installed in each boiler. An additional high limit control is often equipped to shut down the boiler in case the sensed water pressure or temperature exceeds the predetermined limits. A flame detector is often used to monitor the flame and an airflow sensor to verify the combustion airflow. As soon as the flame is extinguished or the combustion airflow is not sensed, the fuel valve closes and the boiler is shut down. ASHRAE/IES Standard 90.1-1989 specifies that gas-fired boilers should have a minimum AFUE of 80%.

Low-Pressure Warm Air Heating Systems

A *low-pressure warm air heating system* is often equipped with an upflow gas-fired furnace having a furnace heat capacity Q_f to air flow $\dot{V}_a$ ratio, $Q_f / \dot{V}_a$, of 50 to 70 Btu/hr.cfm and a temperature rise immediately after the furnace of 50 to 70°F. The external pressure loss of the supply and return duct system is usually not higher than 0.5 in. WG. The supply temperature differential $(T_s - T_r)$ is often 20 to 35°F. The heating system is often integrated with a cooling system, forming a heating/cooling system.

Low-pressure warm air heating systems usually have a heating capacity not exceeding 100,000 Btu/hr. They are often used in residences and sometimes in small commercial buildings.

System Characteristics. A low-pressure warm air heating system is equipped with either supply and return ducts or a supply duct and a return plenum. Recirculating air is then returned from living, dining, bed, and study rooms to the return plenum through door undercuts in case the doors are closed.

The location of the furnace has a significant effect on the efficiency of the heating system. According to Locklin et al. (1987), if the gas furnace of a low-pressure warm air heating system is installed in a closet and its supply duct is mounted inside the conditioned space, its system efficiency may be 20% higher than that of installations whose furnace and supply duct are in the attic or basement.

When a low-pressure warm air heating system is operating, the supply duct leakage in the attic or basement raises its pressure to a positive value and promotes exfiltration. Return duct leakage extracts the ambient air, lowers the attic or basement pressure to a negative value, and induces infiltration. Gammage et al. (1986) reported that both types of leakage increase the whole house infiltration to 0.78 ach when the low-pressure warm air heating system is operating. The infiltration rate is only 0.44 ach when the low-pressure warm air heating system is shut off.

If the supply temperature differential $\Delta T_s = (T_s - T_r)$ exceeds 30°F, or if there is a high ceiling, thermal stratification may form in the conditioned space. Greater supply volume flow rates and suitable locations of the supply and return outlets may reduce thermal stratification and vertical temperature difference.

Part-Load Operation and Control. For a low-pressure warm air heating system, a space thermostat is often used to control the gas valve of the furnace operated in on–off or high–low–off mode. The proportion of on and off times in an operating cycle can be varied to meet the change of space heating load. The time period of an on–off operating cycle is usually 5 to 15 min.

A warm-air heating system that has an external pressure higher than 0.5 in. WG is often integrated with a cooling system and becomes a part of an air-conditioning system.

Low-Temperature Hot Water Heating System Using Fin-Tube Heaters

In a *low-temperature hot water heating system*, the operating temperature is 250°F or less with a maximum working pressure not exceeding 150 psig, usually less than 30 psig. Low-temperature hot water heating systems are widely used for space heating in residences and commercial buildings.

Fin-Tube Heaters. A *fin-tube heater* is a device installed directly inside the conditioned space to add heat to the space through radiant and convective heat transfer. A fin-tube heater consists of a finned-tube element and an outer casing as shown in Figure 10.3(a). The tubes are often made of copper and steel. Copper tubes are generally 0.75, 1, and 1.25 in. in diameter and steel tubes 1.25 and 2 in. in diameter. The fins are usually made of aluminum for copper tubes and of steel for steel tubes. Fin density may vary from 24 to 60 fins per foot. A fin heater may have a maximum length of 12 ft. The outer casing of a finned-tube heater always has a bottom inlet and top outlet for better convection.

The most widely used finned-tube heater is the *baseboard heater*, which is often mounted on cold walls at a level 7 to 10 in. from the floor. It is usually 3 in. deep and has one fin-tube row. A wall finned-tube heater has a greater height. A convector has a cabinet-type enclosure and is often installed under the windowsill.

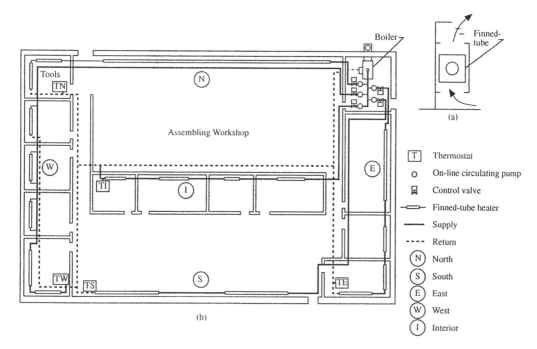

FIGURE 10.3 A two-pipe individual-loop low-temperature hot water heating system: (a) finned-tube heater and (b) piping layout.

Two-Pipe Individual-Loop Systems. Current low-temperature hot water heating systems using finned-tube heaters are often equipped with zone controls. Zone controls can avoid overheating rooms facing south and underheating rooms facing north because of the effects of solar radiation.

Figure 10.3(b) shows the piping layout of a *two-pipe individual-loop system* that is widely used in low-temperature hot water heating systems. Two-pipe means that there are a supply main and a return main pipe instead of one common main for both supply and return. Individual-loop means that there is an individual loop for each control zone. Several finned-tube heaters in a large room can be connected in series, while finned-tube heaters in several small rooms can be connected in reverse return arrangement.

The sizing of low-temperature hot water pipes is usually based on a pressure drop of 1 to 3 ft per 100 ft of pipe length. For a small low-temperature hot water heating system, an open-type expansion tank is often used. A diaphragm tank is often used for a large system. On-line circulating pumps with low head are often used.

Part Load and Control. Usually a hot water sensor located at the exit of the hot water boiler is used to control the firing rate of the boiler at part-load operation. Its set point is usually reset according to the outdoor temperature. Zone control is provided by sensing the return hot water temperature from each individual loop or zone and then varying the water volume flow rate supplied to that zone by modulating the speed of each corresponding on-line circulating pump or its control valve. For hot water heating systems using multiple boilers, on and off for each specific boiler depend not only on the heating demand, but also on minimizing the energy cost.

Infrared Heating

Infrared heating is a process that uses radiant heat transfer from a gas-fired or electrically heated high-temperature device to provide space heating on a localized area for the health and comfort of the occupants or to maintain a suitable indoor environment for a manufacturing process.

An *infrared heater* is a device used to provide infrared heating. Heat radiates from an infrared heater in the form of electromagnetic waves and scatters in all directions. Most infrared heaters have reflectors to focus the radiant beam onto a localized area. Therefore, they are often called *beam radiant heaters*.

Infrared heaters are widely used in high-ceiling supermarkets, factories, warehouses, gymnasiums, skating rinks, and outdoor loading docks.

Gas Infrared Heaters. Infrared heaters can be divided into two categories: gas and electric infrared heaters. *Gas infrared heaters* are again divided into porous matrix gas infrared heaters and indirect gas infrared heaters. In a *porous matrix gas infrared heater*, a gas and air mixture is supplied and distributed evenly through a porous ceramic, a stainless steel panel, or a metallic screen, which is exposed to the ambient air and backed by a reflector. Combustion takes place at the exposed surface with a maximum temperature of about 1600°F. An indirect infrared heater consists of a burner, a radiating tube, and a reflector. Combustion takes place inside the radiating tube at a temperature not exceeding 1200°F.

Gas infrared heaters are usually vented and have a small conversion efficiency. Only 10 to 20% of the input energy of an open combustion gas infrared heater is radiated in the form of infrared radiant energy. Usually 4 cfm of combustion air is required for 1000 Btu/hr gas input. A thermostat often controls a gas valve in on–off mode. For standing pilot ignition, a sensor and a controller are used to cut off the gas supply if the flame is extinguished.

Electric Infrared Heaters. An *electric infrared heater* is usually made of nickel–chromium wire or tungsten filaments mounted inside an electrically insulated metal tube or quartz tube with or without inert gas. The heater also contains a reflector that directs the radiant beam to the localized area requiring heating. Nickel–chromium wires often operate at a temperature of 1200 to 1800°F. A thermostat is also used to switch on or cut off the electric current. An electric infrared heater has a far higher conversion efficiency and is cleaner and more easily managed.

Design Considerations. An acceptable radiative temperature increase $(T_{rad} - T_r)$ of 20 to 25°F is often adopted for normal clothed occupants using infrared heating. The corresponding required watt density for infrared heaters is 30 to 37 W/ft². At a mounting height of 11 ft, two heaters having a spacing of 6.5 ft can provide a watt density of 33 W/ft² and cover an area of 12 × 13 ft. The mounting height of the infrared heaters should not be lower than 10 ft. Otherwise the occupants may feel discomfort from the overhead radiant beam. Refer to Grimm and Rosaler (1990), *Handbook of HVAC Design,* for details.

Gas and electric infrared heaters should not be used in places where there is danger of ignitable gas or materials that may decompose into toxic gases.

11

Refrigeration Systems

Shan K. Wang

Consultant

11   Refrigeration Systems 111
           Classifications of Refrigeration Systems

Refrigeration Systems

Classifications of Refrigeration Systems

Most of the refrigeration systems used for air-conditioning are vapor compression systems. Because of the increase in the energy cost of natural gas in the 1980s, the application of absorption refrigeration systems has dropped sharply. According to Commercial Buildings Characteristics 1992, absorption refrigeration system have a weight of less than 3% of the total amount of refrigeration used in commercial buildings in the United States. Air expansion refrigeration systems are used mainly in aircraft and cryogenics.

Refrigeration systems used for air-conditioning can be classified mainly in the following categories:

- Direct expansion (DX) systems and heat pumps
- Centrifugal chillers
- Screw chillers
- Absorption systems

Each can be either a single-stage or a multistage system.

Direct Expansion Refrigeration Systems

A *direct expansion refrigeration (DX) system*, or simply *DX system*, is part of the packaged air-conditioning system. The DX coil in the packaged unit is used to cool and dehumidify the air directly as shown in Figure 11.1(a). According to EIA Commercial Buildings Characteristics 1992, about 74% of the floor space of commercial buildings in the United States was cooled by DX refrigeration systems.Refrigerants R-22 and R-134a are widely used. Azeotropics and near azeotropics are the refrigerants often used for low-evaporating-temperature systems like those in supermarkets. Because of the limitation of the size of the air system, the refrigeration capacity of DX systems is usually 3 to 100 tons.

Components and Accessories.   In addition to the DX coil, a DX refrigeration system has the following components and accessories:

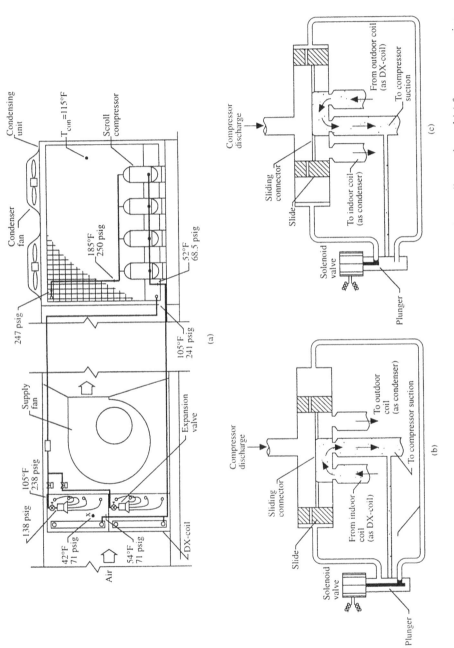

FIGURE 11.1 A DX refrigeration system: (a) schematic diagram; (b) four-way reversing valve, cooling mode; and (c) four-way reversing valve, heating mode.

- *Compressor(s)* — Both reciprocating and scroll compressors are widely used in DX systems. Scroll compressors are gradually replacing reciprocating compressors because they have fewer parts and comparatively higher efficiency. For large DX systems, multiple compressors are adopted.
- *Condensers* — Most DX systems in rooftop packaged units are air cooled. Water-cooled condensers are adopted mainly for DX systems in indoor packaged units due to their compact volume. Evaporative-cooled condensers are also available.
- *Refrigeration feed* — Thermostatic expansion valves are widely used as the throttling and refrigerant flow control devices in medium and large DX systems, whereas capillary tubes are used in small and medium-sized systems.
- *Oil lubrication* — R-22 is partly miscible with mineral oil. Since R-134a is not miscible with mineral oil, synthetic polyolester oil should be used. For medium and large reciprocating compressors, an oil pump of vane, gear, or centrifugal type is used to force the lubricating oil to the bearings and moving surfaces via grooves. For small reciprocating compressors, splash lubrication using the rotation of the crankshaft and the connecting rod to splash oil onto the bearing surface and the cylinder walls is used.

A scroll compressor is often equipped with a centrifugal oil pump to force the oil to lubricate the orbiting scroll journal bearing and motor bearing. For the scroll contact surfaces, lubrication is provided by the small amount of oil entrained in the suction vapor.

- *Refrigerant piping* — Refrigerant piping transports refrigerant through the compressor, condenser, expansion valve, and DX coil to provide the required refrigeration effect. As shown in Figure 11.1(a), from the exit of the DX coil to the inlet of the compressor(s) is the *suction line*. From the outlet of the compressor to the inlet of the air-cooled condenser is the *discharge line*. From the exit of the condenser to the inlet of the expansion valve is the *liquid line*.

Halocarbon refrigerant pipes are mainly made of copper tubes of L type. In a packaged unit, refrigerant pipes are usually sized and connected in the factory. However, the refrigerant pipes in field-built and split DX systems for R-22 are sized on the basis of a pressure drop of 2.91 psi corresponding to a change of saturated temperature ΔT_{suc} of 2°F at 40°F for the suction line and a pressure drop of 3.05 psi corresponding to 1°F at 105°F for the discharge and liquid line. The pressure drop includes pressure losses of pipe and fittings. Refrigerant pipes should also be sized to bring back the entrained oil from the DX coil and condenser through the discharge and suction lines.

Accessories include a filter dryer to remove moisture from the refrigerant, strainer to remove foreign matter, and sight glass to observe the condition of refrigerant flow (whether bubbles are seen because of the presence of flash gas in the liquid line).

Capacity Controls. In DX systems, control of the mass flow rate of refrigerant through the compressor(s) is often used as the primary refrigeration capacity control. Row or intertwined face control at the DX coil is also used in conjunction with the capacity control of the compressor(s).

Three methods of capacity controls are widely used for reciprocating and scroll compressors in DX systems:

- *On–off control* — Starting or stopping the compressor is a kind of step control of the refrigerant flow to the compressor. It is simple and inexpensive, but there is a 100% variation in capacity for DX systems installed with only a single compressor. On–off control is widely used for small systems or DX systems with multiple compressors.
- *Cylinder unloader* — For a reciprocating compressor having multiple cylinders, a cylinder unloader mechanism bypasses the compressed gas from one, two, or three cylinders to the suction chamber to reduce the refrigeration capacity and compressing energy.

- *Speed modulation* — A two-speed motor is often used to drive scroll or reciprocating compressors so that the capacity can be reduced 50% at lower speed.

Safety Controls. In *low- and high-pressure control*, the compressor is stopped when suction pressure drops below a preset value, the cut-in pressure, or the discharge pressure of the hot gas approaches a dangerous level, the cut-out pressure.

In *low-temperature control*, a sensor is mounted on the outer pipe surface of the DX coil. When the surface temperature drops below 32°F, the controller stops the compressor to prevent frosting.

If the pressure of the oil discharged from the pump does not reach a predetermined level within a certain period, a mechanism in *oil pressure failure control* opens the circuit contact and stops the compressor.

In *motor overload control*, a sensor is used to measure the temperature of the electric winding or the electric current to protect the motor from overheating and overloading.

Pump-down control is an effective means of preventing the migration of the refrigerant from the DX coil (evaporator) to the crankcase of the reciprocating compressor. This prevents mixing of refrigerant and oil to form slugs, which may damage the compressor.

When a rise of suction pressure is sensed by a sensor, a DDC controller opens a solenoid valve and the liquid refrigerant enters the DX coil. As the buildup of vapor pressure exceeds the cut-in pressure, the compressor starts. When the DX system needs to shut down, the solenoid valve is closed first; the compressor still pumps the gaseous refrigerant to the condenser. As the suction pressure drops below the cut-in pressure, the compressor stops.

Full- and Part-Load Operations. Consider a DX system in a rooftop packaged unit using four scroll compressors, each with a refrigeration capacity of 10 tons at full load. Performance curves of the condensing unit and the DX coil of this DX system are shown in Figure 11.2(a). A DDC controller actuates on–off for these scroll compressors based on the signal from a discharge air temperature T_{dis} sensor.

On a hot summer day, when the rooftop packaged unit starts with a DX coil load or refrigeration capacity, $q_{rc} \approx 40$ tons, all four scroll compressors are operating. The operating point is at A′ with a suction temperature T_{suc} of about 42°F, and the discharge air temperature T_{dis} is maintained around 53°F. As the space cooling load q_{rc} as well as T_{dis} decreases, the required DX coil load q_{rl} drops to 35 tons. Less evaporation in the DX coil causes a decrease of T_{suc} to about 40°F, and the operating point may move downward to point A with a DX coil refrigeration capacity of 39 tons. Since $q_{rc} > q_{rl}$, T_{dis} drops continually until it reaches 50°F, point A in Figure 11.2(b), and the DDC controller shuts down one of the scroll compressors. The operating point immediately shifts to B′ on the three-compressor curve.

Because the refrigeration capacity at point B′ q_{rc} is 29 tons, which is less than the required $q_{rl} = 35$ tons, both T_{dis} and T_{suc} rise. When the operating point moves up to B* and T_{dis} reaches 56°F, the DDC controller starts all four scroll compressors at operating point A″ with a refrigeration capacity of 42 tons. Since $q_{rc} > q_{rl}$, the operating point again moves downward along the four-compressor curve and forms an operating cycle A″AB′ and B*. The timing of the operating period on four- or three-compressor performance curves balances any required q_{rl} between 29 and 42 tons. Less evaporation at part load in the DX coil results in a greater superheating region and therefore less refrigeration capacity to balance the reduction of refrigeration capacity of the compressor(s) as well as the condensing unit. The condition will be similar when $q_{rl} < 30$ tons, only three- or two-compressor, or two- or one-compressor, or even on–off of one compressor forms an operating cycle.

Main Problems in DX Systems

- *Liquid slugging* is formed by a mixture of liquid refrigerant and oil. It is formed because of the flooding back of liquid refrigerant from the DX coil to the crankcase of the reciprocating compressor due to insufficient superheating. It also may be caused by migration of liquid refrig- erant from the warmer indoor DX coil to the colder outdoor compressor during the shut-down period in a split packaged unit. Liquid slugging dilutes the lubricating oil and causes serious loss

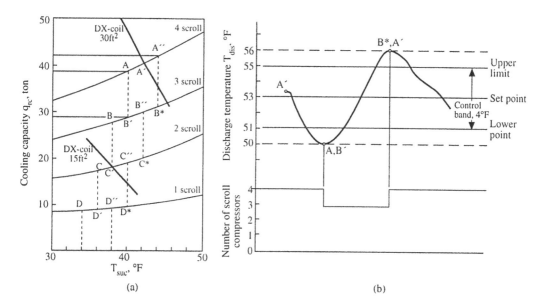

FIGURE 11.2 Capacity control of a DX refrigeration system: (a) performance curves and operating points and (b) locus of control point.

of oil in the crankcase of the reciprocating compressor. Liquid slugging is incompressible. When it enters the compression chamber of a reciprocating compressor, it may damage the valve, piston, and other components. Pump-down control and installation of a crankcase heater are effective means of preventing liquid refrigerant migration and flooding back.

- *Compressor short cycling* — For on–off control, too short a cycle, such as less than 3 min, may pump oil away from the compressor or damage system components. It is due mainly to a too close low-pressure control differential or to reduced air flow.

- *Defrosting* — If the surface of a DX coil is 32°F or lower, frost accumulates on it. Frost blocks the air passage and reduces the rate of heat transfer. It should be removed periodically. The process of removing frost is called defrosting.

   Air at a temperature above 36°F, hot gas inside the refrigerant tubes and an installed electric heating element can be used for defrosting. The defrosting cycle is often controlled by sensing the temperature or pressure difference of air entering the DX coil during a fixed time interval.

- *Refrigerant charge* — Insufficient refrigerant charge causes lower refrigeration capacity, lower suction temperature, and short on–off cycles. Overcharging refrigerant may cause a higher condensing pressure because part of the condensing surface becomes flooded with liquid refrigerant.

Heat Pumps

A *heat pump* in the form of a packaged unit is also a *heat pump system*. A heat pump can either extract heat from a heat source and reject heat to air and water at a higher temperature for heating, or provide refrigeration at a lower temperature and reject condensing heat at a higher temperature for cooling. During summer, the heat extraction, or refrigeration effect, is the useful effect for cooling in a heat pump. In winter, the rejected heat and the heat from a supplementary heater provide heating in a heat pump.

There are three types of heat pumps: air-source, water-source, and ground-coupled heat pumps. Ground-coupled heat pumps have limited applications. Water-source heat pump systems are covered in detail in a later section.

Air-Source Heat Pump. An *air-source heat pump*, or *air-to-air heat pump*, is a DX system with an additional four-way reversing valve to change the refrigerant flow from cooling mode in summer to heating mode in winter and vice versa. The variation in connections between four means of refrigerant flow — compressor suction, compressor discharge, DX coil exit, and condenser inlet — causes the function of the indoor and outdoor coils to reverse. In an air-source heat pump, the coil used to cool or to heat the recirculating/outdoor air is called the *indoor coil.* The coil used to reject heat to or absorb heat from the outside atmosphere is called the *outdoor coil.* A short capillary or restrict tube is often used instead of a thermostatic expansion valve. Both reciprocating and scroll compressors are used in air-source heat pumps. R-22 is the refrigerant widely used. Currently available air-source heat pumps usually have a cooling capacity of $1\frac{1}{2}$ to 40 tons.

Cooling and Heating Mode Operation. In *cooling mode operation,* as shown in Figure 11.1(b), the solenoid valve is deenergized and drops downward. The high-pressure hot gas pushes the sliding connector to the left end. The compressor discharge connects to the outdoor coil, and the indoor coil connects to the compressor inlet.

In *heating mode operation,* as shown in Figure 11.1(c), the solenoid plunger moves upward and pushes the slide connector to the right-hand side. The compressor discharge connects to the indoor coil, and the outdoor coil exit connects to the compressor suction.

System Performance. The performance of an air-source heat pump depends on the outdoor air temperature T_o, in °F as well as the required space heating load q_{rh}. During cooling mode operation, both the refrigeration capacity q_{rc}, in Btu/hr, and EER for the heat pump EER_{hp}, in Btu/hr.W, increase as T_o drops. During heating mode operation, the heating capacity q_{hp}, in Btu/hr, and COP_{hp} decrease, and q_{rh} increases as the T_o drops. When $q_{rh} > q_{hp}$, supplementary heating is required. If COP_{hp} drops below 1, electric heating may be more economical than a heat pump.

If on–off is used for compressor capacity control for an air-source heat pump in a split packaged unit, refrigerant tends to migrate from the warmer outdoor coil to the cooler indoor coil in summer and from the warmer indoor coil to the cooler outdoor coil in winter during the off period. When the compressor starts again, 2 to 5 min of reduced capacity is experienced before the heat pump can be operated at full capacity. Such a loss is called a *cycling loss.*

In winter, most air-source heat pumps switch from the heating mode to cooling mode operation and force the hot gas to the outdoor coil to melt frost. After the frost is melted, the heat pump is switched back to heating mode operation. During defrosting, supplementary electric heating is often necessary to prevent a cold air supply from the air-source heat pump.

Minimum Performance. ASHRAE/IES Standard 90.1-1989 specifies a minimum performance for air-cooled DX systems in packaged units as covered in Section 7. For air-cooled, electrically operated rooftop heat pumps (air-source heat pumps), minimum performance characteristics are:

Cooling EER	8.9
Heating, COP (at $T_oF = 47°F$)	3.0
Heating, COP (at $T_oF = 17°F$)	2.0

Centrifugal Chillers

A *chiller* is a refrigeration machine using a liquid cooler as an evaporator to produce chilled water as the cooling medium in a central air-conditioning system. A *centrifugal chiller,* as shown in Figure 11.3(a), is a refrigeration machine using a centrifugal compressor to produce chilled water. It is often a factory-assembled unit with an integrated DDC control system and sometimes may separate into pieces for transportation. A centrifugal chiller is also a *centrifugal vapor compression refrigeration system.*

Refrigerants. According to Hummel et al. (1991), in 1988 there were about 73,000 centrifugal chillers in the United States. Of these, 80% use R-11, 10% use R-12, and the remaining 10% use R-22 and others. As mentioned in Section 4, production of CFCs, including R-11 and R-12, ceased at the end of 1995 with limited exception for service. R-123 (HCFC) will replace R-11. The chiller's efficiency may drop 2 to 4%, and a capacity reduction of 5% is possible. R-123 has low toxicity. Its allowable exposure limit was raised

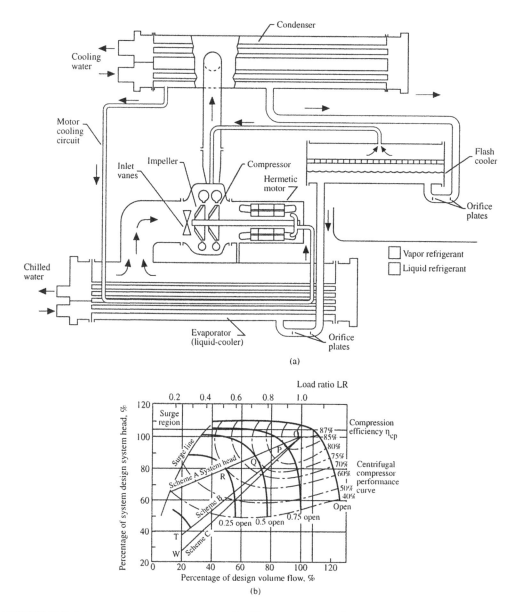

FIGURE 11.3 A two-stage water-cooled centrifugal chiller: (a) schematic diagram and (b) centrifugal compressor performance map at constant speed.

to 50 ppm in 1997 from 30 ppm in 1993 by its manufacturers. A monitor and alarm device to detect R-123 in air should be installed in plant rooms and places where there may be refrigerant leaks.

R-134a (HFC) will replace R-12. According to Lowe and Ares (1995), as a result of the changeout from R-12 to R-134a for a 5000-hp centrifugal chiller in Sears Tower, Chicago, its speed increased from 4878 to 5300 rpm, its cooling capacity is 12 to 24% less, and its efficiency is 12 to 16% worse.

System Components. A centrifugal chiller consists of a centrifugal compressor, an evaporator or liquid cooler, a condenser, a flash cooler, throttling devices, piping connections, and controls. A purge unit is optional.

- Centrifugal compressor — According to the number of internally connected impellers, the centrifugal compressor could have a single, two, or more than two stages. A two-stage impeller with a flash cooler is most widely used because of its higher system performance and comparatively simple construction. Centrifugal compressors having a refrigeration capacity less than 1200 tons are often hermetic. Very large centrifugal compressors are of open type. A gear train is often required to raise the speed of the impeller except for very large impellers using direct drive.

- Evaporator — Usually a liquid cooler of flooded shell-and-tube type evaporator is adopted because of its compact size and high rate of heat transfer.
- Condenser — Water-cooled, horizontal shell-and-tube condensers are widely used.
- Flash cooler — For a two-stage centrifugal compressor, a single-stage flash cooler is used. For a three-stage compressor, a two-stage flash cooler is used.
- Orifice plates and float valves — Both multiple-orifice plates such as that shown in Figure 11.3(a) and float valves are used as throttling devices in centrifugal chillers.
- Purge unit — R-123 has an evaporating pressure p_{ev} = 5.8 psia at 40°F, which is lower than atmospheric pressure. Air and other noncondensable gases may leak into the evaporator through cracks and gaps and usually accumulate in the upper part of the condenser. These noncondensable gases raise the condensing pressure, reduce the refrigerant flow, and lower the rate of heat transfer. A purge unit uses a cooling coil to separate the refrigerant and water from the noncondensable gases and purge the gases by using a vacuum pump.

Performance Ratings. The refrigeration cycle of a typical water-cooled, two-stage centrifugal chiller with a flash cooler was covered in Section 4. Centrifugal chillers have the same refrigeration capacity as centrifugal compressors, 100 to 10,000 tons. According to ARI Standard 550-88, the refrigeration capacity of a centrifugal chiller is rated as follows:

Chilled water temperature leaving evaporator T_{el}:	100% load 44°F
	0% load 44°F
Chilled water flow rate:	2.4 gpm/ton
Condenser water temperature entering condenser T_{cc}:	100% load 85°F
	0% load 60°F
Condenser water flow rate:	3.0 gpm/ton
Fouling factor in evaporator and condenser:	0.00025 hr.ft².°F/Btu

The integrated part-load value (IPLV) of a centrifugal chiller or other chillers at standard rating conditions can be calculated as:

$$IPLV = 0.1(A + B)/2 + 0.5(B + C)/2 + 0.3(C + D)/2 + 0.1D \qquad (11.1)$$

where *A, B, C,* and *D* = kW/ton or COP at 100, 75, 50, and 25% load, respectively. If the operating conditions are different from the standard rating conditions, when T_{el} is 40 to 50°F, for each °F increase or decrease of T_{el}, there is roughly a 1.5% difference in refrigeration capacity and energy use; when T_{cc} is between 80 to 90°F, for each °F of increase or decrease of T_{cc}, there is roughly a 1% increase or decrease in refrigeration capacity and 0.6% in energy use.

ASHRAE/IES Standard 90.1-1989 and ARI Standard 550-88 specify the minimum performance for water-cooled water chillers from January 1, 1992:

	COP	IPLV
≥300 tons	5.2	5.3
≥150 tons < 300 tons	4.2	4.5
<150 tons	3.8	3.9

COP = 5.0 is equivalent to about 0.70 kW/ton. New, installed centrifugal chillers often have an energy consumption of 0.50 kW/ton.

Air-cooled centrifugal chillers have COPs from 2.5 to 2.8. Their energy performance is far poorer than that of water-cooled chillers. Their application is limited to locations where city water is not allowed to be used as makeup water for cooling towers.

Capacity Control. The refrigeration capacity of a centrifugal chiller is controlled by modulating the refrigerant flow at the centrifugal compressor. There are mainly two types of capacity controls: varying the opening and angle of the inlet vanes, and using an adjustable-frequency AC inverter to vary the rotating speed of the centrifugal compressor.

When the opening of the inlet vanes has been reduced, the refrigerant flow is throttled and imparted with a rotation. The result is a new performance curve at lower head and flow. If the rotating speed of a centrifugal compressor is reduced, it also has a new performance curve at lower volume flow and head. Inlet vanes are inexpensive, whereas the AC inverter speed modulation is more energy efficient at part-load operation.

Centrifugal Compressor Performance Map. Figure 11.3(b) shows a single-stage, water-cooled *centrifugal compressor performance map* for constant speed using inlet vanes for capacity modulation. A performance map consists of the compressor's performance curves at various operating conditions. The *performance curve* of a centrifugal compressor shows the relationship of volume flow of refrigerant $\dot{V}_r$ and its head lift Δp or compression efficiency η_{cp} at that volume flow. It is assumed that η_{cp} for a two-stage compressor is equal to the average of the two single-stage impellers having a head of their sum.

On the map, the required *system head curve* indicates the required system head lift at that volume flow of refrigerant. The intersection of the performance curve and the required system head curve is called the *operating point* O, P, Q, R, ... as shown in Figure 11.3(b). One of the important operating characteristics of a centrifugal chiller (a centrifugal vapor compression refrigeration system as well) is that the required system head lift is mainly determined according to the difference in condensing and evaporating pressure $\Delta p_{c\text{-}e} = (p_{con} - p_{ev})$. The pressure losses due to the refrigerant piping, fittings, and components are minor.

In Figure 11.3(b), the abscissa is the percentage of design volume flow of refrigerant, % $\dot{V}_r$, or load ratio; the ordinate is the percentage of design system head $\Delta H_{s.d}$, or percentage of design temperature lift $(T_{con} - T_{ev})$. Here load ratio LR is defined as the ratio of the refrigeration load to the design refrigeration load $q_{rl}/q_{rl.d}$. There are three schemes of required system head curves:

- Scheme A — T_{ce} = constant and T_{el} = constant
- Scheme B — T_{el} = constant and a drop of 2.5°F of T_{ce} for each 0.1 reduction of load ratio
- Scheme C — A reset of T_{el} of 1°F increase and a drop of 2.5°F of T_{ce} for each 0.1 reduction of load ratio

At design $\dot{V}_r$ and system head $H_{s.d}$, $\eta_{cp} = 0.87$. As $\dot{V}_r$, load ratio, and required system head Δp decrease, η_{cp} drops accordingly.

Surge is a unstable operation of a centrifugal compressor or fan resulting in vibration and noise. In a centrifugal chiller, surge occurs when the centrifugal compressor is unable to develop a discharge pressure that satisfies the requirement at the condenser. A centrifugal compressor should never be operated in the surge region.

Part-Load Operation. During part-load operation, if T_{el} and T_{ce} remain constant, the evaporating temperature T_{ev} tends to increase from the design load value because there is a greater evaporating surface and a smaller temperature difference $(T_{el} - T_{ev})$. Similarly, T_{con} tends to decrease.

The ratio of actual compressor power input at part load to the power input at design load may be slightly higher or lower than the load ratios, depending on whether the outdoor wet bulb is constant or varying at part load or whether there is a T_{el} reset; it also depends on the degree of drop of η_{cp} at part load.

Specific Controls. In addition to generic controls, specific controls for a centrifugal chiller include:

- Chilled water leaving temperature T_{el} and reset

- Condenser water temperature T_{cc} control
- On and off of multiple chillers based on actual measured coil load
- Air purge control
- Safety controls like oil pressure, low-temperature freezing protection, high condensing pressure control, motor overheating, and time delaying

Centrifugal Chillers Incorporating Heat Recovery. A *HVAC&R heat recovery system* converts waste heat or waste cooling from any HVAC&R process into useful heat and cooling. A heat recovery system is often subordinate to a parent system, such as a heat recovery system to a centrifugal chiller.

A centrifugal chiller incorporating a heat recovery system often uses a double-bundle condenser in which water tubes are classified as tower bundles and heating bundles. Heat rejected in the condenser may be either discharged to the atmosphere through the tower bundle and cooling tower or used for heating through the heating bundle. A temperature sensor is installed to sense the temperature of return hot water from the heating coils in the perimeter zone. A DDC controller is used to modulate a bypass three-way valve which determines the amount of condenser water supplied to the heating bundle. The tower and heating bundles may be enclosed in the same shell, but baffle sheets are required to guide the water flows.

A centrifugal chiller incorporating a heat recovery system provides cooling for the interior zone and heating for the perimeter zone simultaneously in winter with a higher COP_{hr}. However, it needs a higher condenser water-leaving temperature T_{cl} of 105 to 110°F, compared with 95°F or even lower in a cooling-only centrifugal chiller. An increase of 10 to 12°F of the difference $(T_{con} - T_{ev})$ requires an additional 10 to 15% power input to the compressor. For a refrigeration plant equipped with multiple chillers, it is more energy efficient and lower in first cost to have only part of them equipped with double-bundle condensers.

Screw Chillers

A *screw chiller* or a *helical rotary chiller* is a refrigeration machine using a screw compressor to produce chilled water. A factory-fabricated and assembled screw chiller itself is also a screw vapor compression refrigeration system.

Twin-screw chillers are more widely used than single-screw chillers. A twin-screw chiller consists of mainly a twin-screw compressor, a flooded shell-and-tube liquid cooler as evaporator, a water-cooled condenser, throttling devices, an oil separator, an oil cooler, piping, and controls as shown in Figure 11.4(a). The construction of twin-screw compressors has already been covered. For evaporator, condenser, and throttling devices, they are similar to those in centrifugal chillers. Most twin-screw chillers have a refrigeration capacity of 100 to 1000 tons.

Following are the systems characteristics of screw chillers.

Variable Volume Ratio. The ratio of vapor refrigerant trapped within the interlobe space during the intake process V_{in} to the volume of trapped hot gas discharged V_{dis} is called the *built-in volume ratio* of the twin-screw compressor $V_i = V_{in}/V_{dis}$, or simply *volume ratio*, all in ft³.

There are two types of twin-screw chiller: fixed and variable volume ratio. For a twin-screw chiller of *fixed volume ratio*, the isentropic efficiency η_{isen} becomes maximum when the system required compression ratio $R_{s.com} \approx V_i$. Here $R_{s.com} = p_{con}/p_{ev}$. If $p_{dis} > p_{con}$, overcompression occurs, as shown in Figure 11.4(b). The discharged hot gas reexpands to match the condensing pressure. If $p_{dis} < p_{con}$, undercompression occurs (Figure 11.4[c]). A volume of gas at condensing pressure reenters the trapped volume at the beginning of the discharge process. Both over- and undercompression cause a reduction of η_{isen}.

For a twin-screw chiller of *variable volume ratio*, there are two slides: a sliding valve is used for capacity control and a second slide. By moving the second slide back and forth, the radial discharge port can be relocated. This allows variation of suction and discharge pressure levels and still maintains maximum efficiency.

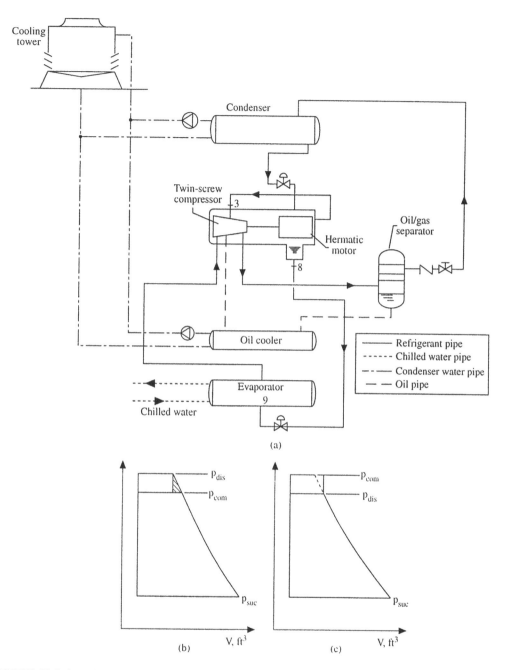

FIGURE 11.4 A typical twin-screw chiller: (a) schematic diagram, (b) over-compression, and (c) under-compression.

Economizer. The hermetic motor shell is connected to an intermediate point of the compression process and maintains an intermediate pressure p_i between p_{con} and p_{ev}. Liquid refrigerant at condensing pressure p_{con} is throttled to p_i, and a portion of the liquid is flashed into vapor. This causes a drop in the temperature of the remaining liquid refrigerant down to the saturated temperature corresponding to p_i. Although the compression in a twin-screw compressor is in continuous progression, the mixing of flashed gas with the compressed gas at the intermediate point actually divides the compression process into two stages. The resulting economizing effect is similar to that of a two-stage compound refrigeration system with a flash cooler: an increase of the refrigeration effect and a saving of the compression power from $(p_{con} - p_{ev})$ to $(p_{con} - p_i)$.

Oil Separation, Oil Cooling, and Oil Injection.   Oil entrained in the discharged hot gas enters an oil separator. In the separator, oil impinges on an internal perforated surface and is collected because of its inertia. Oil drops to an oil sump through perforation. It is then cooled by condenser water in a heat exchanger. A heater is often used to vaporize the liquid refrigerant in the oil sump to prevent dilution of the oil. Since the oil sump is on the high-pressure side of the refrigeration system, oil is forced to the rotor bearings and injected to the rotors for lubrication.

Oil slugging is not a problem for twin-screw compressors. When suction vapor contains a small amount of liquid refrigerant that carries over from the oil separator, often called *wet suction*, it often has the benefit of scavenging the oil from the evaporator.

Twin-screw compressors are positive displacement compressors. They are critical in oil lubrication, sealing, and cooling. They are also more energy efficient than reciprocating compressors. Twin-screw chillers are gaining more applications, especially for ice-storage systems with cold air distribution.

12

Thermal Storage Systems

Shan K. Wang
Consultant

12   Thermal Storage Systems 123
     Thermal Storage Systems and Off-Peak Air-Conditioning
     Systems • Ice-Storage Systems • Chilled-Water Storage
     Systems

Thermal Storage Systems

Thermal Storage Systems and Off-Peak Air-Conditioning Systems

Many electric utilities in the United States have their *on-peak hours* between noon and 8 p.m. during summer weekdays, which include the peak-load hours of air-conditioning. Because the capital cost of a new power plant is so high, from $1200 to $4000 per kW, electric utilities tend to increase their power output by using customers' thermal energy storage (TES) systems, or simply thermal storage systems, which are much less expensive.

A *thermal storage system* as shown in Figure 12.1(a) may have the same refrigeration equipment, like chillers, additional storage tank(s), additional piping, pumps, and controls. The electric-driven compressors are operated during off-peak, partial-peak, and on-peak hours. *Off-peak hours* are often nighttime hours. *Partial-peak hours* are hours between on-peak and off-peak hours in a weekday's 24-hr day-and-night cycle. Chilled water or ice is stored in tanks to provide cooling for buildings during on-peak hours when higher electric demand and electric rates are effective. Although thermal storage systems operate during nighttime when outdoor dry and wet bulbs are lower, they are not necessarily energy saving due to lower evaporating temperature, additional pump power, and energy losses. Thermal storage systems significantly reduce the electric energy cost.

Utilities in the United States often use higher electric demand and rates as well as incentive bonus to encourage the shift of electric load from on-peak to off-peak hours by using thermal storage systems and others. Such a shift not only saves expensive capital cost, but also increases the use of base-load high-efficiency coal and nuclear plants instead of inefficient diesel and gas turbine plants.

0-8493-0057-6/00/$0.00+$.50
© 2000 by CRC Press LLC

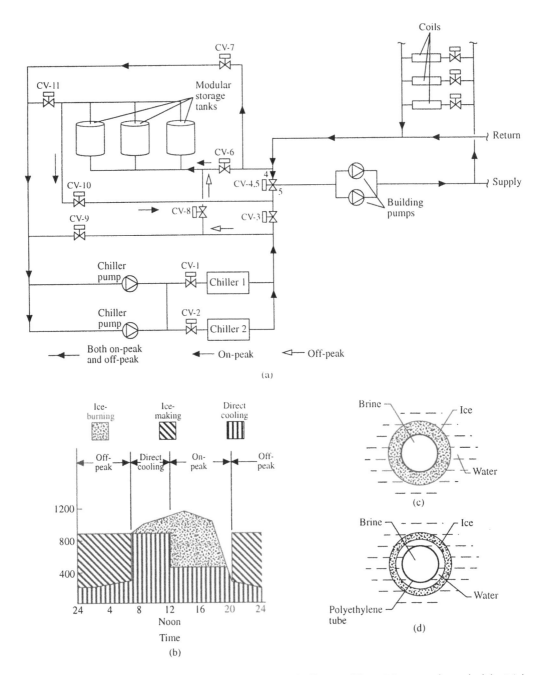

FIGURE 12.1 A brine-coil ice-storage system: (a) schematic diagram, (b) partial-storage time schedule, (c) ice making, and (d) ice burning.

The air-conditioning systems that operate during off-peak and partial-peak hours for thermal storage, or those that use mainly natural gas to provide cooling to avoid higher electric demand and rates during on-peak hours, are called *off-peak air-conditioning systems*. These systems include ice-storage and chilled-water storage systems, desiccant cooling systems, absorption systems, and gas engine chillers.

Absorption chillers and desiccant cooling systems are covered in other sections. Gas engine-driven reciprocating chillers are often a cogeneration plant with heat recovery from engine jacket and exhaust gas, and will not be covered here.

Full Storage and Partial Storage

Ice and chilled-water storage systems are the two most common thermal energy storage systems today. Knebel (1995) estimated that more than 4000 cool storage systems are operated in various commercial buildings.

The unit of stored thermal energy for cooling is ton-hour, or ton.hr. One *ton.hr* is the refrigeration capacity of one refrigeration ton during a 1-hr period, or 12,000 Btu.

In order to achieve minimum life-cycle cost, thermal storage systems could be either full storage or partial storage. For a *full-storage*, or *load shift,* thermal storage system, all refrigeration compressors cease to operate during on-peak hours. Building refrigeration loads are entirely offset by the chilled water or brine from the thermal storage system within an on-peak period. In a *partial storage,* or *load-leveling,* thermal storage system as shown in Figure 12.1(b) all or some refrigeration compressor(s) are operated during on-peak hours.

Direct cooling is the process in which refrigeration compressors produce refrigeration to cool the building directly. During a specific time interval, if the cost of direct cooling is lower than the stored energy, the operation of a thermal storage system is said to be in *chiller priority.* On the contrary, if the cost of direct cooling is higher than the cost of stored energy, the operation is said to be at *storage priority.*

The optimum size of a thermal storage system is mainly determined according to the utility's electric rate structure, especially a time-of-day structure whose electric rates are different between on-peak, partial-peak, and off-peak hours. Not only the design day's instantaneous building peak cooling load is important, but also an hour-by-hour cooling load profile of the design day is required for thermal storage design. A simple payback or a life-cycle cost analysis is usually necessary.

Ice-Storage Systems

System Characteristics

In an *ice-thermal-storage* system, or simply an *ice-storage* system, ice is stored in a tank or other containers to provide cooling for buildings in on-peak hours or on- and partial-peak hours. One pound of ice can store $[(1 \times 144) + (55 - 35)] = 164$ Btu instead of $(60 - 44) = 16$ Btu for chilled water storage. For the same cooling capacity, the storage volume of ice is only about 12% of chilled water. In addition, an air-conditioning system using ice storage often adopts cold air distribution to supply conditioned air at a temperature typically at 44°F. Cold air distribution reduces the size of air-side equipment, ducts, and investment as well as fan energy use. It also improves the indoor air quality of the conditioned space. Since the late 1980s, ice storage has gained more applications than chilled water storage.

Brine is a coolant without freezing and flashing during normal operation. The freezing point of brine, which has a mass fraction of ethylene glycol of 25%, drops to 10°F, and a mass fraction of propylene glycol of 25% drops to 15°F. Glycol-water, when glycol is dissolved in water, is another coolant widely used in ice-storage systems. Ice crystals are formed in glycol-water when its temperature drops below its freezing point during normal operation.

In an ice-storage system, *ice making* or *charging* is a process in which compressors are operated to produce ice. *Ice burning,* or *discharging,* is a process in which ice is melted to cool the brine or glycol-water to offset refrigeration load.

Brine-Coil Ice-Storage Systems. Currently used ice-storage systems include brine-coil, ice-harvester, and ice-on-coil systems. According to Knebel (1995), the brine-coil ice-storage system is most widely used today because of its simplicity, flexibility, and reliability as well as using modular ice-storage tanks.

In a typical brine-coil ice-storage system, ice is charged in multiple modular factory-fabricated storage tanks as shown in Figure 12.1(a). In each storage tank, closely spaced polyethylene or plastic tubes are surrounded by water. Brine, a mixture of 25 to 30% of ethylene glycol by mass and 70 to

75% water, circulates inside the tubes at an entering temperature of 24°F during the ice-making process. The water surrounding the tubes freezes into ice up to a thickness of about 1/2 in. as shown in Figure 12.1(c). Brine leaves the storage tank at about 30°F. Water usually at atmospheric pressure is separated from brine by a plastic tube wall. Plastic tubes occupy about one tenth of the tank volume, and another one tenth remains empty for the expansion of ice. Multiple modular storage tanks are always connected in parallel.

During the ice-melting or -burning process, brine returns from the cooling coils in the air-handling units (AHUs) at a temperature of 46°F or higher. It melts the ice on the outer surface of the tubes and is thus cooled to a temperature of 34 to 36°F, as shown in Figure 12.1(d). Brine is then pumped to the AHUs to cool and dehumidify the air again.

Case Study of a Brine-Coil Ice-Storage System. In Tackett (1989), a brine-coil ice-storage system cools a 550,000-ft² office building near Dallas, TX, as shown in Figure 12.1(a). Ethylene glycol water is used as the brine. There are two centrifugal chillers, each of them having a refrigeration capacity of 568 tons when 34°F brine is produced with a power consumption of 0.77 kW/ton for direct cooling. The refrigeration capacity drops to 425 tons if 24°F brine leaves the chiller with a power consumption of 0.85 kW/ton. A demand-limited partial storage is used, as shown in Figure 12.1(b). During on-peak hours, ice is melted; at the same time one chiller is also operating. The system uses 90 brine-coil modular storage tanks with a full-charged ice-storaged capacity of 7500 ton.hr.

For summer cooling, the weekdays' 24-hr day-and-night cycle is divided into three periods:

- Off peak lasts from 8 p.m. to AHU's start the next morning. Ice is charged at a maximum capacity of 650 tons. The chillers also provide 200 tons of direct cooling for refrigeration loads that operate 24 hr continuously.
- Direct cooling lasts from AHU's start until noon on weekdays. Chillers are operated for direct cooling. If required refrigeration load exceeds the chillers' capacity, some ice storage will be melted to supplement the direct cooling.
- On peak lasts from noon to 8 p.m. Ice is burning with one chiller in operation. During ice burning, water separates the tube and ice. Water has a much lower thermal conductivity (0.35 Btu/hr.ft.°F) than ice (1.3 Btu/hr.ft.°F). Therefore, the capacity of a brine-coil ice-storage system is dominated by the ice burning.

Ice-Harvester Ice-Storage Systems. In an *ice-harvester* system, glycol-water flows on the outside surface of the evaporator and forms ice sheets with a thickness of 0.25 to 0.40 in. within 20 to 30 min. Ice is harvested in the form of flakes when hot gas is flowing inside the tubes of the evaporator during a time interval of 20 to 30 sec. Ice flakes fall to the glycol-water in the storage tank below. The glycol-water at a temperature of 34°F is then supplied to the cooling coils in AHUs for conditioning. After cooling and dehumidifying the air, glycol-water returns to the storage tank at a temperature of 50 to 60°F and is again cooled to 34°F again.

Ice harvesting is an intermittent process. It has a cycle loss due to harvesting of about 15%. In addition, because of its operating complexity and maintenance, its applications are more suitable for large ice-storage systems.

Ice-on-Coil Ice-Storage Systems. In an ice-on-coil system, refrigerant is evaporated in the coils submerged in water in a storage tank. Ice of a thickness not exceeding 0.25 in. builds up on the outer surface of the coils. The remaining water in the storage tank is used for cooling in AHUs. Ice-on-coil systems need large amounts of refrigerant charge and are less flexible in operation. They are usually used in industrial applications.

Ice-in-Containers Ice-Storage Systems. Ice-in-containers ice-storage systems store ice in enclosed containers. Brine circulating over the containers produces the ice inside containers. Complexity in control of the ice inventory inside the containers limits the application of the ice-in-containers systems.

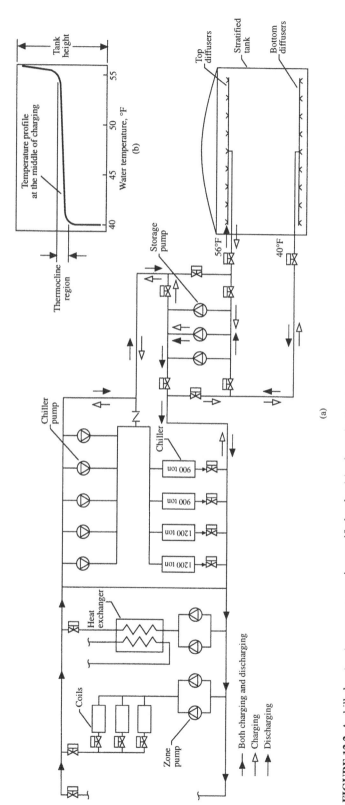

FIGURE 12.2 A chilled-water storage system using stratified tanks: (a) schematic diagram of a chilled-water storage system and (b) thermocline at the middle of charging process.

Chilled-Water Storage Systems

Basics

Chilled-water storage uses the same water chiller and a similar coolant distribution system, except for additional water storage tanks and corresponding piping, additional storage pumps, and controls. The larger the chilled-water storage system, the lower the installation cost per ton.hr storage capacity.

Various types of storage tanks had been used in chilled-water storage systems during the 1970s. A diaphragm tank uses a rubber diaphragm to separate the colder and warmer water. Baffles divide the tank into cells and compartments. Today, stratified tanks have become the most widely used chilled-water storage systems because of their simplicity, low cost, and negligible difference in loss of cooling capacity between stratified tanks and other types of storage tanks.

During the storage of chilled water, the loss in cooling capacity includes direct mixing, heat transfer between warmer return chilled water and colder stored chilled water, and also heat transfer between warmer ambient air and colder water inside the tank. An enthalpy-based easily measured index called *figure of merit (FOM)* is often used to indicate the loss in cooling capacity during chilled-water storage. FOM is defined as:

$$\text{FOM} = q_{\text{dis}}/q_{\text{ch}} \tag{12.1}$$

where q_{dis} = cooling capacity available in the discharge process, Btu/hr
         q_{ch} = theoretical cooling capacity available during charging process, Btu/hr

Charging is the process of filling the storage tank with colder chilled water from the chiller. At the same time, warmer return chilled water is extracted from the storage tank and pumped to the chiller for cooling.

Discharging is the process of discharging the colder stored chilled water from the storage tank to AHUs and terminals. Meanwhile, the warmer return chilled water from the AHUs and terminals fills the tank.

Stratified Tanks. *Stratified tanks* utilize the buoyancy of warmer return chilled water to separate it from the colder stored chilled water during charging and discharging, as shown in Figure 12.2(a). Colder stored chilled water is always charged and discharged from bottom diffusers, and the warmer return chilled water is introduced to and withdrawn from the top diffusers.

Chilled-water storage tanks are usually vertical cylinders and often have a height-to-diameter ratio of 0.25:0.35. Steel is the most commonly used material for above-grade tanks, with a 2-in.-thick spray-on polyurethane foam, a vapor barrier, and a highly reflective coating. Concrete, sometimes precast, pre-stressed tanks are widely used for underground tanks.

A key factor to reduce loss in cooling capacity during chilled water storage is to reduce mixing of colder and warmer water streams at the inlet. If the velocity pressure of the inlet stream is less than the buoyancy pressure, the entering colder stored chilled water at the bottom of tank will stratify. Refer to Wang's handbook (1993) and Knebel (1995) for details.

A *thermocline* is a stratified region in a chilled-water storage tank of which there is a steep temperature gradient as shown in Figure 12.2(b). Water temperature often varies from 42°F to about 60°F. Thermocline separates the bottom colder stored chilled water from the top warmer return chilled water. The thinner the thermocline, the lower the mixing loss.

Diffusers and symmetrical connected piping are used to evenly distribute the incoming water streams with sufficient low velocity, usually lower than 0.9 ft/sec. Inlet stream from bottom diffusers should be downward and from the top diffusers should be upward or horizontal.

Field measurements indicate that stratified tanks have a FOM between 0.85 to 0.9.

13

Air System Basics

Shan K. Wang

Consultant

13   Air System Basics.. 129
     Fan-Duct Systems • System Effect • Modulation of Air Systems
     • Fan Combinations in Air-Handling Units and Packaged Units
     • Fan Energy Use • Year-Round Operation and Economizers •
     Outdoor Ventilation Air Supply

Air System Basics

Fan-Duct Systems

Flow Resistance

Flow resistance is a property of fluid flow which measures the characteristics of a flow passage resisting the fluid flow with a corresponding total pressure loss Δp, in in. WG, at a specific volume flow rate $\mathring{V}$, in cfm:

$$\Delta p = R \mathring{V}^2 \tag{13.1}$$

where R = flow resistance (in. WG/(cfm)^2).

For a duct system that consists of several duct sections connected in series, its flow resistance R_s, in in. WG/(cfm)^2, can be calculated as

$$R_s = R_1 + R_2 + \ldots + R_n \tag{13.2}$$

where R_1, R_2, ... R_n = flow resistance of duct section 1, 2, ... n in the duct system (in. WG/(cfm)^2).

For a duct system that consists of several duct sections connected in parallel, its flow resistance R_p, in in. WG/(cfm)^2, is:

$$1 / \sqrt{R_p} = 1 / \sqrt{R_1} + 1 / \sqrt{R_2} + \ldots + 1 / \sqrt{R_n} \tag{13.3}$$

0-8493-0057-6/00/$0.00+$.50
© 2000 by CRC Press LLC

Fan-Duct System

In a *fan-duct system*, a fan or several fans are connected to ductwork or ductwork and equipment. The volume flow and pressure loss characteristics of a duct system can be described by its performance curve, called *system curve*, and is described by $\Delta p = R\ \mathring{V}^2$.

An *air system* or an air handling system is a kind of fan-duct system. In addition, an outdoor ventilation air system to supply outdoor ventilation air, an exhaust system to exhaust contaminated air, and a smoke control system to provide fire protection are all air systems, that is, fan-duct systems.

Primary, Secondary, and Transfer Air

Primary air is the conditioned air or makeup air. Secondary air is often the induced space air, plenum air, or recirculating air. Transfer air is the indoor air that moves to a conditioned space from an adjacent area.

System-Operating Point

A *system-operating point* indicates the operating condition of an air system or fan-duct system. Since the operating point of a fan must lie on the fan performance curve, and the operating point of a duct system on the system curve, the system operating point of an air system must be the intersection point P_s of the fan performance curve and system curve as shown in Figure 13.1(a).

Fan-Laws

For the same air system operated at speeds n_1 and n_2, both in rpm, their relationship of $\mathring{V}$ volume flow rate, in cfm, system total pressure loss, in in. WG, and fan power input, in hp, can be expressed as

$$\mathring{V}_2 / \mathring{V}_1 = n_2/n_1$$

$$\Delta p_{t2}/\Delta p_{t1} = \left(n_2/n_1\right)^2\left(\rho_2/\rho_1\right) \tag{13.4}$$

$$P_2/P_1 = \left(n_2/n_1\right)^3\left(\rho_2/\rho_1\right)$$

where ρ = air density (lb/ft³). Subscripts 1 and 2 indicate the original and the changed operating conditions. For air systems that are geometrically and dynamically similar:

$$\mathring{V}_2 / \mathring{V}_1 = \left(D_2/D_1\right)^3\left(n_2/n_1\right)$$

$$\Delta p_{t2}/\Delta p_{t1} = \left(D_2/D_1\right)^2\left(n_2/n_1\right)^2\left(\rho_2/\rho_1\right) \tag{13.5}$$

$$P_2/P_1 = \left(D_2/D_1\right)^5\left(n_2/n_1\right)^3\left(\rho_2/\rho_1\right)$$

where D = diameter of the impeller (ft).

Geometrically similar means that two systems are similar in shape and construction. For two systems that are dynamically similar, they must be geometrically similar, and in addition, their velocity distribution or profile of fluid flow should also be similar. When fluid flows in the air systems are at high Reynolds number, such as Re > 10,000, their velocity profiles can be considered similar to each other.

System Effect

The system effect Δp_{se}, in in. WG, of an air system is its additional total pressure loss caused by uneven or nonuniform velocity profile at the fan inlet, or at duct fittings after fan outlet, due to the actual inlet

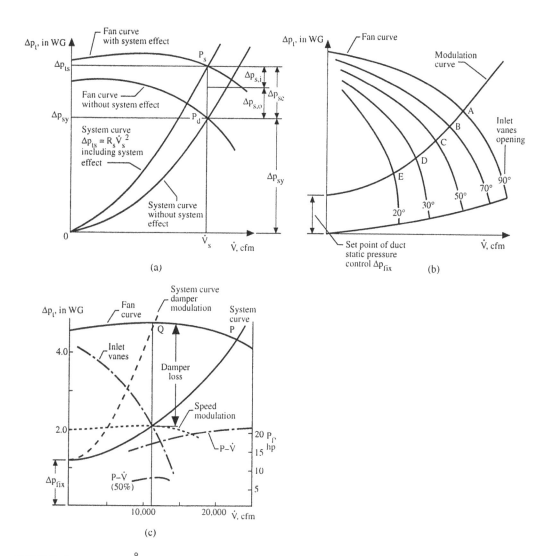

FIGURE 13.1 Air system $\overset{\circ}{V}$- Δp_t performance: (a) system operating point and system effect, (b) modulation curve, and (c) damper, inlet vanes, and fan speed modulation.

and outlet connections as compared with the total pressure loss of the fan test unit during laboratory ratings. The selected fan total pressure which includes the system effect Δp_{ts}, in in. WG, as shown in Figure 13.1(a), can be calculated as

$$\Delta p_{ts} = \Delta p_{sy} + \Delta p_{se} = \Delta p_{sy} + \Delta p_{s.i} + \Delta p_{s.o}$$

$$= \Delta p_{sy} + C_{s.i}\left(v_{fi}/4005\right)^2 + C_{s.o}\left(v_{fo}/4005\right)^2$$

(13.6)

where Δp_{sy} = calculated total pressure loss of the air system, in WG
   $\Delta p_{s.i}$, $\Delta p_{s.o}$ = fan inlet and outlet system effect loss, in WG
   $C_{s.i}$, $C_{s.o}$ = local loss coefficient of inlet and outlet system effect, in WG
   v_{fi}, v_{fo} = velocity at fan inlet (fan collar) and fan outlet, fpm

Both $C_{s,i}$ and $C_{s,o}$ are mainly affected by the length of connected duct between the duct fitting and fan inlet or outlet, by the configuration of the duct fitting, and by the air velocity at the inlet or outlet. Because v_{fi} and v_{fo} are often the maximum velocity of the air system, system effect should not be overlooked. According to AMCA Fan and Systems (1973), a square elbow (height to turning radius ratio R/H = 0.75) connected to a fan inlet with a connected duct length of 2 D_e (equivalent diameter) and v_{fi} = 3000 fpm may have a 0.67 in. WG $\Delta p_{i,o}$ loss. Refer to AMCA Fan and Systems (1973) or Wang's handbook (1993) for details.

Modulation of Air Systems

Air systems can be classified into two categories according to their operating volume flow: constant volume and variable-air-volume systems. The volume flow rate of a *constant volume system* remains constant during all the operating time. Its supply temperature is raised during part load. For a *variable-air-volume (VAV) system*, its volume flow rate is reduced to match the reduction of space load at part-load operation. The system pressure loss of a VAV system can be divided into two parts: variable part Δp_{var} and fixed part Δp_{fix}, which is the set point of the duct static pressure control as shown in Figure 13.1(b) and (c). The *modulation curve* of a VAV air system its its operating curve, or the locus of system operating points when its volume flow rate is modulated at full- and part-load operation.

The volume flow and system pressure loss of an air system can be modulated either by changing its fan characteristics or by varying its flow resistance of the duct system. Currently used types of modulation of volume flow rate of VAV air systems are

1. *Damper modulation* uses an air damper to vary the opening of the air flow passage and therefore its flow resistance.
2. *Inlet vanes modulation* varies the opening and the angle of inlet vanes at the centrifugal fan inlet and then gives different fan performance curves.
3. *Inlet cone modulation* varies the peripheral area of the fan impeller and therefore its performance curve.
4. *Blade pitch modulation* varies the blade angle of the axial fan and its performance curve.
5. *Fan speed modulation* using *adjustable frequency AC drives* varies the fan speed by supplying a variable-frequency and variable-voltage power source. There are three types of AC drives: adjustable voltage, adjustable current, and pulse width modulation (PWM). The PWM is universally applicable.

Damper modulation wastes energy. Inlet vanes are low in cost and are not so energy efficient compared with AC drives and inlet cones. Inlet cone is not expensive and is suitable for backward curved centrifugal fans. Blade pitch modulation is energy efficient and is mainly used for vane and tubular axial fans. AC drive is the most energy-efficient type of modulation; however, it is expensive and often considered cost effective for air systems using large centrifugal fans.

Example 13.1

A multizone VAV system equipped with a centrifugal fan has the following characteristics:

$\mathring{V}$ (cfm)	5,000	10,000	15,000	20,000	25,000
Δp_t, in. WG	4.75	4.85	4.83	4.60	4.20
P, hp		17.0	18.6	20.5	21.2

At design condition, it has a volume flow rate of 22,500 cfm and a fan total pressure of 4.45 in. WG. The set point of duct static pressure control is 1.20 in. WG.

When this VAV system is modulated by inlet vanes to 50% of design flow, its fan performance curves show the following characteristics:

$\overset{\circ}{V}$ (cfm)	5,000	10,000	11,250
Δp_t, in. WG	3.6	2.5	2.1
P, hp		7.5	7.8

Determine the fan power input when damper, inlet vanes, or AC drive fan speed modulation is used. Assume that the fan total efficiency remains the same at design condition when the fan speed is reduced.

Solutions

1. At 50% design flow, the volume flow of this VAV system is 0.5 × 22,500 = 11,250 cfm. The flow resistance of the variable part of this VAV system is

$$R_{var} = \Delta p_{va} = R\overset{\circ}{V}^2 = (4.45 - 1.20)/(22,500)^2 = 6.42 \times 10^{-9} \text{ in. WG}/(\text{cfm})^2$$

When damper modulation is used, the system operating point Q must be the intersection of the fan curve and the system curve that has a higher flow resistance and a $\overset{\circ}{V}$ = 11,250 cfm. From Figure 13.1(c), at point Q, the fan power input is 17.0 hp.

2. From the given information, when inlet vane modulation is used, the fan power input is 7.8 hp.
3. The total pressure loss of the variable part of the VAV system at 50% volume flow is

$$\Delta p_{var} = R_{var} \overset{\circ}{V}^2 = 6.42 \times 10^{-9}(11,250)^2 = 0.81 \text{ in. WG}$$

From Figure 13.1(c), since the fan power input at design condition is 21.2 hp, then its fan total efficiency is:

$$\eta_t = \overset{\circ}{V}\Delta p_{tf}/(6356 P_t) = 22,500 \times 4.45/(6356 \times 21.2) = 74.3\%$$

The fan power input at 50% design volume flow is:

$$P = \overset{\circ}{V}\Delta p_{tf}/(6356\eta_t) = 11,250(0.81 + 1.20)/(6356 \times 0.743) = 4.8 \text{ hp}$$

Damper modulation has a energy waste of (17 − 4.8) = 12.2 hp

Fan Combinations in Air-Handling Units and Packaged Units

Currently used fan combinations in air-handling units (AHUs) and packaged units (PUs) (except dual-duct VAV systems) are shown in Figure 13.2(a), (b), and (c).

Supply and Exhaust Fan/Barometric Damper Combination

An air system equipped with a single supply fan and a constant-volume exhaust fan, or a supply fan and a barometric damper combination as shown in Figure 13.2(a), is often used in buildings where there is no return duct or the pressure loss of the return duct is low. An all-outdoor air economizer cycle is usually not adopted due to the extremely high space pressure. A barometric relief damper is often installed in or connected to the conditioned space to prevent excessively high space pressure. When the space-positive pressure exerted on the barometric damper is greater than the weight of its

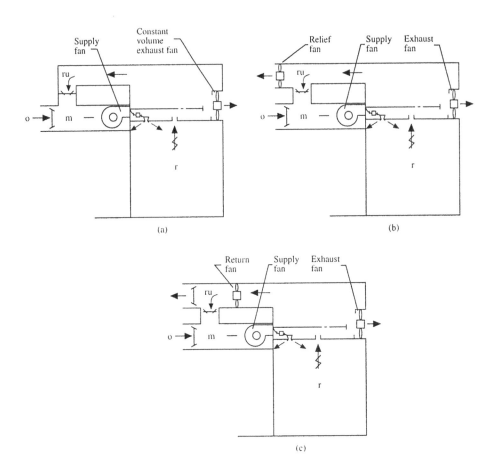

FIGURE 13.2 Fan combinations: (a) supply and exhaust fan, (b) supply and relief fan, and (c) supply and return fan.

damper and/or a spring, the damper is opened and the excessive space pressure is relieved.

Consider a VAV rooftop packaged system using a supply fan and a constant-volume exhaust system to serve a typical floor in an office building. This air system has the following design parameters

During minimum outdoor ventilation air *recirculating mode* at summer design volume flow and at 50% of design volume flow, the outdoor damper is at its minimum opening. The recirculating damper is fully opened. The outdoor air intake at the PU must be approximately equal to the exfiltration at the conditioned space due to the positive space pressure p_r, in in. WG. By using the iteration method, the calculated pressure characteristics and the corresponding volume flow rates are shown below :

Supply volume flow rate:		20,000 cfm
Total pressure loss:	Across the recirculating damper	0.1 in. WG
	Filter and coils	2.5 in. WG
	Supply main duct	0.85 in. WG
	Return system between point r and ru	0.2 in. WG
Effective leakage area on the building shell		3.35 ft²
Minimum outdoor ventilation air		3000 cfm
Space pressure at design flow		+0.05 in. WG

Point	o	m	r	ru
Design flow:				
p_s in. WG	0	−0.20	+0.05	−0.15
$\overset{\circ}{V}$, cfm	3,000	20,000	20,000	17,000
50% design:				
p, in. WG	0	−0.092	+0.0225	−0.052
$\overset{\circ}{V}$, cfm	2,025	10,000	10,000	7,975

Here, o represents outdoor, m the mixing box, r the space, and ru the recirculating air inlet to the PU. When the supply volume flow is reduced from the design volume flow to 50% of design flow during the recirculating mode, the total pressure in the mixing box increases from −0.20 to −0.092 in. WG and the outdoor air intake reduces from 3000 to 2025 cfm. Refer to Wang's (1993) *Handbook of Air Conditioning and Refrigeration* for details.

Supply and Relief Fan Combination

Figure 13.2(b) shows the schematic diagrams of an air system of supply fan and relief fan combination. A relief fan is used to relieve undesirable high positive space pressure by extracting space air and relieving it to the outside atmosphere. A relief fan is always installed in the relief flow passage after the junction of return flow, relief flow, and recirculating flow passage, point ru. It is usually energized only when the air system is operated in air economizer mode. A relief fan is often an axial fan. Since the relief fan is not energized during recirculating mode operation, the volume flow and pressure characteristics of a supply fan and relief fan combination are the same as that in a single supply fan and barometric damper combination when they have the same design parameters.

During air economizer mode, the outdoor air damper(s) are fully opened and the recirculating damper closed. The space pressure p_r = +0.05 in. WG is maintained by modulating the relief fan speed or relief damper opening. The pressure and volume flow characteristics of a supply and relief fan combination at design volume flow and 50% of design flow are as follows:

Point	o	m	r	ru
Design flow:				
p_s in. WG	0	−0.20	+0.05	−0.15
$\overset{\circ}{V}$, cfm	20,000	20,000	20,000	17,000

Point	o	m	r	ru
50% design:				
p, in. WG	0	−0.057	+0.05	+0.007
$\overset{\circ}{V}$, cfm	10,000	10,000	10,000	7,000

Supply Fan and Return Fan Combination

A *return fan* is always installed at the upstream of the junction of return, recirculating, and exhaust flow passage, point ru as shown in Figure 13.2(c). A supply and return fan combination has similar pressure and volume flow characteristics as that of a supply and relief fan combination, except a higher total pressure at point ru. If the return fan is improperly selected and has an excessive fan total pressure, total pressure at point m may be positive. There will be no outdoor intake at the PU or AHU, and at the same time there will also be a negative space pressure and an infiltration to the space.

Comparison of These Three Fan Combination Systems

A supply fan and barometric damper combination is simpler and less expensive. It is suitable for an air system which does not operate at air economizer mode and has a low pressure drop in the return system.

For those air systems whose pressure drop of return system is not exceeding 0.3 in. WG, or there is a considerable pressure drop in relief or exhaust flow passage, a supply and relief fan combination is recommended. For air systems whose return system has a pressure drop exceeding 0.6 in. WG, or those requiring a negative space pressure, a supply and return fan combination seems more appropriate.

Year-Round Operation and Economizers

Consider a typical single-duct VAV reheat system to serve a typical floor whose indoor temperature is 75°F with a relative humidity of 50%, as shown in Figure 13.3(a). During summer, the off-coil temperature is 55°F. The year-round operation of this air system can be divided into four regions on the psychrometric chart, as shown in Figure 13.3(b):

- *Region I — Refrigeration/evaporative cooling.* In this region, the enthalpy of the outdoor air h_o is higher than the enthalpy of the recirculating air h_{ru}, $h_o > h_{ru}$. It is more energy efficient to condition the mixture of recirculating air and minimum outdoor air.

- *Region II — Free cooling and refrigeration.* In this region, $h_o \leq h_{ru}$. It is more energy efficient and also provides better indoor air quality to extract 100% outdoor air.

- *Region III — Free cooling evaporative cooling, and refrigeration.* In this region, extract 100% outdoor air for free cooling because $h_o \leq h_{ru}$. Use evaporative cooling and refrigeration to cool and humidify if necessary.

- *Region IV — Winter heating.* Maintain a 55°F supply temperature by mixing the recirculating air with the outdoor air until the outdoor air is reduced to a minimum value. Provide heating if necessary.

An economizer is a device consisting of dampers and control that uses the free cooling capacity of either outdoor air or evaporatively cooled water from the cooling tower instead of mechanical refrigeration. An air economizer uses outdoor air for free cooling. There are two kinds of air economizers: enthalpy-based, in which the enthalpy of outdoor and recirculating air is compared, and temperature-based, in which temperature is compared. A water economizer uses evaporatively cooled water.

Fan Energy Use

For an air system, fan energy use for each cfm of conditioned air supplied from the AHUs and PUs to the conditioned space within a certain time period, in W/cfm, can be calculated as

$$W/cfm = 0.1175 \Delta p_{sy} / (\eta_f \eta_m) \tag{13.7}$$

where Δp_{sy} = mean system total pressure loss during a certain time period, in. WG
η_f = fan total efficiency
η_m = combined motor and drive (direct drive or belt drive) efficiency

For an air system using a separate outdoor ventilation system, its fan energy use, in W/cfm, is then calculated as

$$W/cfm = (1 + R_{o.s}) [0.1175 \Delta p_{sy} / (\eta_f \eta_m)] \tag{13.8}$$

where $R_{o.s}$ = ratio of outdoor air volume flow rate to supply volume flow rate.

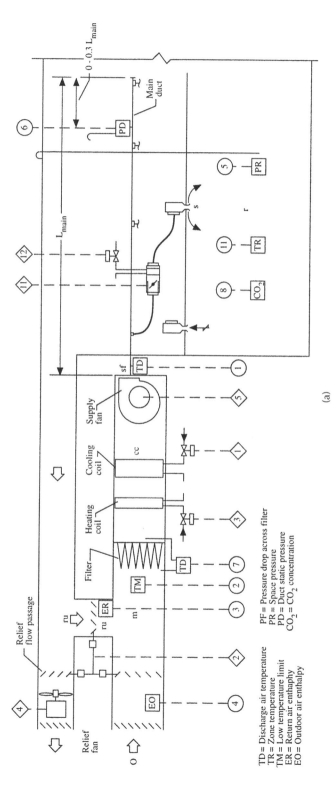

TD = Discharge air temperature
TR = Zone temperature
TM = Low temperature limit
ER = Return air enthaphy
EO = Outdoor air enthalpy

PF = Pressure drop across filter
PR = Space pressure
PD = Duct static pressure
CO_2 = CO_2 concentration

(a)

FIGURE 13.3 Year-round operation, discharge air temperature, and duct static pressure control for a VAV reheat system: (a) control diagram, (b) year-round operation, and (c) discharge air temperature control output diagram.

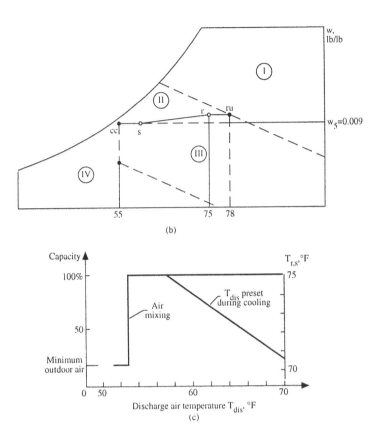

FIGURE 13.3 (continued)

Outdoor Ventilation Air Supply

Basics

- An adequate amount of outdoor ventilation air supply is the key factor to provide acceptable indoor air quality (IAQ) for a conditioned space. Although an inadequate amount of outdoor ventilation air supply causes poor IAQ, an oversupply of outdoor ventilation air other than in an air economizer cycle is often a waste of energy.

- According to local codes and ANSI/ASHRAE Standard 62-1989, the minimum outdoor ventilation rate for each person must be provided at the outdoor air intake of AHU or PU, or by an outdoor air ventilation system. If the minimum outdoor ventilation air rate is reduced by using higher efficiency filters to remove air contaminants in the recirculating air, then indoor air contaminant concentration must be lower than the specified level in ANSI/ASHRAE Standard 62-1989.

- For a multizone air system, although the ratio of outdoor ventilation air rate to supply air volume flow rate required may be varied from zone to zone, the excessive outdoor air supply to a specified zone will increase the content of unused outdoor air in the recirculating air in AHU or PU. This helps to solve the problem in any zone that needs more outdoor air.

- Since the occupancy in many buildings is often variable and intermittent, a demand-based variable amount of outdoor ventilation air control should be used instead of time-based constant volume outdoor ventilation air control, except during the air economizer cycle.

- Carbon dioxide (CO_2) is a gaseous body effluent. CO_2 is an indicator of representative odor and an indicator of adequate outdoor ventilation rate at specific outdoor and indoor air concentration in a control zone at steady state. For most of the comfort air-conditioning systems, it is suitable to use CO_2 as a key parameter to control the intake volume flow rate of outdoor ventilation air to maintain an indoor CO_2 concentration not exceeding 800 to 1000 ppm in a critical or representative zone. As mentioned in Section 5, Persily (1993) showed that the actual measured indoor daily maximum CO_2 concentration levels in five buildings were all within 400 to 820 ppm.

If a field survey finds that a specific indoor air contaminant exceeds a specified indoor concentration, then a gas sensor for this specific contaminant or a mixed gas sensor should be used to control this specific indoor concentration level.

Types of Minimum Outdoor Ventilation Air Control.   There are four types of minimum outdoor ventilation air control that are currently used:

- Type I uses a CO_2 sensor or a mixed gas sensor and a DDC controller to control the volume flow rate of outdoor ventilation air for a separate outdoor ventilation air system on the demand-based principle.

- Type II uses a CO_2 or mixed gas sensor and a DDC controller to control the ratio of the openings between outdoor and recirculating dampers and, therefore, the volume flow rates of outdoor air and recirculating air in AHUs or PUs on the demand-based principle.

- Type III uses a flow sensor or a pressure sensor and a DDC controller to control the openings of outdoor and recirculating dampers to provide a nearly constant volume outdoor air intake in VAV AHUs or VAV PUs.

- Type IV adjusts the opening of the outdoor damper manually to provide a constant volume of outdoor air in constant-volume AHUs and PUs. If the outdoor intake is mounted on the external wall without a windshield, the volume flow of outdoor ventilation air intake will be affected by wind force and direction.

Type I is the best minimum outdoor ventilation air control for the air system. For a VAV system, it is expensive. Type II is a better choice. Type III is more complicated and may cause energy waste. Type IV has the same result as Type III and is mainly used in constant-volume systems.

Outdoor intake must be located in a position away from the influence of exhaust outlets. Fans, control dampers, and filters should be properly operated and maintained in order to provide a proper amount of outdoor ventilation air as well as an acceptable IAQ.

14

Absorption Systems

Shan K. Wang

Individual Consultant

14   Absorption Systems 139
      Double-Effect Direct-Fired Absorption Chillers • Absorption
      Cycles, Parallel-, Series-, and Reverse-Parallel Flow

Absorption System

Absorption systems use heat energy to produce refrigeration as well as heating if it is required. Water is the refrigerant and aqueous lithium bromide (LiBr) is widely used as the carrier to absorb the refrigerant and provide a higher coefficient of performance.

The mixture of water and anhydrous LiBr is called *solution*. The composition of a solution is usually expressed by its mass fraction, or percentage of LiBr, often called *concentration*. When the water vapor has boiled off from the solution, it is called *concentration solution*. If the solution has absorbed the water vapor, it is called *diluted solution*.

Absorption systems can be divided into the following categories:

- *Absorption chillers* use heat energy to produce refrigeration.
- *Absorption chiller/heaters* use direct-fired heat input to provide cooling or heating separately.
- *Absorption heat pumps* extract heat energy from the evaporator, add to the heat input, and release them both to the hot water for heating.
- *Absorption heat transformers* raise the temperature of the waste heat source to a required level.

Most recently installed absorption chillers use direct-fired natural gas as the heat source in many locations in the United States where there are high electric demand and electric rate at on-peak hours. Absorption chillers also are free from CFC and HCFC. An energy cost analysis should be done to determine whether an electric chiller or a gas-fired absorption chiller is the suitable choice.

Absorption heat pumps have only limited applications in district heating. Most absorption heat transformers need industrial waste heat. Both of them will not be covered here.

Double-Effect Direct-Fired Absorption Chillers

Figure 14.1(a) shows a double-effect direct-fired absorption chiller. *Double effect* means that there are two generators. *Direct fired* means that gas is directly fired at the generator instead of using steam or hot water. A single-effect absorption chiller using steam as the heat input to its single generator has a COP only from 0.7 to 0.8, whereas a double-effect direct-fired absorption chiller has a COP approximately equal to 1 and therefore is the mot widely used absorption chiller in the United States for new and retrofit projects today. The refrigeration capacity of double-effect direct-fired absorption chillers varies from 100 to 1500 tons.

A double-effect direct-fired absorption chiller mainly consists of the following components and controls:

- Evaporator — An *evaporator* is comprised of a tube bundle, spray nozzles, a water trough, a refrigerant pump, and an outer shell. Chilled water flows inside the tubes. A refrigerant pump sprays the liquid refrigerant over the outer surface of the tube bundle for a higher rate of evaporation. A water trough is located at the bottom to maintain a water level for recirculation.

- Absorber — In an *absorber*, there are tube bundles in which cooling water flows inside the tubes. Solution is sprayed over the outer surface of the tube bundle to absorb the water vapor. A solution pump is used to pump the diluted solution to the heat exchanger and low-temperature generator.

- Heat exchangers — There are two heat exchangers: *low-temperature heat exchanger* in which the temperature of hot concentrated solution is lower, and *high-temperature heat exchanger* in which the temperature of hot concentrated solution is higher. In both heat exchangers, heat is transferred from the hot concentrated solution to the cold diluted solution. Shell-and-tube or plate-and-frame heat exchangers are most widely used for their higher effectiveness.

- Generators — *Generators* are also called *desorbers*. In the *direct-fired generator*, there are the fire tube, flue tube, vapor/liquid separator, and flue-gas economizer. Heat is supplied from the gas burner or other waste heat source. The *low-temperature generator* is often of the shell-and-tube type. The water vapor vaporized in the direct-fired generator is condensed inside the tubes. The latent heat of condensation thus released is used to vaporize the dilute solution in the low-temperature generator.

- Condenser — A condenser is usually also of the shell-and-tube type. Cooling water from the absorber flows inside the tubes.

- Throttling devices — Orifices and valves are often used as throttling devices to reduce the pressure of refrigerant and solution to the required values.

- Air purge unit — Since the pressure inside the absorption chiller is below atmospheric pressure, air and other noncondensable gases will leak into it from the ambient air. An *air purge unit* is used to remove these noncondensable gases from the chiller. A typical air purge unit is comprised of a pickup tube, a purge chamber, a solution spray, cooling water tubes, a vacuum pump, a solenoid valve, and a manual shut-off valve.

When noncondensable gases leak into the system, they tend to migrate to the absorber where pressure is lowest. Noncondensable gases and water vapor are picked from the absorber through the pickup tube. Water vapor is absorbed by the solution spray and returned to the absorber through a liquid trap at the bottom of the purge chamber. Heat of absorption is removed by the cooling water inside the tubes. Noncondensable gases are then evacuated from the chamber periodically by a vacuum pump to the outdoor atmosphere.

Palladium cells are used to continuously remove a small amount of hydrogen that is produced due to corrosion. Corrosion inhibitors like lithium chromate are needed to protect the machine parts from the corrosive effect of the absorbent when air is present.

Absorption Cycles, Parallel-, Series-, and Reverse-Parallel Flow

An *absorption* cycle shows the properties of the solution and its variation in concentrations, temperature, and pressure during absorbing, heat exchanging, and concentration processes on an equilibrium chart as shown in Figure 14.1(b). The ordinate of the equilibrium chart is the saturated temperature and pressure

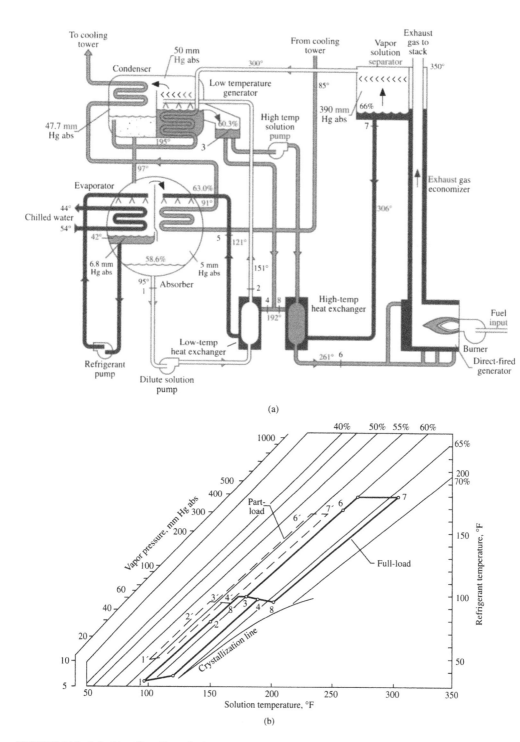

FIGURE 14.1 A double-effect direct-fired reverse-parallel-flow absorption chiller: (a) schematic diagram (reprinted by permission from the Trane catalog) and (b) absorption cycle.

of water vapor, in °F and mm Hg abs. The abscissa is the temperature of the solution, in °F. Concentration lines are incline lines. At the bottom of the concentration lines, there is a *crystallization line* or *saturation line*. If the mass of fraction of LiBr in a solution which remains at constant temperature is higher than the saturated condition, that part of LiBr exceeding the saturation condition tends to form solid crystals.

Because there are two generators, the flow of solution from the absorber to generators can be in series flow, parallel flow, or reverse-parallel flow. In a series-flow system, the diluted solution from the absorber is first pumped to the direct-fired generator and then to the low-temperature generator. In a parallel-flow system, diluted solution is pumped to both direct-fired and low-temperature generators in parallel. In a reverse-parallel-flow system as shown in Figure 14.1(a), diluted solution is first pumped to the low-temperature generator. After that, the partly concentrated solution is then sent to the direct-fired generator as well as to the intermediate point 4 between high- and low-temperature heat exchangers in parallel. At point 4, partly concentrated solution mixes with concentrated solution from a direct-fired generator. A reverse-parallel-flow system is more energy efficient.

Solution and Refrigerant Flow

In a typical double-effect direct-fired reverse-parallel-flow absorption chiller operated at design full load, water is usually evaporated at a temperature of 42°F and a saturated pressure of 6.8 mm Hg abs in the evaporator. Chilled water returns from the AHUs or fan coils at a temperature typically 54°F, cools, and leaves the evaporator at 44°F. A refrigeration effect is produced due to the vaporization of water vapor and the removal of latent heat of vaporization from the chilled water.

Water vapor in the evaporator is then extracted to the absorber due to its lower vapor pressure. It is absorbed by the concentrated LiBr solution at a pressure of about 5 mm Hg abs. After absorption, the solution is diluted to a concentration of 58.6% and its temperature increases to 95°F (point 1). Most of the heat of absorption and the sensible heat of the solution is removed by the cooling water inside the tube bundle. Diluted solution is then pumped by a solution pump to the low-temperature generator through a low-temperature heat exchanger.

In the low-temperature generator, the dilute solution is partly concentrated to 60.3% at a solution temperature of 180°F (point 3). It then divides into two streams: one of them is pumped to the direct-fired generator through a high-temperature heat exchanger, and the other stream having a slightly greater mass flow rate is sent to the intermediate point 4. In the direct-fired generator, the concentrated solution leaves at a concentration of 66% and a solution temperature of 306°F (point 7).

The mixture of concentrated and partly concentrated solution at point 4 has a concentration of 63% and a temperature of 192°F. It enters the low-temperature heat exchanger. Its temperature drops to 121°F before entering the absorber (point 5).

In the direct-fired generator, water is boiled off at a pressure of about 390 mm Hg abs. The boiled-off water vapor flows through the submerged tube in the low-temperature generator. The release of latent heat of condensation causes the evaporation of water from the dilution solution at a vapor pressure of about 50 mm Hg abs. The boiled-off water vapor in the low-temperature generator flows to the condenser through the top passage and is condensed into liquid water at a temperature of about 99°F and a vapor pressure of 47.7 mm Hg abs. This condensed liquid water is combined with the condensed water from the submerged tube at the trough. Both of them return to the evaporator after its pressure is throttled by an orifice plate.

Part-Load Operation and Capacity Control

During part-load operation, a double-effect direct-fired reverse-parallel-flow absorption chiller adjusts its capacity by reducing the heat input to the direct-fired generator through the burner. Lower heat input results at less water vapor boiled off from the solution in the generators. This causes the drop in solution concentration, the amount of water vapor extracted, the rate of evaporation, and the refrigeration capacity. Due to less water vapor being extracted, both evaporating pressure and temperature will rise. Since the amount of water vapor to be condensed is greater than that boiled off from the generators, both the condensing pressure and condensing temperature decrease.

Coefficient of Performance (COP)

The COP of an absorption chiller can be calculated as

$$COP = 12,000/q_{1g} \qquad (14.1)$$

where q_{1g} = heat input to the direct-fired generator per ton of refrigeration output (Btu/hr.ton).

Safety Controls

Safety controls in an absorption chiller include the following:

* Crystallization controls are devices available to prevent crystallization and dissolve crystals. Absorption chillers are now designed to operate in a region away from the crystallization line. It is no longer a serious problem in newly developed absorption systems. One such device uses a bypass valve to permit refrigerant to flow to the concentration solution line when crystallization is detected. Condenser water temperature is controlled by using a three-way bypass valve to mix the recirculating water with the evaporated cooled water from the tower to avoid the sudden drop of the temperature of concentrated solution in the absorber.
* Low-temperature cut-out control shuts down the absorption chiller if the temperature of the refrigerant in the evaporator falls below a preset limit to protect the evaporator from freezing.
* Chilled and cooling water flow switches stop the absorption chiller when the mass flow rate of chilled water or the supply of cooling water falls below a preset value.
* A high-pressure relief valve is often installed on the shell of the direct-fired generator to prevent its pressure from exceeding a predetermined value.
* Monitoring of low and high pressure of gas supply and flame ignition are required for direct-fired burner(s).
* Interlocked controls between absorption chiller and chilled water pumps, cooling water pumps, and cooling tower fans are used to guarantee that they are in normal operation before the absorption chiller starts.

Absorption Chiller/Heater

A double-effect direct-fired reverse-parallel-flow absorption chiller/heater has approximately the same system components, construction, and flow process as the absorption chiller. The cooling mode operation is the same as in an absorption chiller. During the heating mode of an absorption chiller/heater, its evaporator becomes the condenser and is used to condense the water vapor that has been boiled off from the direct-fired generator. At design condition, hot water is supplied at a temperature of 130 to 140°F. The condenser and the low-temperature generator are not in operation. (See Figure 14.1.)

In order to increase the coefficient of performance of the absorption chiller, a triple-effect cycle with three condensers and three desorbers has been proposed and is under development. A triple-effect absorption chiller is predicted to have a coefficient of performance of around 1.5. Its initial cost is also considerably increased due to a greater number of condensers and desorbers.

15

Air-Conditioning Systems and Selection

Shan K. Wang

Consultant

15 Air-Conditioning Systems and Selection........................ 147
    Basics in Classification • Individual Systems • Packaged
    Systems • Central Systems • Air-Conditioning System
    Selection • Comparison of Various Systems • Subsystems •
    Energy Conservation Recommendations

Air-Conditioning Systems and Selection

Basics in Classification

The purpose of classifing air-conditioning or HVAC&R systems is to distinguish one type from another so that an optimum air-conditioning system can be selected according to the requirements. Proper classification of air-conditioning systems also will provide a background for using knowledge-based expert systems to help the designer to select an air-conditioning system and its subsystems.

Since air system characteristics directly affect the space indoor environmental parameters and the indoor air quality, the characteristics of an air system should be clearly designated in the classification.

The system and equipment should be compatible with each other. Each system has its own characteristics which are significantly different from others.

Individual Systems

As described in Section 1, air conditoning or HVAC&R systems can be classified as individual, space, packaged, and central systems.

Individual systems usually have no duct and are installed only in rooms that have external walls and external windows. Individual systems can again be subdivided into the following.

Room Air-Conditioner Systems

A room air conditioner is the sole factory-fabricated self-contained equipment used in the room air-conditioning system. It is often mounted on or under the window sill or on a window frame as shown in Figure 1.1. A room air-conditioner consists mainly of an indoor coil, a small forward-curved centrifugal fan for indoor coil, a capillary tube, a low-efficiency dry and reusable filter, grilles, a thermostat or other controls located in the indoor compartment, and a rotary, scroll, or reciprocating compressor, an outdoor coil, and a propeller fan for the outdoor coil located in the outdoor compartment. There is an outdoor ventilation air intake opening and a manually operated damper on the casing that divides the indoor and outdoor compartments. Room air-conditioners have a cooling capacity between 1/2 to 2 tons.

The following are system characteristics of a room air-conditioner system:

Room heat pump system is a room air-conditioner plus a four-way reversing valve which provides
    both the summer cooling and winter heating.
Air system: single supply fan
    Fan, motor, and drive combined efficiency: 25%
    Fan energy use: 0.3 to 0.4 W/cfm
    Fan speed: HI-LO 2-speed or HI-MED-LO 3-speed
    Outdoor ventilation air system: type IV
Cooling system: DX system, air-cooled condenser
    EER 7.5 to 9.5 Btu/hr.W
    Evaporating temperature T_{ev} at design load: typically 45°F
Heating system: electric heating (if any)
Part-load: on–off of refrigeration compressor
Sound level: indoor NC 45 to 50
Maintenance: More maintenance work is required.

Summer and winter mode air-conditioning cycles of a room air-conditioning system are similar to that shown in Figure 3.4.

All fan, motor, and drive combined efficiencies for various air-conditioning systems are from data in ASHRAE Standard 90.1-1989.

Packaged Terminal Air-Conditioner (PTAC) Systems

A packaged terminal air-conditioner is the primary equipment in a PTAC system. A PTAC system is similar to a room air-conditioner system. Their main differences are

- A PTAC uses a wall sleeve and is intended to be mounted through the wall.
- Heating is available from hot water, steam, heat pump, electric heater, and sometimes even direct-fired gas heaters.

PTACs are available in cooling capacity between 1/2 to 1 1/2 tons and a heating capacity of 2500 to 35,000 Btu/hr.

Space (Space-Conditioning) Systems

Most space conditioning air-conditioning systems cool, heat, and filtrate their recirculating space air above or in the conditioned space. Space conditioning systems often incorporate heat recovery by transferring the heat rejected from the interior zone to the perimeter zone through the condenser(s). Space systems often have a separate outdoor ventilation air system to supply the required outdoor ventilation air.

Space systems can be subdivided into four-pipe fan-coil systems and water-source heat pump systems.

Four-Pipe Fan-Coil Systems

In a four-pipe fan-coil unit system, space recirculating air is cooled and heated at a fan coil by using four pipes: one chilled water supply, one hot water supply, one chilled water return, and one hot water return. Outdoor ventilation air is conditioned at a make-up AHU or primary AHU. It is then supplied to the fan coil where it mixes with the recirculating air, as shown in Figure 15.1(a), or is supplied to the conditioned space directly.

A *fan-coil unit* or a *fan coil* is a *terminal* as shown in Figure 15.1(b). Fan-coil units are available in standard sizes 02, 03, 04, 06, 08, 10, and 12 which correspond to 200 cfm, 400 cfm, and so on in volume flow

The following are system characteristics of a four-pipe fan-coil system:

A *two-pipe fan-coil system* has a supply and a return pipe only. Because of the problems of changeover from chilled water to hot water and vice versa, its applications are limited.

A *water-cooling electric heating fan-coil system* uses chilled water for cooling and an electric heater for heating as shown in Figure 1.2. This system is often used in a location that has a mild winter.

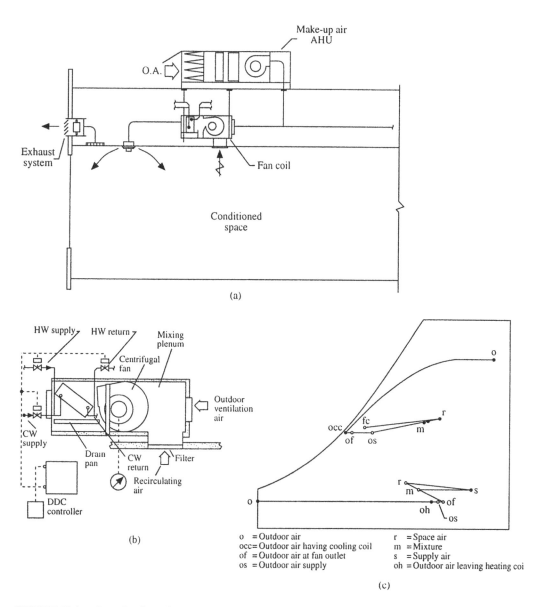

o   = Outdoor air
occ= Outdoor air having cooling coil
of = Outdoor air at fan outlet
os = Outdoor air supply

r   = Space air
m = Mixture
s   = Supply air
oh = Outdoor air leaving heating coi

(c)

FIGURE 15.1 A four-pipe fan-coil system: (a) schematic diagram, (b) fan-coil unit, and (c) air-conditioning cycle.

Air system:

    Fan-coil, space air recirculating

    Fan, motor, and drive combined efficiency: 25%

    Fan speed: HI-LO 2-speed and HI-MED-LO 3-speed

    External pressure for fan coil: 0.06 to 0.2 in. WG

    System fan(s) energy use: 0.45 to 0.5 W/cfm

    No return air and return air duct

    Outdoor ventilation air system: type I

    An exhaust system to exhaust part of the outdoor ventilation air

Cooling system: chilled water from centrifugal or screw chiller

    Water-cooled chiller energy use: 0.4 to 0.65 kW/ton

Heating system: hot water from boiler, electric heater

Part load: control the flow rate of chilled and hot water supplied to the coil. Since air leaving coil

    temperature T_{cc} rises during summer mode part load, space relative humidity will be higher.

Sound level: indoor NC 40 to 45

Maintenance: higher maintenance cost

System fan(s) energy use: 0.45 to 0.55 W/cfm (includes all fans in the four-pipe fan-coil system)

An air-conditioning cycle for a four-pipe fan-coil system with outdoor ventilation air delivered to the suction side of the fan coil is shown in Figure 15.1(c). A part of the space cooling and dehumidifying load is usually taken care by the conditioned outdoor ventilation air from the make-up AHU. A double-bundle condenser is often adopted in a centrifugal chiller to incorporate heat recovery for providing winter heating.

Water-Source Heat Pump Systems

Water-source heat pumps (WSHPs) are the primary equipment in a water-source heat pump system as shown in Figure 15.2(a). A *water-source heat pump* usually consists of an air coil to cool and heat the air; a water coil to reject and extract heat from the condenser water; a forward-curved centrifugal fan; reciprocating, rotary, or scroll compressor(s); a short capillary tube; a reversing valve; controls; and an outer casing. WSHPs could be either a horizontal or vertical unit. WSHPs usually have cooling capacities

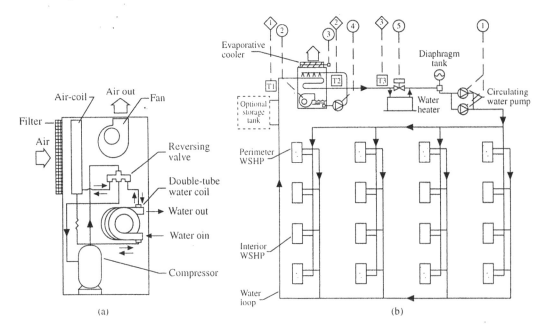

FIGURE 15.2 A water-source heat pump system: (a) vertical system and (b) system schematic diagram..

between 1/2 to 26 tons. Small-capacity WSHPs of 3 tons or less without ducts are used in perimeter zones, whereas large-capacity WSHPs with ductwork are used only in interior zones.

In addition to the WSHPs, a WSHP system usually is also comprised of an evaporative cooler or cooling tower to cool the condenser water; a boiler to provide the supplementary heat for the condenser water if necessary; two circulating pumps, one of them being standby; and controls, as shown in Figure 15.2(b). A separate outdoor ventilation air system is required to supply outdoor air to the WSHP or directly to the space.

During hot weather, such as outdoor wet bulb at 78°F, all the WSHPs are operated in cooling mode. Condenser water leaves the evaporative cooler at a temperature typically 92°F and absorbs condensing heat rejected from the condensers — the water coils in WSHPs. Condenser water is then raised to 104°F and enters the evaporative cooler. In an evaporative cooler, condenser water is evaporatively cooled indirectly by atmospheric air, so that it would not foul the inner surface of water coils in WSHPs.

During moderate weather, the WSHPs serving the shady side of a building may be in heating mode, and while serving the sunny side of the building and the interior space in cooling mode. During cold weather, most of the WSHPs serving perimeter zones are in heating mode, while serving interior spaces are in cooling mode except morning warm-up. Cooling WSHPs reject heat to the condenser water loop; meanwhile heating WSHPs absorb heat from the condenser water loop. The condenser water is usually maintained at 60 to 90°F. If its temperature rises above 90°F, the evaporative cooler is energized. If it drops below 60°F, the boiler or electric heater is energized. A WSHP system itself is a combination of WSHP and a heat recovery system to transfer the heat from the interior space and sunny side of the building to the perimeter zone and shady side of building for heating in winter, spring, and fall.

System characteristics of air, cooling, and heating in a WSHP system are similar to a room conditioner heat pump system. In addition:

Outdoor ventilating air system: type I and IV
Water system: two-pipe, close circuit
Centrifugal water circulating pump
Water storage tank is optional

To prevent freezing in locations where outdoor temperature may drop below 32°F, isolate the outdoor portion of the water loop, outdoor evaporative cooler, and the pipe work from the indoor portion by using a plate-and-frame heat exchanger. Add ethylene or propylene glycol to the outdoor water loop for freezing protection.

There is another space system called a panel heating and cooling system. Because of its higher installation cost and dehumidification must be performed in the separate ventilation systems, its applications are very limited.

A space conditioning system has the benefit of a separate demand-based outdoor ventilation air system. A WSHP system incorporates heat recovery automatically. However, its indoor sound level is higher; only a low-efficiency air filter is used for recirculating air, and more space maintenance is required than central and packaged systems. Because of the increase of the minimum outdoor ventilation air rate to 15 cfm/person recently, it may gain more applications in the future.

Packaged Systems

In packaged systems, air is cooled directly by a DX coil and heated by direct-fired gas furnace or electric heater in a packaged unit (PU) instead of chilled and hot water from a central plant in a central system. Packaged systems are different from space conditioning systems since variable-air-volume supply and air economizer could be features in a packaged system. Packaged systems are often used to serve two or more rooms with supply and return ducts instead of serving individual rooms only in an individual system.

As mentioned in Section 7, packaged units are divided according to their locations into rooftop, split, or indoor units. Based on their operating characteristics, packaged systems can be subdivided into the following systems:

Single-Zone Constant-Volume (CV) Packaged Systems

Although a single-zone CV packaged system may have duct supplies to and returns from two or more rooms, there is only a single zone sensor located in the representative room or space. A CV system has a constant supply volume flow rate during operation except the undesirable reduction of volume flow due to the increase of pressure drop across the filter.

A single-zone CV packaged system consists mainly of a centrifugal fan, a DX coil, a direct-fired gas furnace or an electric heater, a low or medium efficiency filter, mixing box, dampers, DDC controls, and an outer casing. A relief or a return fan is equipped for larger systems.

A single-zone CV packaged system serving a church is shown in Figure 1.3. This system operates on basic air-conditioning cycles as shown in Figure 3.4 during cooling and heating modes.

The system characteristics of a single-zone CV packaged system are

Air system: single supply fan, a relief or return fan for a large system
    Fan, motor, and drive combined efficiency: 40 to 45%
    Fan total pressure: 1.5 to 3 in. WG
    Fan(s) energy use: 0.45 to 0.8 W/cfm
    Outdoor ventilation air system: type IV and II
    Enthalpy or temperature air economizer
Cooling systems: DX system, air cooled
    Compressor: reciprocating or scroll
    EER: 8.9 to 10.0 Btu/hr.W
Heating system: direct-fired gas furnace, air-source heat pump, or electric heating
Part load: on–off or step control of the compressor capacity, DX-coil effective area, and the gas flow to
    the burner
Sound level: indoor NC 35 to 45
Maintenance: higher maintenance cost than central systems

Single-zone, CV packaged systems are widely used in residences, small retail stores, and other commercial buildings.

Constant-Volume Zone-Reheat Packaged Systems

System construction and system characteristics of a CV zone-reheat system are similar to the single-zone CV packaged systems except:

1. It serves multizones and has a sensor and a DDC controller for each zone.
2. There is a reheating coil or electric heater in the branch duct for each zone.

A CV zone-reheat packaged system cools and heats simultaneously and therefore wastes energy. It is usually used for the manufacturing process and space needs control of temperature and humidity simultaneously.

Variable-Air-Volume Packaged Systems

A variable-air-volume (VAV) system varies its volume flow rate to match the reduction of space load at part load. A VAV packaged system, also called a *VAV cooling packaged system*, is a multizone system and uses a VAV box in each zone to control the zone temperature at part load during summer cooling mode operation, as shown in Figure 15.3(a).

A *VAV box* is a terminal in which the supply volume flow rate of the conditioned supply air is modulated by varying the opening of the air passage by means of a single blade damper, as shown in Figure 15.3(b), or a moving disc against a cone-shaped casing.

The following are the system characteristics of a VAV packaged system:

Single-zone VAV packaged system which serves a single zone without VAV boxes. A DDC controller modulates the position of the inlet vanes or the fan speed according to the signal of the space temperature sensor.

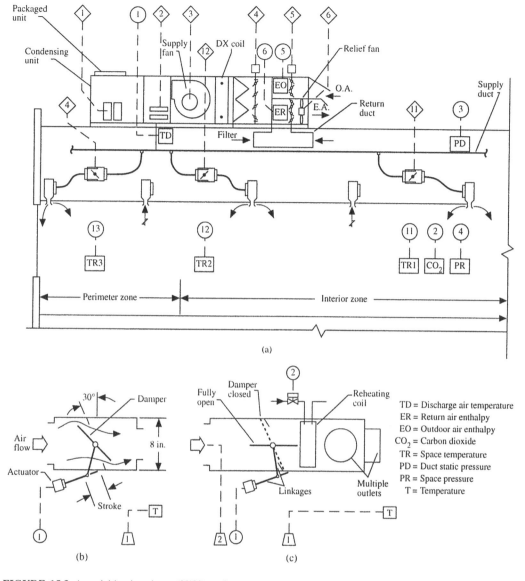

FIGURE 15.3 A variable-air-volume (VAV) package system: (a) schematic diagram, (b) VAV box, (c) reheating box, (d) parallel-flow fan-powered VAV box.

Air system: a supply/relief fan or supply/return fan combination. Space pressurization control by a relief/return fan

    Fan, motor, and drive combined efficiency: 45%

    Supply fan total pressure: 3.75 to 4.5 in. WG

    Fan(s) energy use at design condition: 1 to 1.25 W/cfm

    VAV box minimum setting: 30% of peak supply volume flow

    Outdoor ventilation air system: type II and III

Economizer: enthalpy air economizer or water economizer

Cooling system: DX coil, air-, water-, or evaporative-cooled condenser

    Compressor: reciprocating, scroll, and screw

    EER: 8.9 to 12 Btu/hr.W

Capacity: 20 to 100 tons

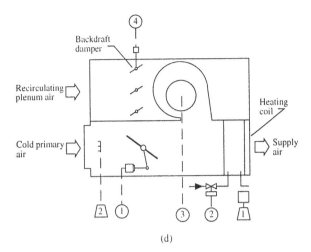

(d)

FIGURE 15.3 (continued)

Part load: zone volume flow modulation by VAV box; step control of compressor capacity; modulation
   of gas flow to burner; and discharge air temperature reset

Smoke control: exhausts smoke on the fire floor, and supplies air and pressurizes the floors immediately
   above or below the fire floor

Diagnostics: a diagnostic module displays the status and readings of various switches, dampers,
   sensors, etc. and the operative problems by means of expert system

Maintenance: higher than central system

Sound level: indoor NC 30 to 45

Heating system characteristics as well as the air-conditioning cycles are similar as that in a single-
zone CV packaged system.

VAV Reheat Packaged Systems

A VAV reheat packaged system has its system construction and characteristics similar to that in a VAV
packaged system except in each VAV box there is an additional reheating coil. Such a VAV box is called
a *reheating VAV box*, as shown in Figure 15.2(a) and 15.3(c). VAV reheat packaged systems are used to
serve perimeter zones where winter heating is required.

Fan-Powered VAV Packaged Systems

A fan-powered VAV packaged system is similar to that of a VAV packaged system except *fan-powered
VAV boxes* as shown in Figure 15.3(d) are used instead of VAV boxes.

   There are two types of fan-powered VAV boxes: parallel-flow and series-flow boxes. In a *parallel-
flow* fan-powered box, the plenum air flow induced by the fan is parallel with the cold primary air flow
through the VAV box. These two air streams are then combined and mixed together. In a *series-flow*
box, cold primarily from the VAV box is mixed with the induced plenum air and then flows through the
small fan. The parallel-flow fan-powered VAV box is more widely used.

   In a fan-powered VAV box, volume flow dropping to minimum setting, extracting of ceiling plenum
air, and energizing of reheating coil will actuate in sequence to maintain the space temperature during
part-load/heating mode operation. A fan-powered VAV box can also mix the cold primary air from cold
air distribution with the ceiling plenum air and provides greater space air movements during minimum
space load.

   Packaged systems are lower in installation cost and occupy less space than central systems. During
the past two decades, DDC-controlled packaged systems have evolved into sophisticated equipment and
provide many features that only a built-up central system could provide before.

Central Systems

Central systems use chilled and hot water that comes from the central plant to cool and heat the air in the air-handling units (AHUs). Central systems are built-up systems. The most clean, most quiet thermal-storage systems, and the systems which offer the most sophisticated features, are always central systems. Central systems can be subdivided into the following.

Single-Zone Constant-Volume Central Systems

A single-zone CV central system uses a single controller to control the flow of chilled water, hot water, or the opening of dampers to maintain a predetermined indoor temperature, relative humidity, or air contaminants. They are often used in manufacturing factories. The system characteristics of a single-zone CV central system are

Single-zone CV air washer central system uses air washer to control both space relative humidity and temperature. This system is widely used in textile mills. The reason to use constant volume is to dilute the fiber dusts produced during manufacturing. A rotary filter is often used for high dust-collecting capacity.

Air system: supply and return fan combination
    Fan, motor, and drive combined efficiency: 45 to 50%
    Outdoor ventilation air system: type II and IV
Economizer: air or water economizer
Smoke control: exhaust smoke on the fire floor, and pressurize adjacent floor(s) or area
Cooling system: centrifugal or screw chiller, water-cooled condenser
    Cooling energy use: 0.4 to 0.65 kW/ton
Heating system: hot water from boiler or from heat recovery system
Part load: modulate the water mass flow to cooling and heating coils in AHUs, and discharge air
    temperature reset
Sound level: indoor NC 30 to 45. Silencers are used both in supply and return air systems if they are
    required
Maintenance: in central plant and fan rooms, lower maintenance cost

Single-Zone CV Clean Room Systems

This is the central system which controls the air cleanliness, temperature, and relative humidity in Class 1, 10, 100, 1000, and 10,000 clean rooms for electronic, pharmaceutical, and precision manufacturing and other industries. Figure 15.4(a) shows a schematic diagram of this system. The recirculating air unit (RAU) uses prefilter, HEPA filters, and a water cooling coil to control the space air cleanliness and required space temperature, whereas a make-up air unit (MAU) supplies conditioned outdoor air, always within narrow dew point limits to the RAU at any outside climate, as shown in Figure 15.4(b). A unidirectional air flow of 90 fpm is maintained at the working area. For details, refer to *ASHRAE Handbook 1991 HVAC Applications* and Wang's *Handbook of Air Conditioning and Refrigeration*.

CV Zone-Reheat Central Systems

These systems have their system construction and characteristics similar to that for a single-zone CV central system, except they serve multizone spaces and there is a reheating coil, hot water, or electric heating in each zone. CV zone-reheat central systems are often used for health care facilities and in industrial applications.

VAV Central Systems

A VAV central system is used to serve multizone space and is also called *VAV cooling central system.* Its schematic diagram is similar to that of a VAV packaged system (Figure 15.3) except air will be cooled or heated by water cooling or heating coils in the AHUs. The same VAV box shown in Figure 15.3(b) will be used in a VAV central system. The system characteristics of VAV central systems are as follows:

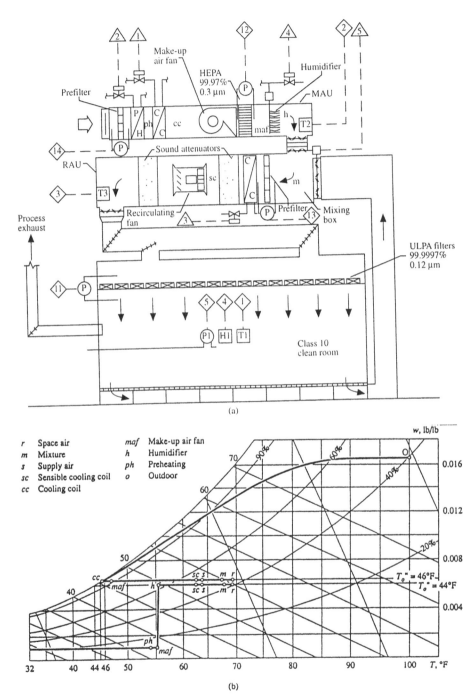

r   Space air   maf   Make-up air fan
m   Mixture   h   Humidifier
s   Supply air   ph   Preheating
sc   Sensible cooling coil   o   Outdoor
cc   Cooling coil

FIGURE 15.4 A single-zone CV clean room system: (a) schematic diagram and (b) air-conditioning cycle. (Source: Wang, S. K., *Handbook of Air Conditioning and Refrigeration*, McGraw-Hill, 1993. Reprinted by permission.)

Single-zone VAV central system differs from a VAV central system only because it serves a single zone, and therefore there is no VAV box in the branch ducts. Supply volume flow is modulated by inlet vanes and AC inverter.

> Air system: supply and relief/return fan combination
>> Fan, motor, and drive combined efficiency for airfoil centrifugal fan with AC inverter fan speed modulation: 55%
>> Fan(s) energy use: 0.9 to 1.2 W/cfm
>> VAV box: minimum setting 30% of peak supply volume flow
>> Outdoor ventilation air system: type I, II, and III
> Cooling system: centrifugal, screw, and reciprocating chillers, water-cooled condenser, with energy use 0.4 to 0.65 kW/ton; or sometimes absorption chiller
> Heating system: hot water from boiler or electric heating at the terminals
> Economizer: air and water economizer
> Part load: zone volume flow modulation, discharge air temperature reset, and chilled water temperature reset
> Smoke control: exhausts smoke from the fire floor and pressurizes the immediate floors above and below
> Sound level: indoor NC 20 to 40. Silencers are often used both in supply and return systems.
> Maintenance: less space maintenance

VAV central systems are widely used for interior zone in buildings.

VAV Reheat Central Systems

A VAV reheat system is similar in system construction and characteristics to that in a VAV central system except that reheating boxes are used instead of VAV boxes in a VAV central system.

Fan-Powered VAV Central Systems

A fan-powered VAV central system is similar in system construction and characteristics to that in a VAV central system except that fan-powered VAV boxes are used instead of VAV boxes.

Dual-Duct VAV Central Systems

A dual-duct VAV system uses a warm air duct and a cold air duct to supply both warm and cold air to each zone in a multizone space, as shown in Figure 15.5(a). Warm and cold air are first mixed in a mixing VAV box, and are then supplied to the conditioned space. Warm air duct is only used for perimeter zones.

A *mixing VAV box* consists of two equal air passages, one for warm air and one for cold air, arranged in parallel. Each of them has a single blade damper and its volume flow rate is modulated. Warm and cold air are then combined, mixed, and supplied to the space.

A dual-duct VAV system is usually either a single supply fan and a relief/return fan combination, or a warm air supply fan, a cold air supply fan, and a relief/return fan. A separate warm air fan and cold air supply fan are beneficial in discharge air temperature reset and fan energy use.

During summer cooling mode operation, the mixture of recirculating air and outdoor air is used as the warm air supply. The heating coil is not energized. During winter heating mode operation, mixture of outdoor and recirculating air or 100% outdoor air is used as the cold air supply; the cooling coil is not energized.

Because there is often air leakage at the dampers in the mixing VAV box, more cold air supply is needed to compensate for the leaked warm air or leaked cold air.

Other system characteristics of a dual-duct VAV central system are similar to a VAV central system.

Dual-Duct CV Central System

This is another version of a dual-duct VAV central system and is similar in construction to a dual-duct VAV system, except that a mixing box is used instead of a mixing VAV box. The supply volume flow rate from a mixing box is nearly constant. Dual-duct CV central systems have only limited applications, like health care facilities, etc.

An ice- or chilled-water storage system is always a central system plus a thermal storage system. The thermal storage system does not influence the system characteristics of the air distribution, and air cooling and heating — except for a greater head lift for a refrigeration compressor — is needed for ice-storage systems. Therefore, the following central systems should be added:

- VAV ice-storage or chilled-water systems
- VAV reheat ice-storage or chilled-water storage systems
- Fan-powered VAV ice-storage systems

Some of the air-conditioning systems are not listed because they are not effective or are a waste of energy, and therefore rarely used in new and retrofit projects such as:

- High-velocity induction space conditioning systems which need a higher pressure drop primary air to induce recirculating air in the induction unit and use more energy than fan-coil systems
- Multizone central systems which mix warm and cool air at the fan room and use a supply duct from fan room to each control zone
- Air skin central systems which use a warm air heating system to offset transmission loss in the perimeter zone and overlook the effect of the solar radiation from variation building orientations

In the future, there will be newly developed systems added to this classification list.

Air-Conditioning System Selection

As described in Section 1, the goal of an air-conditioning or HVAC&R system is to provide a healthy, comfortable, manufacturable indoor environment at acceptable indoor air quality, keeping the system energy efficient. Various occupancies have their own requirements for their indoor environment. The basic considerations to select an air-conditioning system include:

1. The selection of an air-conditioning system must satisfy the required space temperature, relative humidity, air cleanliness, sound level, and pressurization. For a Class 100 clean room, a single-zone CV clean room system is always selected. A four-pipe fan-coil space conditioning system is usually considered suitable for guest rooms in hotels for operative convenience, better privacy, and a guaranteed outdoor ventilation air system. A concert hall needs a very quiet single-zone VAV central system for its main hall and balcony.

2. The size of the project has a considerable influence on the selection. For a small-size residential air-conditioning system, a single-zone constant-volume packaged system is often the first choice.

3. Energy-efficient measures are specified by local codes. Comparison of alternatives by annual energy-use computer programs for medium and large projects is often necessary. Selection of energy source includes electricity or gas, and also using electrical energy at off-peak hours, like thermal storage systems is important to achieve minimum energy cost.

   For a building whose sound level requirement is not critical and conditioned space is comprised of both perimeter and interior zones, a WSHP system incorporating heat recovery is especially suitable for energy saving.

4. First cost or investment is another critical factor that often determines the selection.

5. Selection of an air-conditioning system is the result of synthetical assessment. It is difficult to combine the effect of comfort, reliability, safety, and cost. Experience and detailed computer program comparisons are both important.

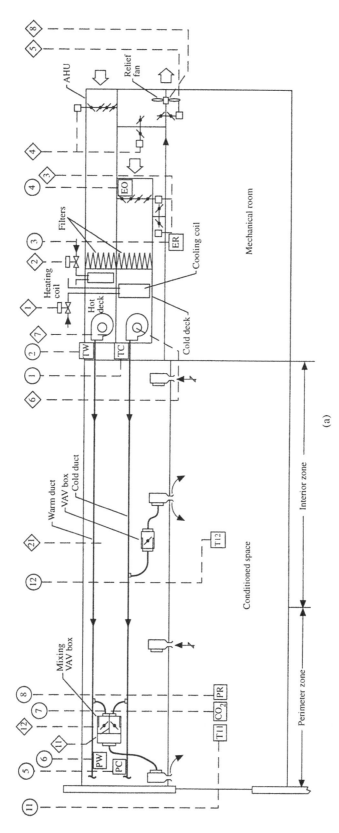

FIGURE 15.5 A dual-duct VAV central system: (a) schematic diagram.

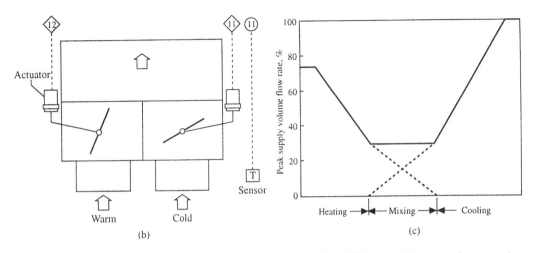

FIGURE 15.5 (continued) A dual-duct VAV central system: (b) mixing VAV box and (c) volume flow-operating mode diagram.

The selection procedure usually begins whether an individual, space conditioning, packaged, central system, or CV, VAV, VAV reheat, fan-powered VAV, dual-duct VAV, or thermal storage system is selected. Then the air, refrigeration, heating, and control subsystems will be determined. After that, choose the option, the feature, the construction, etc. in each subsystem.

Comparison of Various Systems

The sequential order of system performance — excellent, very good, good, satisfactory — regarding temperature and relative humidity control (T&HC), outdoor ventilation air (OA), sound level, energy use, first cost, and maintenance for individual, space conditioning (SC), packaged, and central systems is as follows:

	Excellent (low or less)	Very good	Good	Satisfactory
T&HC	Central	Packaged	Space	Individual
IAQ	Space	Central	Packaged	Individual
Sound	Central	Packaged	Space	Individual
Energy use	Individual	Space	Packaged	Central
First cost	Individual	Packaged	Space	Central
Maintenance	Central	Packaged	Space	Individual

Among the packaged and central systems, VAV cooling systems are used only for interior zones. VAV reheat, fan-powered VAV, and dual-duct VAV central systems are all for perimeter zones. VAV reheat systems are simple and effective, but have a certain degree of simultaneous cooling and heating when their volume flow has been reduced to minimum setting. Fan-powered VAV systems have the function of mixing cold primary air with ceiling plenum air. They are widely used in ice-storage systems with cold air distribution. Fan-powered VAV is also helpful to create a greater air movement at minimum cold primary air flow. Dual-duct VAV systems are effective and more flexible in operation. They are also more complicated and expensive.

Subsystems

Air Systems

The economical size of an air system is often 10,000 to 25,000 cfm. A very large air system always has higher duct pressure loss and is more difficult to balance. For highrise buildings of four stories and higher, floor-by-floor AHU(s) or PU(s) (one or more AHU or PU per floor) are often adopted. Such an arrangement is beneficial for the balance of supply and return volume flow in VAV systems and also for fire protection. A fan-powered VAV system using a riser to supply less cold primary to the fan-powered VAV box at various floors may have a larger air system. Its risers can be used as supply and exhaust ducts for a smoke-control system during a building fire.

In air systems, constant-volume systems are widely used in small systems or to dilute air contaminants in health care facilities and manufacturing applications. VAV systems save fan energy and have better operating characteristics. They are widely used in commercial buildings and in many factories.

Refrigeration Systems

For comfort air-conditioning systems, the amounts of required cooling capacity and energy saving are dominant factors in the selection of the refrigeration system. For packaged systems having cooling capacity less than 100 tons, reciprocating and scroll vapor compression systems with air-cooled condensers are most widely used. Evaporative-cooled condensers are available in many packaged units manufactured for their lower energy use. Scroll compressors are gradually replacing the reciprocating compressors for their simple construction and energy saving. For chillers of cooling capacity of 100 tons and greater, centrifugal chillers are still most widely used for effective operation, reliability, and energy efficiency. Screw chillers have become more popular in many applications, especially for ice-storage systems.

Heating Systems

For locations where there is a cold and long winter, a perimeter baseboard hot water heating system or dual-duct VAV systems are often a suitable choice. For perimeter zones in locations where winter is mild, winter heating is often provided by using warm air supply from AHU or PU from terminals with electric or hot water heaters. Direct-fired furnace warm air supply may be used for morning warm-up. For interior or conditioned zones, a cold air supply during occupied periods in winter and a warm air supply from the PUs or AHUs during morning warm-up period is often used.

Control Systems

Today, DDC microprocessor-based control with open data communication protocol is often the choice for medium- and large-size HVAC&R projects. For each of the air, cooling, and heating systems, carefully select the required generic and specific control systems. If a simple control system and a more complicated control system can provide the same required results, the simple one is always the choice.

Energy Conservation Recommendations

1. Turn off electric lights, personal computers, and office appliances when they are not needed. Shut down AHUs, PUs, fan coils, VAV boxes, compressors, fans, and pumps when the space or zone they serve is not occupied or working.
2. Provide optimum start and stop for the AHUs and PUs and terminals daily.
3. Temperature set point should be at its optimum value. For comfort systems, provide a dead band between summer and winter mode operation. Temperature of discharged air from the AHU or PU and chilled water leaving the chiller should be reset according to space or outdoor temperature or the system load.
4. Reduce air leakages from ducts and dampers. Reduce the number of duct fittings and pipe fittings and their pressure loss along the design path if this does not affect the effectiveness of the duct system. The maximum design velocity in ducts for comfort systems should not exceed 3000 fpm, except that a still higher velocity is extremely necessary.
5. Adopt first the energy-efficient cooling methods: air and water economizer, evaporative cooler, or ground water instead of refrigeration.

6. Use cost-effective high-efficiency compressors, fans, pumps, and motors as well as evaporative-cooled condensers in PUs. Use adjustable-frequency fan speed modulation for large centrifugal fans. Equipment should be properly sized. Over-sized equipment will not be energy efficient.

7. Use heat recovery systems and waste heat for winter heating or reheating. Use a heat-pump system whenever its COP_{hp} is greater than 1.

8. For medium- and large-size air-conditioning systems, use VAV systems instead of CV systems except for health care or applications where dilution of air contaminant is needed. Use variable flow for building-loop and distribution-loop water systems.

9. Use double- and triple-pane windows with low emissive coatings. Construct low U-value roofs and external walls.

References

AMCA. 1973. Fan and Systems Publication 201. AMCA, Arlington Heights, IL.

Amistadi, H. 1993. Design and drawing software review, *Eng. Syst.* 6:18–29.

ANSI/ASHRAE. 1992. ANSI/ASHRAE Standard 34-1992, *Numbering Designation and Safety Classification of Refrigerants*. ASHRAE, Atlanta, GA.

ASHRAE. 1991. *ASHRAE Handbook, HVAC Applications*. ASHRAE, Atlanta, GA.

ASHRAE. 1992. *ASHRAE Handbook, HVAC Systems and Equipment*. ASHRAE, Atlanta, GA.

ASHRAE. 1993. *ASHRAE Handbook, Fundamentals*. ASHRAE, Atlanta, GA.

ASHRAE. 1994. *ASHRAE Handbook, Refrigeration*. ASHRAE, Atlanta, GA.

Bayer, C.W. and Black, M.S. 1988. IAQ evaluations of three office buildings. *ASHRAE J.* 7:48–52.

Bushby, S.T. and Newman, H.M. 1994. BACnet: a technical update, *ASHRAE J.* 1:S72–84.

Carlson, G.F. 1968. Hydronic systems: analysis and evaluation, I. *ASHRAE J.* 10:2–11.

DOE. 1981. DOE-2 Reference Material (Version 2.1A). National Technical Information Service, Springfield, VA.

Dorgan, C.E. and Elleson, J.S. 1988. Cold air distribution. *ASHRAE Trans.* I:2008–2025.

Durkin, J. 1994. *Expert Systems Design and Development*. Macmillan, New York.

EIA. 1994. Commercial Buildings Characteristics 1992. U.S. Government Printing Office, Washington, D.C.

Elyashiv, T. 1994. Beneath the surface: BACnet™ data link and physical layer options. *ASHRAE J.* 11:32–36.

EPA/CPSC. 1988. The Inside Story: A Guide to Indoor Air Quality. Environmental Protection Agency, Washington, D.C.

Fanger, P.O., Melikow, A.K., Hanzawa, H., and Ring, J. 1989. Turbulence and draft. *ASHRAE J.* 4:18–25.

Fiorino, D.P. 1991. Case study of a large, naturally stratified, chilled-water thermal storage system. *ASHRAE Trans.* II:1161–1169.

Gammage, R.B., Hawthorne, A.R., and White, D.A. 1986. Parameters Affecting Air Infiltration and Air Tightness in Thirty-One East Tennessee Homes, Measured Air Leakage in Buildings, ASIM STP 904. American Society of Testing Materials, Philadelphia.

Goldschmidt, I.G. 1994. A data communucations introduction to BACnet™. *ASHRAE J.* 11:22–29.

Gorton, R.L. and Sassi, M.M. 1982. Determination of temperature profiles and loads in a thermally stratified air-conditioning system. I. Model studies. *ASHRAE Trans.* II:14–32.

Grimm, N.R. and Rosaler, R.C. 1990. *Handbook of HVAC Design*. McGraw-Hill, New York.

Hartman, T.B. 1989. TRAV — a new HVAC concept. *Heating/Piping/Air Conditioning*. 7:69–73.

Hayner, A.M. 1994. Engineering in quality. *Eng. Syst.* 1:28–33.

Heyt, H.W. and Diaz, M.J. 1975. Pressure drop in spiral air duct. *ASHRAE Trans.* II:221–232.

Huebscher, R.G. 1948. Friction equivalents for round, square, and rectangular ducts. *ASHRAE Trans.* 101–144.

Hummel, K.E., Nelson, T.P., and Tompson, P.A. 1991. Survey of the use and emissions of chlorofluorocarbons from large chillers. *ASHRAE Trans.* II:416–421.

Jakob, F.E., Locklin, D.W., Fisher, R.D., Flanigan, L.G., and Cudnik, L.A. 1986. SP43 evaluation of system options for residential forced-air heating. *ASHRAE Trans.* IIB:644–673.

Kimura, K. 1977. *Scientific Basis of Air Conditioning*. Applied Science Publishers, London.

Knebel, D.E. 1995. Current trends in thermal storage. *Eng. Syst.* 1:42–58.

Korte, B. 1994. The health of the industry. *Heating/Piping/Air Conditioning*. 1:111–112.

Locklin, D.W., Herold, K.E., Fisher, R.D., Jakob, F.E., and Cudnik, R.A. 1987. Supplemental information from SP43 evaluation of system options for residential forced-air heating. *ASHRA Trans.* II:1934–1958.

Lowe, R. and Ares, R. 1995. From CFC-12 to HFC-134a: an analysis of a refrigerant retrofit project. *Heating/Piping/Air Conditioning*. 1:81–89.

McQuiston, F.C. and Spitler, J.D. 1992. *Cooling and Heating Load Calculating Manual*, 2nd ed. ASHRAE, Atlanta, GA.

Mitalas, G.P. 1972. Transfer function method of calculating cooling loads, heat extraction rate and space temperature, *ASHRAE J.* 12:52–56.

Mitalas, G.P. and Stephenson, D.G. 1967. Room thermal response factors. *ASHRAE Trans.* 2, III.2.1.

Modera, M.P. 1989. Residential duct system leakage: magnitude, impact, and potential for reduction. *ASHRAE Trans.* II:561–569.

Molina, M.J. and Rowland, S. 1974. Stratospheric sink for chloromethanes: chlorine atom catalyzed destruction of ozone. *Nature*. 249:810–812.

NIOSH. 1989. Congressional Testimony of J. Donald Miller, M.D., before the Subcommittee of Superfund, Ocean, and Water Protection, May 26, 1989. NIOSH, Cincinnati, Cleveland.

Parsons, B.K., Pesaran, A.A., Bharathan, D., and Shelpuk, B. 1989. Improving gas-fired heat pump capacity and performance by adding a desiccant dehumidification subsystem. *ASHRAE Trans.* I:835–844.

Persily, A.K. 1993. Ventilation, carbon dioxide, and ASHRAE Standard 62-1989. *ASHRAE J.* 7:40–44.

Reynolds, S. 1994. CFD modeling optimizes contaminant elimination. *Eng. Syst.* 2:35–37.

Rowland, S. 1992. The CFC controversy: issues and answers. *ASHRAE J.* 12:20–27.

Rudoy, W. and Duran, F. 1975. Development of an improved cooling load calculation method. *ASHRAE Trans.* II:19–69.

Scofield, C.M. and DesChamps, N.H. 1984. Indirect evaporative cooling using plate-type heat exchangers. *ASHRAE Trans.* I B:148–153.

Shinn, K.E. 1994. A specifier's guide to BACnet™. *ASHRAE J.* 4:54–58.

Sowell, E.F. 1988. Classification of 200,640 parametric zones for cooling load calculations. *ASHRAE Trans.* II:754–777.

Spitler, J.D., McQuiston, F.C., and Lindsey, K.L. 1993. The CLTD/SCL/CLF Cooling Calculation Method. *ASHRAE Trans.* I:183–192.

Straub, H.E. and Cooper, J.G. 1991. Space heating with ceiling diffusers. *Heating/Piping/Air Conditioning*. May:49–55.

Tackett, R.K. 1989. Case study: office building use ice storage, heat recovery, and cold air distribution. *ASHRAE Trans.* I:1113–1121.

Threlkeld, J.L. 1970. *Thermal Environmental Engineering*. Prentice-Hall, Englewood Cliffs, NJ.

The Trane Company. 1992. TRANE TRACE 600, Engineering Manual. The Trane Co., Lacrosse, WI.

Tinsley, W.E., Swindler, B., and Huggins, D.R. 1992. Rooftop HVAC system offers optimum energy efficiency. *ASHRAE J.* 3:24–28.

Tsal, R.J., Behls, H.F., and Mangel, R. 1988. T-method duct design. I. Optimizing theory. *ASHRAE Trans.* II:90–111.

Tsal, R.J., Behls, H.F., and Mangel, R. 1988. T-method duct design. II. Calculation procedure and economic analysis. *ASHRAE Trans.* II:112–150.

Vaculik, F. and Plett, E.G. 1993. Carbon dioxide concentration-based ventilation control. *ASHRAE Trans.* I:1536–1547.

Van Horn, M. 1986. *Understanding Expert Systems*. Bantam Books, Toronto.

Wang, S.K. 1993. *Handbook of Air Conditioning and Refrigeration.* McGraw-Hill, New York.

Wang, S.K., Leung, K.L., and Wong, W.K. 1984. Sizing a rectangular supply duct with transversal slots by using optimum cost and balanced total pressure principle. *ASHRAE Trans.* II A:414–429.

Williams, P.T., Baker, A.J., and Kelso, R.M. 1994. Numerical calculation of room air motion. III. Three-dimensional CFD simulation of a full scale experiment. *ASHRAE Trans.* I:549–564.

Wong, S.P.W. and Wang, S.K. 1990. Fundamentals of simultaneous heat and moisture transfer between the building envelope and the conditioned space air. *ASHRAE Trans.* II:73–83.

Wright, D.K. 1945. A new friction chart for round ducts. *ASHRA Trans.* 303–316.

16

Desiccant Dehumidification and Air-Conditioning

Zalman Lavan

Professor Emeritus, Illinois Institute of Technology

16 Desiccant Dehumidification and
    Air-Conditioning 165
    Introduction • Sorbents and Desiccants • Dehumidification •
    Liquid Spray Tower • Solid Packed Tower • Rotary Desiccant
    Dehumidifiers • Hybrid Cycles • Solid Desiccant Air-
    Conditioning • Conclusions

Desiccant Dehumidification and Air-Conditioning

Introduction

Desiccant air-conditioning is a promising emerging technology to supplement electrically driven vapor compression systems that rely almost exclusively on R22 refrigerant that causes depletion of the ozone layer. To date, this technology has only a limited market, e.g., in supermarkets where the latent heat loads are very high, in specialized manufacturing facilities that require very dry air, and in hospitals where maximum clean air is required. However, recent emphasis on increased air change requirements (see ASHRAE standards, ANSI 62-1989), improved indoor air quality, and restriction on use of CFC refrigerants (see The Montreal Protocol Agreement, as amended in Copenhagen in 1992, United Nations Environmental Programme, 1992) may stimulate wider penetration of desiccant-based air-conditioning which can be used as stand-alone systems or in combination with conventional systems. (See Table 4.1 for properties of some refrigerants.)

Sorbents and Desiccants

Sorbents are materials which attract and hold certain vapor or liquid substances. The process is referred to **absorption** if a chemical change takes place and as **adsorption** if no chemical change occurs. **Desiccants**, in both liquid and solid forms, are a subset of sorbents that have a high affinity to water molecules. Liquid desiccants *absorb* water molecules, while solid desiccants *adsorb* water molecules and hold them on their vast surfaces (specific surface areas are typically hundreds of square meters per gram).

While desiccants can sorb water in both liquid and vapor forms, the present discussion is limited to **sorption** of water vapor from adjacent air streams. The sorption driving force for both liquid and solid desiccants is a vapor pressure gradient. Adsorption (in solid desiccants) and absorption (in liquid desiccants) occur when the water vapor partial pressure of the surrounding air is larger than that at the desiccant surface. When an air stream is brought in contact with a desiccant, water vapor from the air is attracted by the desiccant, the air is **dehumidified**, and the water content of the desiccant rises. As the water sorbed by the desiccant increases, the sorption rate decreases and finally stops when *sorption equilibrium* is reached. For dehumidification to be resumed, water must be removed from the desiccant by heating. This process is referred to as **desorption, reactivation,** or **regeneration**. The heat of sorption (or desorption) is generally higher than the latent heat of vaporization of water; it approaches the latter as sorption equilibrium is reached.

Some typical *liquid desiccants* are water solutions of calcium chloride (CaCl), lithium chloride (LiCl), lithium bromide (LiBr), and triethylene glycol. The equilibrium water vapor pressure at the solution surface as a function of temperature and water content is shown in Figure 16.1 for water-lithium chloride solution. The surface vapor pressure (and dew point) increases with increasing solution temperature and decreases with increasing moisture content.

Common *solid desiccants* are silica gel, molecular sieves (zeolites), activated alumina, and activated carbon. The equilibrium sorption capacity (or moisture content) at a constant temperature, referred to as an **isotherm**, is usually presented as percent water (mass of water divided by mass of dry desiccant) vs. percent relative humidity (vapor pressure divided by saturation vapor pressure). Sorption capacity decreases with increasing temperature, but the spread of isotherms is relatively small (especially for concave down isotherms). Figure 16.2 shows normalized loading (sorption capacity divided by sorption capacity at 100% relative humidity) vs. relative humidity for silica gel, molecular sieve, and a generic desiccant, type 1 (modified) or simply 1-M (Collier et al., 1986).

Dehumidification

Dehumidification by vapor compression systems is accomplished by cooling the air below the dew point and then reheating it. The performance is greatly hindered when the desired outlet dew point is below 40°F due to frost formation on the cooling coils (*ASHRAE, Systems and Equipment Handbook,* 1992).

Desiccant dehumidification is accomplished by direct exchange of water vapor between an air stream and a desiccant material due to water vapor pressure difference. Figure 16.3 shows the cyclic operation of a desiccant dehumidification system.

In *sorption (1–2)*, dry and cold desiccant (point 1) sorbs moisture since the vapor pressure at the surface is lower than that of the air stream. During this process the moisture content (loading or uptake) increases, the surface vapor pressure increases, and the liberated heat of sorption raises the desiccant temperature. During *desorption (2–3)*, the desiccant is subjected to a hot air stream, and moisture is removed and transferred to the surrounding air. The surface vapor pressure is increased and the desiccant temperature rises due to the added heat. The cycle is closed by *cooling (3–1)*. The desiccant is cooled while its moisture content is constant and the surface vapor pressure is lowered. The above cycle of sorption, desorption, and cooling can be modified by combining the sorption process with cooling to approach isothermal rather than adiabatic sorption.

Desirable Characteristics for High-Performance Liquid and Solid Desiccant Dehumidifiers

High equilibrium moisture sorption capacity
High heat and mass transfer rates
Low heat input for regeneration
Low pressure drop
Large contact transfer surface area per unit volume

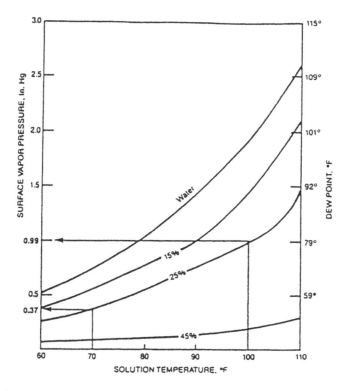

FIGURE 16.1 Surface vapor pressure of water-lithium chloride solutions. (Source: ASHRAE 1993, *Fundamentals Handbook*, chap. 19. With permission.)

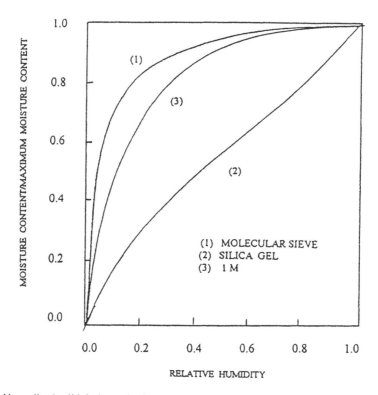

FIGURE 16.2 Normalized solid desiccant isotherms.

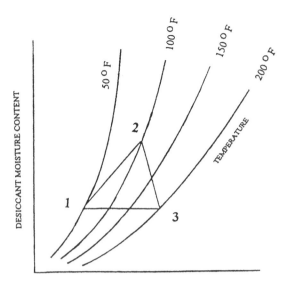

FIGURE 16.3 Cyclic dehumidification processes.

Compatible desiccant/contact materials
Inexpensive materials and manufacturing techniques
Minimum deterioration and maintenance

Additional Requirements for Liquid Desiccant Dehumidifiers

Small liquid side resistance to moisture diffusion
Minimum crystallization

Additional Requirements for Solid Desiccant Dehumidifiers

The desiccant should not deliquesce even at 100% relative humidity.
The airflow channels should be uniform.
The desiccant should be bonded well to the matrix.
The material should not be carciogenic or combustible.

Liquid Spray Tower

Figure 16.4 is a schematic of a liquid spray tower. A desiccant solution from the sump is continuously sprayed downward in the absorber, while air, the process stream, moves upward. The air is dehumidified and the desiccant solution absorbs moisture and is weakened. In order to maintain the desired solution concentration, a fraction of the solution from the sump is passed through the regenerator, where it is heated by the heating coil and gives up moisture to the desorbing air stream. The strong, concentrated solution is then returned to the sump. The heat liberated in the absorber during dehumidification is removed by the cooling coil to facilitate continuous absorption (see Figures 16.1 and 16.3). The process air stream exits at a relatively low temperature. If sufficiently low water temperature is available (an underground well, for example), the process stream could provide both sensible and latent cooling.

The heating and cooling coils, shown in Figure 16.4, are often eliminated and the liquid solutions are passed through heating and cooling heat exchangers before entering the spray towers.

Advantages

The system is controlled to deliver the desired level of dry air by adjusting the solution concentration.

Uniform exit process stream conditions can be maintained.

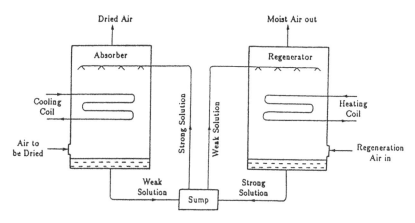

FIGURE 16.4 Liquid desiccant dehumidifier with heating and cooling coils.

A concentrated solution can be economically stored for subsequent drying use.

The system can serve as a humidifier when required by simply weakening the solution.

When used in conjunction with conventional A/C systems, humidity control is improved and energy is conserved.

Disadvantages

Some desiccants are corrosive.

Response time is relatively large.

Maintenance can be extensive.

Crystallization may be a problem.

Solid Packed Tower

The dehumidification system, shown in Figure 16.5, consists of two side-by-side cylindrical containers filled with solid desiccant and a heat exchanger acting as a desiccant cooler. The air stream to be processed is passed through dry desiccant in one of the containers, while a heated air stream is passed over the moist desiccant in the other. Adsorption (1–2) takes place in the first container, desorption (2–3) in the other container, and cooling (3–1) occurs in the desiccant cooler. The function of the two containers is periodically switched by redirecting the two air streams.

Advantages

No corrosion or crystallization

Low maintenance

Very low dew point can be achieved

Disadvantages

The air flow velocity must be low in order to maintain uniform velocity through the containers and to avoid dusting.

Uniform exit process stream dew point cannot be maintained due to changing moisture content in the adsorbing desiccant.

Rotary Desiccant Dehumidifiers

A typical rotary solid desiccant dehumidifier is shown in Figure 16.6. Unlike the intermittent operation of packed towers, rotary desiccant dehumidifiers use a wheel (or drum) that rotates continuously and delivers air at constant humidity levels.

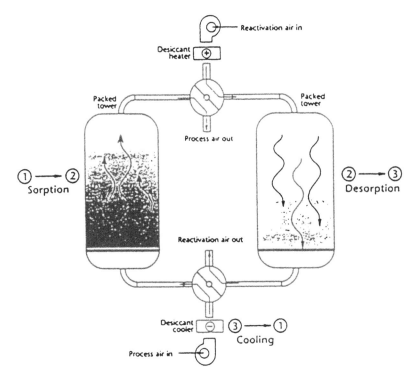

FIGURE 16.5 Solid packed tower dehumidification. (From Harriman, L. G., III. 1990. *The Dehumidification Handbook*, 2nd ed. Munters Cargocaire. With permission.)

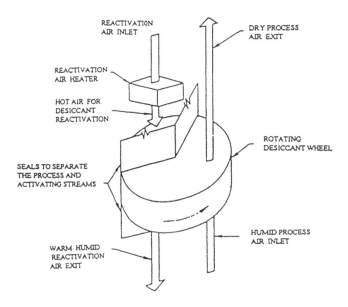

FIGURE 16.6 Rotary desiccant dehumidification wheel. (Source: ASHRAE 1992, *Systems and Equipment Handbook*. chap. 22. With permission.)

Desiccant wheels typically consist of very fine desiccant particles dispersed and impregnated with a fibrous or ceramic medium shaped like a honeycomb or fluted corrugated paper. The wheel is divided into two segments. The process stream flows through the channels in one segment, while the regenerating (or reactivating) stream flows through the other segment.

Desiccant Material

The desired desiccant properties for optimum dehumidification performance are a suitable isotherm shape and a large moisture sorption capacity. The isotherms of silica gel are almost linear. The moisture sorption capacity is high; the desiccant is reactivated at relatively low temperatures and is suitable for moderate dehumidification. Molecular sieves have very steep isotherms at low relative humidity. The desiccant is reactivated at relatively high temperatures and is used for deep dehumidification. The isotherm of the type 1-M yields optimum dehumidification performance (Collier et al., 1986), especially when used in conjunction with high regeneration temperatures.

The Desiccant Wheel

Some considerations for selection of desiccant wheels are:

Appropriate desiccant materials
Large desiccant content
Wheel depth and flute size (for large contact surface area and low pressure drop)
Size and cost

The actual performance depends on several additional factors that must be addressed. These include:

Inlet process air temperature and humidity
Desired exit process air humidity
Inlet reactivating air temperature and humidity
Face velocity of the two air streams
Size of reactivation segment

It should be noted that:

Higher inlet process air humidity results in higher exit humidity and temperature (more heat of sorption is released).
Lower face velocity of the process stream results in lower exit humidity and higher temperature.
Higher regeneration temperatures result in deeper drying, hence lower exit process air humidity and higher temperature.
When lower exit air temperature is required, the exit process air should be cooled by a heat exchanger.
Final cooling of the exit process air can be achieved by partial humidification (this counteracts in part previous dehumidification).

The following is a range of typical parameters for rotary desiccant wheels:

Rotation speed: 4 to 10 rpm
Desiccant fraction: 70 to 80%
Flute size: 1 to 2 mm
Reactivation segment: 25 to 30% of wheel
Face velocity: 300 to 700 fpm
Reactivating temperature: 100 to 300°F

Hybrid Cycles

A limited number of hybrid systems consisting of desiccant dehumidifiers and electrically driven vapor compression air-conditioners are presently in use in supermarkets. This application is uniquely suited for this purpose since the latent heat loads are high due to the large number of people and frequent traffic through doors. Also, low relative humidity air is advantageous for open-case displays.

Vapor compression systems are inefficient below a dew point of 45 to 50°F. When used in supermarkets, they require high airflow rates, the air must be reheated for comfort, and the evaporator coils must be defrosted frequently. Hybrid systems offer improved performance and lower energy cost in these cases.

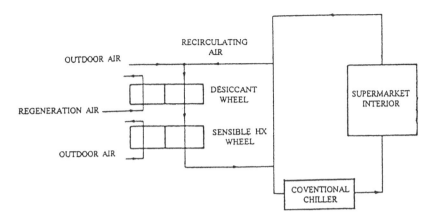

FIGURE 16.7 Hybrid air-conditioning system for supermarkets.

Figure 16.7 shows a typical hybrid air-conditioning system for supermarkets. A mixture of outdoor and recirculated air is first passed through the desiccant and sensible heat exchanger wheels, where it is dehumidified and precooled. It then enters the conventional chiller before it is introduced to the interior of the supermarket. The sensible heat exchanger wheel is cooled by outdoor air and the desiccant wheel is regenerated by air heated with natural gas. Energy cost can be further reduced by preheating the reactivating air stream with waste heat rejected from the condenser of the refrigeration and/or air-conditioning systems.

The advantages of these hybrid systems are

Air-conditioning requirement is reduced by up to 20%.
The vapor compression system operates at a higher coefficient of performance (COP) since the evaporator coils are at a higher temperature.
Airflow requirements are reduced; electric fan energy is saved and duct sizes are reduced.
The refrigeration cases run more efficiently since the frequency of defrost cycles is greatly reduced.

Solid Desiccant Air-Conditioning

Several stand-alone desiccant air-conditioning systems were suggested and extensively studied. These systems consist of a desiccant wheel, a sensible heat exchanger wheel, and evaporating pads. Sorption can be adiabatic or cooled (if cooling is combined with sorption). When room air is dehumidified and recirculated, the system is said to operate in the **recirculating** mode. When 100% outside air is used as the process stream, the system operates in the **ventilating mode**.

Ventilation Mode

In the *adsorption path* the process air stream drawn from the outdoors is passed through the dry section of the desiccant wheel where it is dehumidified and heated by the liberated heat of sorption. It then passes through the sensible heat exchanger wheel and exits as dry but slightly warm air. The hot and dry air leaving the dehumidifier enters the heat exchanger, where it is sensibly cooled down to near room temperature. It is then passed through the evaporative cooler, where it is further cooled and slightly humidified as it enters the conditioned space.

In the *desorption path,* air is drawn from the conditioned space; it is humidified (and thus cooled) in the evaporative cooler. The air stream enters the sensible heat exchanger, where it is preheated, and it is then heated to the desired regeneration temperature by a suitable heat source (natural gas, waste heat, or solar energy), passed through the desiccant wheel (regenerating the desiccant material), and discharged out of doors.

Performance. In order to achieve high performance, the maximum moisture content of the desiccant should be high and the isotherm should have the optimum shape (1 M). In addition, Zheng et al. (1993)

showed that the optimum performance is very sensitive to the rotational speed of the desiccant wheel. Glav (1966) introduced stage regeneration. He showed that performance is improved when the reactivation segment of the wheel is at a temperature which increases in the direction of rotation. Collier (Collier et al., 1986) showed that well-designed open-cycle desiccant cooling systems can have a thermal COP of 1.3. This, however, would require the use of high-effectiveness sensible heat exchangers, which would be large and expensive. Smaller and more affordable heat exchangers should yield system COPs in the order of unity. An extensive review of the state-of-the-art assessment of desiccant cooling is given by Pesaran et al. (1992).

Conclusions

Desiccant-based air-conditioning offers significant advantages over conventional systems. Desiccant systems are already successfully used in some supermarkets. It is expected that these systems will gradually attain wider market penetration due to environmental requirements and potential energy savings.

The advantages of desiccant air-conditioning are summarized below:

No CFC refrigerants are used.
Indoor air quality is improved.
Large latent heat loads and dry air requirements are conveniently handled.
Individual control of temperature and humidity is possible.
The energy source may be natural gas and/or waste heat.
Less circulated air is required.
Summer electric peak is reduced.

Defining Terms

Absorb, absorption: When a chemical change takes place during sorption.
Adsorb, adsorption: When no chemical change occurs during sorption.
Dehumidification: Process of removing water vapor from air.
Desiccant: A subset of sorbents that has a particular affinity to water.
Desorb, desorption: Process of removing the sorbed material from the sorbent.
Isotherm: Sorbed material vs. relative humidity at a constant temperature.
Reactivation: Process of removing the sorbed material from the sorbent.
Recirculation: Indoor air only is continuously processed.
Regeneration: Process of removing the sorbed material from the sorbent.
Sorbent: A material that attracts and holds other gases or liquids.
Sorption: Binding of one substance to another.
Staged regeneration: When the temperature of the regeneration segment of the desiccant wheel is not uniform.
Ventilation mode: 100% of outdoor air is processed.

References

ANSI/ASHRAE 62-1989. *Ventilation for Acceptable Indoor Air Quality*. American Society of Heating, Refrigeration and Air-Conditioning Engineers, Atlanta, GA.

ASHRAE. 1992. *HVAC Systems and Equipment Handbook*, Chap. 22. American Society of Heating, Refrigeration and Air Conditioning Engineers, Atlanta, GA.

ASHRAE. 1993. *Fundamentals Handbook*, Chap. 19. American Society of Heating, Refrigeration and Air Conditioning Engineers, Atlanta, GA.

Collier, R.K. 1989. Desiccant properties and their effect on cooling system performance. *ASHRAE Trans.* 95(1):823–827.

Collier, R.K, Cale, T.S., and Lavan, Z. 1986. *Advanced Desiccant Materials Assessment,* pb-87-172805/XAB. Gas Research Institute, Chicago, IL.

Glav, B.O. 1966. Air Conditioning Apparatus, U.S. Patent No. 3251402.

Harriman, L.G. III. 1990. *The Dehumidification Handbook Second Edition.* Munters Cargocaire, Amesbury, MA.

Pesaran, A.A., Penny, T.R., and Czanderna. 1992. *Desiccant Cooling: State-of-the-Art Assessment.* National Renewable Energy Laboratory, Golden, CO.

United Nations Environmental Programme. 1992. Report of the fourth meeting of the parties to the Montreal protocol on substances that deplete the ozone layer, November 23–25, 1992, Copenhagen.

Zheng, W., Worek, W.M., and Novosel, D. 1993. Control and optimization of rotational speeds for rotary dehumidifiers. *ASHRAE Trans.* 99(1).

Appendices

Paul Norton

National Renewable Energy Laboratory

A. Properties of Gases and Vapors A-2
B. Properties of Liquids B-35
C. Properties of Solids C-38
D. SI Units D-74
E. Miscellaneous E-75

Appendix A. Properties of Gases and Vapors

TABLE A.1 Properties of Dry Air at Atmospheric Pressure

Symbols and Units:

K = absolute temperature, degrees Kelvin

deg C = temperature, degrees Celsius

deg F = temperature, degrees Fahrenheit

ρ = density, kg/m^3

c_p = specific heat capacity, kJ/kg·K

c_p/c_v = specific heat capacity ratio, dimensionless

μ = viscosity, N·s/m^2 × 10^6 (For N·s/m^2 (= kg/m·s) multiply tabulated values by 10^{-6})

k = thermal conductivity, W/m·k × 10^3 (For W/m·K multiply tabulated values by 10^{-3})

Pr = Prandtl number, dimensionless

h = enthalpy, kJ/kg

V_s = sound velocity, m/s

Temperature			Properties							
K	deg C	deg F	ρ	c_p	c_p/c_v	μ	k	Pr	h	V_s
100	−173.15	−280	3.598	1.028		6.929	9.248	.770	98.42	198.4
110	−163.15	−262	3.256	1.022	1.420 2	7.633	10.15	.768	108.7	208.7
120	−153.15	−244	2.975	1.017	1.416 6	8.319	11.05	.766	118.8	218.4
130	−143.15	−226	2.740	1.014	1.413 9	8.990	11.94	.763	129.0	227.6
140	−133.15	−208	2.540	1.012	1.411 9	9.646	12.84	.761	139.1	236.4
150	−123.15	−190	2.367	1.010	1.410 2	10.28	13.73	.758	149.2	245.0
160	−113.15	−172	2.217	1.009	1.408 9	10.91	14.61	.754	159.4	253.2
170	−103.15	−154	2.085	1.008	1.407 9	11.52	15.49	.750	169.4	261.0
180	−93.15	−136	1.968	1.007	1.407 1	12.12	16.37	.746	179.5	268.7
190	−83.15	−118	1.863	1.007	1.406 4	12.71	17.23	.743	189.6	276.2
200	−73.15	−100	1.769	1.006	1.405 7	13.28	18.09	.739	199.7	283.4
205	−68.15	−91	1.726	1.006	1.405 5	13.56	18.52	.738	204.7	286.9
210	−63.15	−82	1.684	1.006	1.405 3	13.85	18.94	.736	209.7	290.5
215	−58.15	−73	1.646	1.006	1.405 0	14.12	19.36	.734	214.8	293.9
220	−53.15	−64	1.607	1.006	1.404 8	14.40	19.78	.732	219.8	297.4
225	−48.15	−55	1.572	1.006	1.404 6	14.67	20.20	.731	224.8	300.8
230	−43.15	−46	1.537	1.006	1.404 4	14.94	20.62	.729	229.8	304.1
235	−38.15	−37	1.505	1.006	1.404 2	15.20	21.04	.727	234.9	307.4
240	−33.15	−28	1.473	1.005	1.404 0	15.47	21.45	.725	239.9	310.6
245	−28.15	−19	1.443	1.005	1.403 8	15.73	21.86	.724	244.9	313.8
250	−23.15	−10	1.413	1.005	1.403 6	15.99	22.27	.722	250.0	317.1
255	−18.15	−1	1.386	1.005	1.403 4	16.25	22.68	.721	255.0	320.2
260	−13.15	8	1.359	1.005	1.403 2	16.50	23.08	.719	260.0	323.4
265	−8.15	17	1.333	1.005	1.403 0	16.75	23.48	.717	265.0	326.5
270	−3.15	26	1.308	1.006	1.402 9	17.00	23.88	.716	270.1	329.6
275	+1.85	35	1.285	1.006	1.402 6	17.26	24.28	.715	275.1	332.6
280	6.85	44	1.261	1.006	1.402 4	17.50	24.67	.713	280.1	335.6
285	11.85	53	1.240	1.006	1.402 2	17.74	25.06	.711	285.1	338.5
290	16.85	62	1.218	1.006	1.402 0	17.98	25.47	.710	290.2	341.5
295	21.85	71	1.197	1.006	1.401 8	18.22	25.85	.709	295.2	344.4
300	26.85	80	1.177	1.006	1.401 7	18.46	26.24	.708	300.2	347.3
305	31.85	89	1.158	1.006	1.401 5	18.70	26.63	.707	305.3	350.2
310	36.85	98	1.139	1.007	1.401 3	18.93	27.01	.705	310.3	353.1
315	41.85	107	1.121	1.007	1.401 0	19.15	27.40	.704	315.3	355.8
320	46.85	116	1.103	1.007	1.400 8	19.39	27.78	.703	320.4	358.7

**Condensed and computed from: "Tables of Thermal Properties of Gases," National Bureau of Standards Circular 564, U.S. Government Printing Office. November 1955.

TABLE A.1 (continued) Properties of Dry Air at Atmospheric Pressure

Temperature			Properties							
K	deg C	deg F	ρ	c_p	c_p/c_v	μ	k	Pr	h	V_s
325	51.85	125	1.086	1.008	1.400 6	19.63	28.15	.702	325.4	361.4
330	56.85	134	1.070	1.008	1.400 4	19.85	28.53	.701	330.4	364.2
335	61.85	143	1.054	1.008	1.400 1	20.08	28.90	.700	335.5	366.9
340	66.85	152	1.038	1.008	1.399 9	20.30	29.28	.699	340.5	369.6
345	71.85	161	1.023	1.009	1.399 6	20.52	29.64	.698	345.6	372.3
350	76.85	170	1.008	1.009	1.399 3	20.75	30.03	.697	350.6	375.0
355	81.85	179	0.994 5	1.010	1.399 0	20.97	30.39	.696	355.7	377.6
360	86.85	188	0.980 5	1.010	1.398 7	21.18	30.78	.695	360.7	380.2
365	91.85	197	0.967 2	1.010	1.398 4	21.38	31.14	.694	365.8	382.8
370	96.85	206	0.953 9	1.011	1.398 1	21.60	31.50	.693	370.8	385.4
375	101.85	215	0.941 3	1.011	1.397 8	21.81	31.86	.692	375.9	388.0
380	106.85	224	0.928 8	1.012	1.397 5	22.02	32.23	.691	380.9	390.5
385	111.85	233	0.916 9	1.012	1.397 1	22.24	32.59	.690	386.0	393.0
390	116.85	242	0.905 0	1.013	1.396 8	22.44	32.95	.690	391.0	395.5
395	121.85	251	0.893 6	1.014	1.396 4	22.65	33.31	.689	396.1	398.0
400	126.85	260	0.882 2	1.014	1.396 1	22.86	33.65	.689	401.2	400.4
410	136.85	278	0.860 8	1.015	1.395 3	23.27	34.35	.688	411.3	405.3
420	146.85	296	0.840 2	1.017	1.394 6	23.66	35.05	.687	421.5	410.2
430	156.85	314	0.820 7	1.018	1.393 8	24.06	35.75	.686	431.7	414.9
440	166.85	332	0.802 1	1.020	1.392 9	24.45	36.43	.684	441.9	419.6
450	176.85	350	0.784 2	1.021	1.392 0	24.85	37.10	.684	452.1	424.2
460	186.85	368	0.767 7	1.023	1.391 1	25.22	37.78	.683	462.3	428.7
470	196.85	386	0.750 9	1.024	1.390 1	25.58	38.46	.682	472.5	433.2
480	206.85	404	0.735 1	1.026	1.389 2	25.96	39.11	.681	482.8	437.6
490	216.85	422	0.720 1	1.028	1.388 1	26.32	39.76	.680	493.0	442.0
500	226.85	440	0.705 7	1.030	1.387 1	26.70	40.41	.680	503.3	446.4
510	236.85	458	0.691 9	1.032	1.386 1	27.06	41.06	.680	513.6	450.6
520	246.85	476	0.678 6	1.034	1.385 1	27.42	41.69	.680	524.0	454.9
530	256.85	494	0.665 8	1.036	1.384 0	27.78	42.32	.680	534.3	459.0
540	266.85	512	0.653 5	1.038	1.382 9	28.14	42.94	.680	544.7	463.2
550	276.85	530	0.641 6	1.040	1.381 8	28.48	43.57	.680	555.1	467.3
560	286.85	548	0.630 1	1.042	1.380 6	28.83	44.20	.680	565.5	471.3
570	296.85	566	0.619 0	1.044	1.379 5	29.17	44.80	.680	575.9	475.3
580	306.85	584	0.608 4	1.047	1.378 3	29.52	45.41	.680	586.4	479.2
590	316.85	602	0.598 0	1.049	1.377 2	29.84	46.01	.680	596.9	483.2
600	326.85	620	0.588 1	1.051	1.376 0	30.17	46.61	680	607.4	486.9
620	346.85	656	0.569 1	1.056	1.373 7	30.82	47.80	.681	628.4	494.5
640	366.85	692	0.551 4	1.061	1.371 4	31.47	48.96	.682	649.6	502.1
660	386.85	728	0.534 7	1.065	1.369 1	32.09	50.12	.682	670.9	509.4
680	406.85	764	0.518 9	1.070	1.366 8	32.71	51.25	.683	692.2	516.7
700	426.85	800	0.504 0	1.075	1.364 6	33.32	52.36	.684	713.7	523.7
720	446.85	836	0.490 1	1.080	1.362 3	33.92	53.45	.685	735.2	531.0
740	466.85	872	0.476 9	1.085	1.360 1	34.52	54.53	.686	756.9	537.6
760	486.85	908	0.464 3	1.089	1.358 0	35.11	55.62	.687	778.6	544.6
780	506.85	944	0.452 4	1.094	1.355 9	35.69	56.68	.688	800.5	551.2
800	526.85	980	0.441 0	1.099	1.354	36.24	57.74	.689	822.4	557.8
850	576.85	1 070	0.415 2	1.110	1.349	37.63	60.30	.693	877.5	574.1
900	626.85	1 160	0.392 0	1.121	1.345	38.97	62.76	.696	933.4	589.6
950	676.85	1 250	0.371 4	1.132	1.340	40.26	65.20	.699	989.7	604.9
1 000	726.85	1 340	0.352 9	1.142	1.336	41.53	67.54	.702	1 046	619.5
1 100	826.85	1 520	0.320 8	1.161	1.329	43.96			1 162	648.0
1 200	926.85	1 700	0.294 1	1.179	1.322	46.26			1 279	675.2
1 300	1 026.85	1 880	0.271 4	1.197	1.316	48.46			1 398	701.0
1 400	1 126.85	2 060	0.252 1	1.214	1.310	50.57			1 518	725.9
1 500	1 220.85	2 240	0.235 3	1.231	1.304	52.61			1 640	749.4
1 600	1 326.85	2 420	0.220 6	1.249	1.299	54.57			1 764	772.6
1 800	1 526.85	2 780	0.196 0	1.288	1.288	58.29			2 018	815.7
2 000	1 726.85	3 140	0.176 4	1.338	1.274				2 280	855.5
2 400	2 126.85	3 860	0.146 7	1.574	1.238				2 853	924.4
2 800	2 526.85	4 580	0.124 5	2.259	1.196				3 599	983.1

TABLE A.2 Ideal Gas Properties of Nitrogen, Oxygen, and Carbon Dioxide

Symbols and Units:

T = absolute temperature, degrees Kelvin
$\bar{h}$ = enthalpy, kJ/kmol
$\bar{u}$ = internal energy, kJ/kmol
$\bar{s}°$ = absolute entropy at standard reference pressure, kJ/kmol K
[$\bar{h}$ = enthalpy of formation per mole at standard state = 0 kJ/kmol]

Part a. Ideal Gas Properties of Nitrogen, N_2

T	$\bar{h}$	$\bar{u}$	$\bar{s}°$	T	$\bar{h}$	$\bar{u}$	$\bar{s}°$
0	0	0	0	600	17,563	12,574	212.066
220	6,391	4,562	182.639	610	17,864	12,792	212.564
230	6,683	4,770	183.938	620	18,166	13,011	213.055
240	6,975	4,979	185.180	630	18,468	13,230	213.541
250	7,266	5,188	186.370	640	18,772	13,450	214.018
260	7,558	5,396	187.514	650	19,075	13,671	214.489
270	7,849	5,604	188.614	660	19,380	13,892	214.954
280	8,141	5,813	189.673	670	19,685	14,114	215.413
290	8,432	6,021	190.695	680	19,991	14,337	215.866
298	8,669	6,190	191.502	690	20,297	14,560	216.314
300	8,723	6,229	191.682	700	20,604	14,784	216.756
310	9,014	6,437	192.638	710	20,912	15,008	217.192
320	9,306	6,645	193.562	720	21,220	15,234	217.624
330	9,597	6,853	194.459	730	21,529	15,460	218.059
340	9,888	7,061	195.328	740	21,839	15,686	218.472
350	10,180	7,270	196.173	750	22,149	15,913	218.889
360	10,471	7,478	196.995	760	22,460	16,141	219.301
370	10,763	7,687	197.794	770	22,772	16,370	219.709
380	11,055	7,895	198.572	780	23,085	16,599	220.113
390	11,347	8,104	199.331	790	23,398	16,830	220.512
400	11,640	8,314	200.071	800	23,714	17,061	220.907
410	11,932	8,523	200.794	810	24,027	17,292	221.298
420	12,225	8,733	201.499	820	24,342	17,524	221.684
430	12,518	8,943	202.189	830	24,658	17,757	222.067
440	12,811	9,153	202.863	840	24,974	17,990	222.447
450	13,105	9,363	203.523	850	25,292	18,224	222.822
460	13,399	9,574	204.170	860	25,610	18,459	223.194
470	13,693	9,786	204.803	870	25,928	18,695	223.562
480	13,988	9,997	205.424	880	26,248	18,931	223.927
490	14,285	10,210	206.033	890	26,568	19,168	224.288
500	14,581	10,423	206.630	900	26,890	19,407	224.647
510	14,876	10,635	207.216	910	27,210	19,644	225.002
520	15,172	10,848	207.792	920	27,532	19,883	225.353
530	15,469	11,062	208.358	930	27,854	20,122	225.701
540	15,766	11,277	208.914	940	28,178	20,362	226.047
550	16,064	11,492	209.461	950	28,501	20,603	226.389
560	16,363	11,707	209.999	960	28,826	20,844	226.728
570	16,662	11,923	210.528	970	29,151	21,086	227.064
580	16,962	12,139	211.049	980	29,476	21,328	227.398
590	17,26_	12,356	211.562	990	29,803	21,571	227.728

Source: Adapted from M.J. Moran and H.N. Shapiro, *Fundamentals of Engineering Thermodynamics*, 3rd. ed., Wiley, New York, 1995, as presented in K. Wark. *Thermodynamics*, 4th ed., McGraw-Hill, New York, 1983, based on the *JANAF Thermochemical Tables*, NSRDS-NBS-37, 1971.

TABLE A.2 (continued) Ideal Gas Properties of Nitrogen, Oxygen, and Carbon Dioxide

T	$\bar{h}$	$\bar{u}$	$\bar{s}^\circ$	T	$\bar{h}$	$\bar{u}$	$\bar{s}^\circ$
1000	30,129	21,815	228.057	1760	56,227	41.594	247.396
1020	30,784	22,304	228.706	1780	56,938	42,139	247.798
1040	31,442	22,795	229.344	1800	57,651	42,685	248.195
1060	32,101	23,288	229.973	1820	58,363	43,231	248.589
1080	32,762	23,782	230.591	1840	59,075	43,777	248.979
1100	33,426	24,280	231.199	1860	59,790	44,324	249.365
1120	34,092	24,780	231.799	1880	60,504	44,873	249.748
1140	34,760	25,282	232.391	1900	61,220	45,423	250.128
1160	35,430	25,786	232.973	1920	61,936	45,973	250.502
1180	36,104	26,291	233.549	1940	62,654	46,524	250.874
1200	36,777	26,799	234.115	1960	63,381	47,075	251.242
1220	37,452	27,308	234.673	1980	64,090	47,627	251.607
1240	38,129	27,819	235.223	2000	64,810	48,181	251.969
1260	38,807	28,331	235.766	2050	66,612	49,567	252.858
1280	39,488	28,845	236.302	2100	68,417	50,957	253.726
1300	40,170	29,361	236.831	2150	70,226	52,351	254.578
1320	40,853	29,878	237.353	2200	72,040	53,749	255.412
1340	41,539	30,398	237.867	2250	73,856	55,149	256.227
1360	42,227	30,919	238.376	2300	75,676	56,553	257.027
1380	42,915	31,441	238.878	2350	77,496	57,958	257.810
1400	43,605	31,964	239.375	2400	79,320	59,366	258.580
1420	44,295	32,489	239.865	2450	81,149	60,779	259.332
1440	44,988	33,014	240.350	2500	82,981	62,195	260.073
1460	45,682	33,543	240.827	2550	84,814	63,613	260.799
1480	46,377	34,071	241.301	2600	86,650	65,033	261.512
1500	47,073	34,601	241.768	2650	88,488	66,455	262.213
1520	47,771	35,133	242.228	2700	90,328	67,880	262.902
1540	48,470	35,665	242.685	2750	92,171	69,306	263.577
1560	49,168	36,197	243.137	2800	94,014	70,734	264.241
1580	49,869	36,732	243.585	2850	95,859	72,163	264.895
1600	50,571	37,268	244.028	2900	97,705	73,593	265.538
1620	51,275	37,806	244.464	2950	99,556	75,028	266.170
1640	51,980	38,344	244.896	3000	101,407	76,464	266.793
1660	52,686	38,884	245.324	3050	103,260	77,902	267.404
1680	53,393	39,424	245.747	3100	105,115	79,341	268.007
1700	54,099	39,965	246.166	3150	106,972	80,782	268.601
1720	54,807	40,507	246.580	3200	108,830	82,224	269.186
1740	55,516	41,049	246.990	3250	110,690	83,668	269.763

TABLE A.2 (continued) Ideal Gas Properties of Nitrogen, Oxygen, and Carbon Dioxide

Part b. Ideal Gas Properties of Oxygen, O_2

T	$\bar{h}$	$\bar{u}$	$\bar{s}°$	T	$\bar{h}$	$\bar{u}$	$\bar{s}°$
0	0	0	0	600	17,929	12,940	226.346
220	6,404	4,575	196.171	610	18,250	13,178	226.877
230	6,694	4,782	197.461	620	18,572	13,417	227.400
240	6,984	4,989	198.696	630	18,895	13,657	227.918
250	7,275	5,197	199.885	640	19,219	13,898	228.429
260	7,566	5,405	201.027	650	19,544	14,140	228.932
270	7,858	5,613	202.128	660	19,870	14,383	229.430
280	8,150	5,822	203.191	670	20,197	14,626	229.920
290	8,443	6,032	204.218	680	20,524	14,871	230.405
298	8,682	6,203	205.033	690	20,854	15,116	230.885
300	8,736	6,242	205.213	700	21,184	15,364	231.358
310	9,030	6,453	206.177	710	21,514	15,611	231.827
320	9,325	6,664	207.112	720	21,845	15,859	232.291
330	9,620	6,877	208.020	730	22,177	16,107	232.748
340	9,916	7,090	208.904	740	22,510	16,357	233.201
350	10,213	7,303	209.765	750	22,844	16,607	233.649
360	10,511	7,518	210.604	760	23,178	16,859	234.091
370	10,809	7,733	211.423	770	23,513	17,111	234.528
380	11,109	7,949	212.222	780	23,850	17,364	234.960
390	11,409	8,166	213.002	790	24,186	17,618	235.387
400	11,711	8,384	213.765	800	24,523	17,872	235.810
410	12,012	8,603	214.510	810	24,861	18,126	236.230
420	12,314	8,822	215.241	820	25,199	18,382	236.644
430	12,618	9,043	215.955	830	25,537	18,637	237.055
440	12,923	9,264	216.656	840	25,877	18,893	237.462
450	13,228	9,487	217.342	850	26,218	19,150	237.864
460	13,535	9,710	218.016	860	26,559	19,408	238.264
470	13,842	9,935	218.676	870	26,899	19,666	238.660
480	14,151	10,160	219.326	880	27,242	19,925	239.051
490	14,460	10,386	219.963	890	27,584	20,185	239.439
500	14,770	10,614	220.589	900	27,928	20,445	239.823
510	15,082	10,842	221.206	910	28,272	20,706	240.203
520	15,395	11,071	221.812	920	28,616	20,967	240.580
530	15,708	11,301	222.409	930	28,960	21,228	240.953
540	16,022	11,533	222.997	940	29,306	21,491	241.323
550	16,338	11,765	223.576	950	29,652	21,754	241.689
560	16,654	11,998	224.146	960	29,999	22,017	242.052
570	16,971	12,232	224.708	970	30,345	22,280	242.411
580	17,290	12,467	225.262	980	30,692	22,544	242.768
590	17,609	12,703	225.808	990	31,041	22,809	243.120

TABLE A.2 (continued)   Ideal Gas Properties of Nitrogen, Oxygen, and Carbon Dioxide

T	$\bar{h}$	$\bar{u}$	$\bar{s}°$	T	$\bar{h}$	$\bar{u}$	$\bar{s}°$
1000	31,389	23,075	243.471	1760	58,880	44,247	263.861
1020	32,088	23,607	244.164	1780	59,624	44,825	264.283
1040	32,789	24,142	244.844	1800	60,371	45,405	264.701
1060	33,490	24,677	245.513	1820	61,118	45,986	265.113
1080	34,194	25,214	246.171	1840	61,866	46,568	265.521
1100	34,899	25,753	246.818	1860	62,616	47,151	265.925
1120	35,606	26,294	247.454	1880	63,365	47,734	266.326
1140	36,314	26,836	248.081	1900	64,116	48,319	266.722
1160	37,023	27,379	248.698	1920	64,868	48,904	267.115
1180	37,734	27,923	249.307	1940	65,620	49,490	267.505
1200	38,447	28,469	249.906	1960	66,374	50,078	267.891
1220	39,162	29,018	250.497	1980	67,127	50,665	268.275
1240	39,877	29,568	251.079	2000	67,881	51,253	268.655
1260	40,594	30,118	251.653	2050	69,772	52,727	269.588
1280	41,312	30,670	252.219	2100	71,668	54,208	270.504
1300	42,033	31,224	252.776	2150	73,573	55,697	271.399
1320	42,753	31,778	253.325	2200	75,484	57,192	272.278
1340	43,475	32,334	253.868	2250	77,397	58,690	273.136
1360	44,198	32,891	254.404	2300	79,316	60,193	273.981
1380	44,923	33,449	254.932	2350	81,243	61,704	274.809
1400	45,648	34,008	255.454	2400	83,174	63,219	275.625
1420	46,374	34,567	255.968	2450	85,112	64,742	276.424
1440	47,102	35,129	256.475	2500	87,057	66,271	277.207
1460	47,831	35,692	256.978	2550	89,004	67,802	277.979
1480	48,561	36,256	257.474	2600	90,956	69,339	278.738
1500	49,292	36,821	257.965	2650	92,916	70,883	279.485
1520	50,024	37,387	258.450	2700	94,881	72,433	280.219
1540	50,756	37,952	258.928	2750	96,852	73,987	280.942
1560	51,490	38,520	259.402	2800	98,826	75,546	281.654
1580	52,224	39,088	259.870	2850	100,808	77,112	282.357
1600	52,961	39,658	260.333	2900	102,793	78,682	283.048
1620	53,696	40,227	260.791	2950	104,785	80,258	283.728
1640	54,434	40,799	261.242	3000	106,780	81,837	284.399
1660	55,172	41,370	261.690	3050	108,778	83,419	285.060
1680	55,912	41,944	262.132	3100	110,784	85,009	285.713
1700	56,652	42,517	262.571	3150	112,795	86,601	286.355
1720	57,394	43,093	263.005	3200	114,809	88,203	286.989
1740	58,136	43,669	263.435	3250	116,827	89,804	287.614

TABLE A.2 (continued) Ideal Gas Properties of Nitrogen, Oxygen, and Carbon Dioxide

Part c. Ideal Gas Properties of Carbon Dioxide, CO_2

T	$\bar{h}$	$\bar{u}$	$\bar{s}^\circ$	T	$\bar{h}$	$\bar{u}$	$\bar{s}^\circ$
0	0	0	0	600	22,280	17,291	243.199
220	6,601	4,772	202.966	610	22,754	17,683	243.983
230	6,938	5,026	204.464	620	23,231	18,076	244.758
240	7,280	5,285	205.920	630	23,709	18,471	245.524
250	7,627	5,548	207.337	640	24,190	18,869	246.282
260	7,979	5,817	208.717	650	24,674	19,270	247.032
270	8,335	6,091	210.062	660	25,160	19,672	247.773
280	8,697	6,369	211.376	670	25,648	20,078	248.507
290	9,063	6,651	212.660	680	26,138	20,484	249.233
298	9,364	6,885	213.685	690	26,631	20,894	249.952
300	9,431	6,939	213.915	700	27,125	21,305	250.663
310	9,807	7,230	215.146	710	27,622	21,719	251.368
320	10,186	7,526	216.351	720	28,121	22,134	252.065
330	10,570	7,826	217.534	730	28,622	22,552	252.755
340	10,959	8,131	218.694	740	29,124	22,972	253.439
350	11,351	8,439	219.831	750	29,629	23,393	254.117
360	11,748	8,752	220.948	760	30,135	23,817	254.787
370	12,148	9,068	222.044	770	30,644	24,242	255.452
380	12,552	9,392	223.122	780	31,154	24,669	256.110
390	12,960	9,718	224.182	790	31,665	25,097	256.762
400	13,372	10,046	225.225	800	32,179	25,527	257.408
410	13,787	10,378	226.250	810	32,694	25,959	258.048
420	14,206	10,714	227.258	820	33,212	26,394	258.682
430	14,628	11,053	228.252	830	33,730	26,829	259.311
440	15,054	11,393	229.230	840	34,251	27,267	259.934
450	15,483	11,742	230.194	850	34,773	27,706	260.551
460	15,916	12,091	231.144	860	35,296	28,125	261.164
470	16,351	12,444	232.080	870	35,821	28,588	261.770
480	16,791	12,800	233.004	880	36,347	29,031	262.371
490	17,232	13,158	233.916	890	36,876	29,476	262.968
500	17,678	13,521	234.814	900	37,405	29,922	263.559
510	18,126	13,885	235.700	910	37,935	30,369	264.146
520	18,576	14,253	236.575	920	38,467	30,818	264.728
530	19,029	14,622	237.439	930	39,000	31,268	265.304
540	19,485	14,996	238.292	940	39,535	31,719	265.877
550	19,945	15,372	239.135	950	40,070	32,171	266.444
560	20,407	15,751	239.962	960	40,607	32,625	267.007
570	20,870	16,131	240.789	970	41,145	33,081	267.566
580	21,337	16,515	241.602	980	41,685	33,537	268.119
590	21,807	16,902	242.405	990	42,226	33,995	268.670

TABLE A.2 (continued) Ideal Gas Properties of Nitrogen, Oxygen, and Carbon Dioxide

T	$\bar{h}$	$\bar{u}$	$\bar{s}^\circ$	T	$\bar{h}$	$\bar{u}$	$\bar{s}^\circ$
1000	42,769	34,455	269.215	1760	86,420	71,787	301.543
1020	43,859	35,378	270.293	1780	87,612	72,812	302.271
1040	44,953	36,306	271.354	1800	88,806	73,840	302.884
1060	46,051	37,238	272.400	1820	90,000	74,868	303.544
1080	47,153	38,174	273.430	1840	91,196	75,897	304.198
1100	48,258	39,112	274.445	1860	92,394	76,929	304.845
1120	49,369	40,057	275.444	1880	93,593	77,962	305.487
1140	50,484	41,006	276.430	1900	94,793	78,996	306.122
1160	51,602	41,957	277.403	1920	95,995	80,031	306.751
1180	52,724	42,913	278.362	1940	97,197	81,067	307.374
1200	53,848	43,871	279.307	1960	98,401	82,105	307.992
1220	54,977	44,834	280.238	1980	99,606	83,144	308.604
1240	56,108	45,799	281.158	2000	100,804	84,185	309.210
1260	57,244	46,768	282.066	2050	103,835	86,791	310.701
1280	58,381	47,739	282.962	2100	106,864	89,404	312.160
1300	59,522	48,713	283.847	2150	109,898	92,023	313.589
1320	60,666	49,691	284.722	2200	112,939	94,648	314.988
1340	61,813	50,672	285.586	2250	115,984	97,277	316.356
1360	62,963	51,656	286.439	2300	119,035	99,912	317.695
1380	64,116	52,643	287.283	2350	122,091	102,552	319.011
1400	65,271	53,631	288.106	2400	125,152	105,197	320.302
1420	66,427	54,621	288.934	2450	128,219	107,849	321.566
1440	67,586	55,614	289.743	2500	131,290	110,504	322.808
1460	68,748	56,609	290.542	2550	134,368	113,166	324.026
1480	69,911	57,606	291.333	2600	137,449	115,832	325.222
1500	71,078	58,606	292.114	2650	140,533	118,500	326.396
1520	72,246	59,609	292.888	2700	143,620	121,172	327.549
1540	73,417	60,613	292.654	2750	146,713	123,849	328.684
1560	74,590	61,620	294.411	2800	149,808	126,528	329.800
1580	76,767	62,630	295.161	2850	152,908	129,212	330.896
1600	76,944	63,741	295.901	2900	156,009	131,898	331.975
1620	78,123	64,653	296.632	2950	159,117	134,589	333.037
1640	79,303	65,668	297.356	3000	162,226	137,283	334.084
1660	80,486	66,592	298.072	3050	165,341	139,982	335.114
1680	81,670	67,702	298.781	3100	168,456	142,681	336.126
1700	82,856	68,721	299.482	3150	171,576	145,385	337.124
1720	84,043	69,742	300.177	3200	174,695	148,089	338.109
1740	85,231	70,764	300.863	3250	177,822	150,801	339.069

TABLE A.3 Psychrometric Table: Properties of Moist Air at 101 325 N/m²

Symbols and Units:

P_s = pressure of water vapor at saturation, N/m²

W_s = humidity ratio at saturation, mass of water vapor associated with unit mass of dry air

V_a = specific volume of dry air, m³/kg

V_s = specific volume of saturated mixture, m³/kg dry air

h_a^a = specific enthalpy of dry air, kJ/kg

h_s = specific enthalpy of saturated mixture, kJ/kg dry air

s_s = specific entropy of saturated mixture, J/K·kg dry air

Temperature			Properties						
C	K	F	P_s	W_s	V_a	V_s	h_a	h_s	s_s
− 40	233.15	− 40	12.838	0.000 079 25	0.659 61	0.659 68	− 22.35	− 22.16	− 90.659
− 30	243.15	− 22	37.992	0.000 234 4	0.688 08	0.688 33	− 12.29	− 11.72	− 46.732
− 25	248.15	− 13	63.248	0.000 390 3	0.702 32	0.702 75	− 7.265	− 6.306	− 24.706
− 20	253.15	− 4	103.19	0.000 637 1	0.716 49	0.717 24	− 2.236	− 0.6653	− 2.2194
− 15	258.15	+ 5	165.18	0.001 020	0.730 72	0.731 91	+ 2.794	5.318	21.189
− 10	263.15	14	259.72	0.001 606	0.744 95	0.746 83	7.823	11.81	46.104
− 5	268.15	23	401.49	0.002 485	0.759 12	0.762 18	12.85	19.04	73.365
0	273.15	32	610.80	0.003 788	0.773 36	0.778 04	17.88	27.35	104.14
5	278.15	41	871.93	0.005 421	0.787 59	0.794 40	22.91	36.52	137.39
10	283.15	50	1 227.2	0.007 658	0.801 76	0.811 63	27.94	47.23	175.54
15	288.15	59	1 704.4	0.010 69	0.816 00	0.829 98	32.97	59.97	220.22
20	293.15	68	2 337.2	0.014 75	0.830 17	0.849 83	38.00	75.42	273.32
25	298.15	77	3 167.0	0.020 16	0.844 34	0.871 62	43.03	94.38	337.39
30	303.15	86	4 242.8	0.027 31	0.858 51	0.896 09	48.07	117.8	415.65
35	308.15	95	5 623.4	0.036 73	0.872 74	0.924 06	53.10	147.3	512.17
40	313.15	104	7 377.6	0.049 11	0.886 92	0.956 65	58.14	184.5	532.31
45	318.15	113	9 584.8	0.065 36	0.901 15	0.995 35	63.17	232.0	783.06
50	323.15	122	12 339	0.086 78	0.915 32	1.042 3	68.21	293.1	975.27
55	328.15	131	15 745	0.115 2	0.929 49	1.100 7	73.25	372.9	1 221.5
60	333.15	140	19 925	0.153 4	0.943 72	1.174 8	78.29	478.5	1 543.5
65	338.15	149	25 014	0.205 5	0.957 90	1.272 1	83.33	621.4	1 973.6
70	343.15	158	31 167	0.278 8	0.972 07	1.404 2	88.38	820.5	2 564.8
75	348.15	167	38 554	0.385 8	0.986 30	1.592 4	93.42	1 110	3 412.8
80	353.15	176	47 365	0.551 9	1.000 5	1.879 1	98.47	1 557	4 710.9
85	358.15	185	57 809	0.836 3	1.014 6	2.363 2	103.5	2 321	6 892.6
90	363.15	194	70 112	1.416	1.028 8	3.340 9	108.6	3 876	11 281

Note: The P_s column in this table gives the vapor pressure of pure water at temperature intervals of five degrees Celsius. For the latest data on vapor pressures at intervals of 0.1 deg C, see "Vapor Pressure Equation for Water," A. Wexler and L. Greenspan, *J. Res. Nat. Bur. Stand.*, 75A(3):213-229, May-June 1971.

* Fpr very low barometric pressures and high wet-bulb temperatures, the values of h_s in this table are somewhat low; for corrections, see "ASHRAE Handbook of Fundamentals."

*Computed from: Psychrometric Tables, in "ASHRAE Handbook of Fundamentals," American Society of Heating, refrigerating and Air-Conditioning Engineers, 1972.

TABLE A.4 Water Vapor at Low Pressures: Perfect Gas Behavior $pv/T = R = 0.461\ 51$ kJ/kg·K

Symbols and Units:

t = thermodynamic temperature, deg C

T = thermodynamic temperature, K

$pv = RT$, kJ/kg

u_o = specific internal energy at zero pressure, kJ/kg

h_o = specific enthalpy at zero pressure, kJ/kg

s_l = specific entropy of semiperfect vapor at 0.1 MN/m², kJ/kg·K

ψ_l = specific Helmholtz free energy of semiperfect vapor at 0.1 MN/m², kJ/kg

ψ_l = specific Helmholtz free energy of semiperfect vapor at 0.1 MN/m², kJ/kg

ζ_l = specific Gibbs free energy of semiperfect vapor at 0.1 MN/m², kJ/kg

p_r = relative pressure, pressure of semiperfect vapor at zero entropy, TN/m²

v_r = relative specific volume, specific volume of semiperfect vapor at zero entropy, mm³/kg

c_{po} = specific heat capacity at constant pressure for zero pressure, kJ/kg·K

c_{vo} = specific heat capacity at constant volume for zero pressure, kJ/kg·K

$k = c_{po}/c_{vo}$ = isentropic exponent, $-(\partial\log p/\partial\log v)_s$

t	T	pv	u_o	h_o	s_l	ψ_l	ζ_l	p_r	v_r	c_{po}	c_{vo}	k
0	273.15	126.06	2 375.5	2 501.5	6.804 2	516.9	643.0	.252 9	498.4	1.858 4	1.396 9	1.330 4
10	283.15	130.68	2 389.4	2 520.1	6.871 1	443.9	574.6	.292 3	447.0	1.860 1	1.398 6	1.330 0
20	293.15	135.29	2 403.4	2 538.7	6.935 7	370.2	505.5	.336 3	402.4	1.862 2	1.400 7	1.329 5
30	303.15	139.91	2 417.5	2 557.4	6.998 2	296.0	435.9	.385 0	363.4	1.864 7	1.403 1	1.328 9
40	313.15	144.52	2 431.5	2 576.0	7.058 7	221.1	365.6	.439 0	329.2	1.867 4	1.405 9	1.328 3
50	323.15	149.14	2 445.6	2 594.7	7.117 5	145.6	294.7	.498 6	299.1	1.870 5	1.409 0	1.327 5
60	333.15	153.75	2 459.7	2 613.4	7.174 5	69.5	223.2	.564 2	272.5	1.873 8	1.412 3	1.326 8
70	343.15	158.37	2 473.8	2 632.2	7.230 0	−7.2	151.2	.636 3	248.9	1.877 4	1.415 9	1.325 9
80	353.15	162.98	2 488.0	2 651.0	7.284 0	−84.3	78.6	.715 2	227.9	1.881 2	1.419 7	1.325 1
90	363.15	167.60	2 502.2	2 669.8	7.336 6	−162.1	5.5	.801 5	209.1	1.885 2	1.423 7	1.324 2
100	373.15	172.21	2 516.5	2 688.7	7.387 8	−240.3	−68.1	.895 7	192.26	1.889 4	1.427 9	1.323 2
120	393.15	181.44	2 545.1	2 726.6	7.486 7	−398.3	−216.8	1.109 7	163.50	1.898 3	1.436 7	1.321 2
140	413.15	190.67	2 573.9	2 764.6	7.581 1	−558.2	−367.5	1.361 7	140.03	1.907 7	1.446 2	1.319 1
160	433.15	199.90	2 603.0	2 802.9	7.671 5	−720.0	−520.1	1.656 4	120.69	1.917 7	1.456 2	1.316 9
180	453.15	209.13	2 632.2	2 841.3	7.758 3	−883.5	−674.4	1.999 1	104.61	1.928 1	1.466 6	1.314 7
200	473.15	218.4	2 661.6	2 880.0	7.841 8	−1 048.7	−830.4	2.396	91.15	1.938 9	1.477 4	1.312 4
300	573.15	264.5	2 812.3	3 076.8	8.218 9	−1 898.4	−1 633.9	5.423	48.77	1.997 5	1.536 0	1.300 5
400	673.15	310.7	2 969.0	3 279.7	8.545 1	−2 783.1	−2 472.5	10.996	28.25	2.061 4	1.599 9	1.288 5
500	773.15	356.8	3 132.4	3 489.2	8.835 2	−3 699	−3 342	20.61	17.310	2.128 7	1.667 2	1.276 8
600	873.15	403.0	3 302.5	3 705.5	9.098 2	−4 642	−4 239	36.45	11.056	2.198 0	1.736 5	1.265 8
700	973.15	449.1	3 479.7	3 928.8	9.340 3	−5 610	−5 161	61.58	7.293	2.268 3	1.806 8	1.255 4
800	1 073.15	495.3	3 663.9	4 159.2	9.565 5	−6 601	−6 106	100.34	4.936	2.338 7	1.877 1	1.245 9
900	1 173.15	541.4	3 855.1	4 396.5	9.776 9	−7 615	−7 073	158.63	3.413	2.407 8	1.946 2	1.237 1
1 000	1 273.15	587.6	4 053.1	4 640.6	9.976 6	−8 649	−8 061	244.5	2.403	2.474 4	2.012 8	1.299 3
1 100	1 373.15	633.7	4 257.5	4 891.2	10.166 1	−9 702	−9 068	368.6	1.719	2.536 9	2.075 4	1.222 4
1 200	1 473.15	679.9	4 467.9	5 147.8	10.346 4	−10 774	−10 094	544.9	1.248	2.593 8	2.132 3	1.216 4
1 300	1 573.15	726.0	4 683.7	5 409.7	10.518 4	−11 863	−11 137	791.0	.918	2.643 1	2.181 6	1.211 5

*Adapted from: "Steam Tables," J. H. Keenan, F. G. Keyes, P. G. Hill, and J. G. Moore, John Wiley & Sons, Inc. 1969 (International Edition - Metric Units).

REFERENCE

For other steam tables in metric units, see "Steam Tables in SI Units," Ministry of Technology, London, 1970.

TABLE A.5 Properties of Saturated Water and Steam

Part a. Temperature Table

Temp. °C	Press. bars	Specific Volume m³/kg		Internal Energy kJ/kg		Enthalpy kJ/kg			Entropy kJ/kg · K		Temp. °C
		Sat. Liquid $v_f \times 10^3$	Sat. Vapor v_g	Sat. Liquid u_f	Sat. Vapor u_g	Sat. Liquid h_f	Evap. h_{fg}	Sat. Vapor h_g	Sat. Liquid s_f	Sat. Vapor s_g	
.01	0.00611	1.0002	206.136	0.00	2375.3	0.01	2501.3	2501.4	0.0000	9.1562	.01
4	0.00813	1.0001	157.232	16.77	2380.9	16.78	2491.9	2508.7	0.0610	9.0514	4
5	0.00872	1.0001	147.120	20.97	2382.3	20.98	2489.6	2510.6	0.0761	9.0257	5
6	0.00935	1.0001	137.734	25.19	2383.6	25.20	2487.2	2512.4	0.0912	9.0003	6
8	0.01072	1.0002	120.917	33.59	2386.4	33.60	2482.5	2516.1	0.1212	8.9501	8
10	0.01228	1.0004	106.379	42.00	2389.2	42.01	2477.7	2519.8	0.1510	8.9008	10
11	0.01312	1.0004	99.857	46.20	2390.5	46.20	2475.4	2521.6	0.1658	8.8765	11
12	0.01402	1.0005	93.784	50.41	2391.9	50.41	2473.0	2523.4	0.1806	8.8524	12
13	0.01497	1.0007	88.124	54.60	2393.3	54.60	2470.7	2525.3	0.1953	8.8285	13
14	0.01598	1.0008	82.848	58.79	2394.7	58.80	2468.3	2527.1	0.2099	8.8048	14
15	0.01705	1.0009	77.926	62.99	2396.1	62.99	2465.9	2528.9	0.2245	8.7814	15
16	0.01818	1.0011	73.333	67.18	2397.4	67.19	2463.6	2530.8	0.2390	8.7582	16
17	0.01938	1.0012	69.044	71.38	2398.8	71.38	2461.2	2532.6	0.2535	8.7351	17
18	0.02064	1.0014	65.038	75.57	2400.2	75.58	2458.8	2534.4	0.2679	8.7123	18
19	0.02198	1.0016	61.293	79.76	2401.6	79.77	2456.5	2536.2	0.2823	8.6897	19
20	0.02339	1.0018	57.791	83.95	2402.9	83.96	2454.1	2538.1	0.2966	8.6672	20
21	0.02487	1.0020	54.514	88.14	2404.3	88.14	2451.8	2539.9	0.3109	8.6450	21
22	0.02645	1.0022	51.447	92.32	2405.7	92.33	2449.4	2541.7	0.3251	8.6229	22
23	0.02810	1.0024	48.574	96.51	2407.0	96.52	2447.0	2543.5	0.3393	8.6011	23
24	0.02985	1.0027	45.883	100.70	2408.4	100.70	2444.7	2545.4	0.3534	8.5794	24
25	0.03169	1.0029	43.360	104.88	2409.8	104.89	2442.3	2547.2	0.3674	8.5580	25
26	0.03363	1.0032	40.994	109.06	2411.1	109.07	2439.9	2549.0	0.3814	8.5367	26
27	0.03567	1.0035	38.774	113.25	2412.5	113.25	2437.6	2550.8	0.3954	8.5156	27
28	0.03782	1.0037	36.690	117.42	2413.9	117.43	2435.2	2552.6	0.4093	8.4946	28
29	0.04008	1.0040	34.733	121.60	2415.2	121.61	2432.8	2554.5	0.4231	8.4739	29
30	0.04246	1.0043	32.894	125.78	2416.6	125.79	2430.5	2556.3	0.4369	8.4533	30
31	0.04496	1.0046	31.165	129.96	2418.0	129.97	2428.1	2558.1	0.4507	8.4329	31
32	0.04759	1.0050	29.540	134.14	2419.3	134.15	2425.7	2559.9	0.4644	8.4127	32
33	0.05034	1.0053	28.011	138.32	2420.7	138.33	2423.4	2561.7	0.4781	8.3927	33·
34	0.05324	1.0056	26.571	142.50	2422.0	142.50	2421.0	2563.5	0.4917	8.3728	34
35	0.05628	1.0060	25.216	146.67	2423.4	146.68	2418.6	2565.3	0.5053	8.3531	35
36	0.05947	1.0063	23.940	150.85	2424.7	150.86	2416.2	2567.1	0.5188	8.3336	36
38	0.06632	1.0071	21.602	159.20	2427.4	159.21	2411.5	2570.7	0.5458	8.2950	38
40	0.07384	1.0078	19.523	167.56	2430.1	167.57	2406.7	2574.3	0.5725	8.2570	40
45	0.09593	1.0099	15.258	188.44	2436.8	188.45	2394.8	2583.2	0.6387	8.1648	45

TABLE A.5 (continued) Properties of Saturated Water and Steam

Temp. °C	Press. bars	Specific Volume m³/kg		Internal Energy kJ/kg		Enthalpy kJ/kg			Entropy kJ/kg · K		Temp. °C
		Sat. Liquid $v_f \times 10^3$	Sat. Vapor v_g	Sat. Liquid u_f	Sat. Vapor u_g	Sat. Liquid h_f	Evap. h_{fg}	Sat. Vapor h_g	Sat. Liquid s_f	Sat. Vapor s_g	
50	.1235	1.0121	12.032	209.32	2443.5	209.33	2382.7	2592.1	.7038	8.0763	50
55	.1576	1.0146	9.568	230.21	2450.1	230.23	2370.7	2600.9	.7679	7.9913	55
60	.1994	1.0172	7.671	251.11	2456.6	251.13	2358.5	2609.6	.8312	7.9096	60
65	.2503	1.0199	6.197	272.02	2463.1	272.06	2346.2	2618.3	.8935	7.8310	65
70	.3119	1.0228	5.042	292.95	2469.6	292.98	2333.8	2626.8	.9549	7.7553	70
75	.3858	1.0259	4.131	313.90	2475.9	313.93	2321.4	2635.3	1.0155	7.6824	75
80	.4739	1.0291	3.407	334.86	2482.2	334.91	2308.8	2643.7	1.0753	7.6122	80
85	.5783	1.0325	2.828	355.84	2488.4	355.90	2296.0	2651.9	1.1343	7.5445	85
90	.7014	1.0360	2.361	376.85	2494.5	376.92	2283.2	2660.1	1.1925	7.4791	90
95	.8455	1.0397	1.982	397.88	2500.6	397.96	2270.2	2668.1	1.2500	7.4159	95
100	1.014	1.0435	1.673	418.94	2506.5	419.04	2257.0	2676.1	1.3069	7.3549	100
110	1.433	1.0516	1.210	461.14	2518.1	461.30	2230.2	2691.5	1.4185	7.2387	110
120	1.985	1.0603	0.8919	503.50	2529.3	503.71	2202.6	2706.3	1.5276	7.1296	120
130	2.701	1.0697	0.6685	546.02	2539.9	546.31	2174.2	2720.5	1.6344	7.0269	130
140	3.613	1.0797	0.5089	588.74	2550.0	589.13	2144.7	2733.9	1.7391	6.9299	140
150	4.758	1.0905	0.3928	631.68	2559.5	632.20	2114.3	2746.5	1.8418	6.8379	150
160	6.178	1.1020	0.3071	674.86	2568.4	675.55	2082.6	2758.1	1.9427	6.7502	160
170	7.917	1.1143	0.2428	718.33	2576.5	719.21	2049.5	2768.7	2.0419	6.6663	170
180	10.02	1.1274	0.1941	762.09	2583.7	763.22	2015.0	2778.2	2.1396	6.5857	180
190	12.54	1.1414	0.1565	806.19	2590.0	807.62	1978.8	2786.4	2.2359	6.5079	190
200	15.54	1.1565	0.1274	850.65	2595.3	852.45	1940.7	2793.2	2.3309	6.4323	200
210	19.06	1.1726	0.1044	895.53	2599.5	897.76	1900.7	2798.5	2.4248	6.3585	210
220	23.18	1.1900	0.08619	940.87	2602.4	943.62	1858.5	2802.1	2.5178	6.2861	220
230	27.95	1.2088	0.07158	986.74	2603.9	990.12	1813.8	2804.0	2.6099	6.2146	230
240	33.44	1.2291	0.05976	1033.2	2604.0	1037.3	1766.5	2803.8	2.7015	6.1437	240
250	39.73	1.2512	0.05013	1080.4	2602.4	1085.4	1716.2	2801.5	2.7927	6.0730	250
260	46.88	1.2755	0.04221	1128.4	2599.0	1134.4	1662.5	2796.6	2.8838	6.0019	260
270	54.99	1.3023	0.03564	1177.4	2593.7	1184.5	1605.2	2789.7	2.9751	5.9301	270
280	64.12	1.3321	0.03017	1227.5	2586.1	1236.0	1543.6	2779.6	3.0668	5.8571	280
290	74.36	1.3656	0.02557	1278.9	2576.0	1289.1	1477.1	2766.2	3.1594	5.7821	290
300	85.81	1.4036	0.02167	1332.0	2563.0	1344.0	1404.9	2749.0	3.2534	5.7045	300
320	112.7	1.4988	0.01549	1444.6	2525.5	1461.5	1238.6	2700.1	3.4480	5.5362	320
340	145.9	1.6379	0.01080	1570.3	2464.6	1594.2	1027.9	2622.0	3.6594	5.3357	340
360	186.5	1.8925	0.006945	1725.2	2351.5	1760.5	720.5	2481.0	3.9147	5.0526	360
374.14	220.9	3.155	0.003155	2029.6	2029.6	2099.3	0	2099.3	4.4298	4.4298	374.14

TABLE A.5 (continued)   Properties of Saturated Water and Steam

Part b. Pressure Table

Press. bars	Temp. °C	Specific Volume m³/kg Sat. Liquid $v_f \times 10^3$	Specific Volume m³/kg Sat. Vapor v_g	Internal Energy kJ/kg Sat. Liquid u_f	Internal Energy kJ/kg Sat. Vapor u_g	Enthalpy kJ/kg Sat. Liquid h_f	Enthalpy kJ/kg Evap. h_{fg}	Enthalpy kJ/kg Sat. Vapor h_g	Entropy kJ/kg·K Sat. Liquid s_f	Entropy kJ/kg·K Sat. Vapor s_g	Press. bars
0.04	28.96	1.0040	34.800	121.45	2415.2	121.46	2432.9	2554.4	0.4226	8.4746	0.04
0.06	36.16	1.0064	23.739	151.53	2425.0	151.53	2415.9	2567.4	0.5210	8.3304	0.06
0.08	41.51	1.0084	18.103	173.87	2432.2	173.88	2403.1	2577.0	0.5926	8.2287	0.08
0.10	45.81	1.0102	14.674	191.82	2437.9	191.83	2392.8	2584.7	0.6493	8.1502	0.10
0.20	60.06	1.0172	7.649	251.38	2456.7	251.40	2358.3	2609.7	0.8320	7.9085	0.20
0.30	69.10	1.0223	5.229	289.20	2468.4	289.23	2336.1	2625.3	0.9439	7.7686	0.30
0.40	75.87	1.0265	3.993	317.53	2477.0	317.58	2319.2	2636.8	1.0259	7.6700	0.40
0.50	81.33	1.0300	3.240	340.44	2483.9	340.49	2305.4	2645.9	1.0910	7.5939	0.50
0.60	85.94	1.0331	2.732	359.79	2489.6	359.86	2293.6	2653.5	1.1453	7.5320	0.60
0.70	89.95	1.0360	2.365	376.63	2494.5	376.70	2283.3	2660.0	1.1919	7.4797	0.70
0.80	93.50	1.0380	2.087	391.58	2498.8	391.66	2274.1	2665.8	1.2329	7.4346	0.80
0.90	96.71	1.0410	1.869	405.06	2502.6	405.15	2265.7	2670.9	1.2695	7.3949	0.90
1.00	99.63	1.0432	1.694	417.36	2506.1	417.46	2258.0	2675.5	1.3026	7.3594	1.00
1.50	111.4	1.0528	1.159	466.94	2519.7	467.11	2226.5	2693.6	1.4336	7.2233	1.50
2.00	120.2	1.0605	0.8857	504.49	2529.5	504.70	2201.9	2706.7	1.5301	7.1271	2.00
2.50	127.4	1.0672	0.7187	535.10	2537.2	535.37	2181.5	2716.9	1.6072	7.0527	2.50
3.00	133.6	1.0732	0.6058	561.15	2543.6	561.47	2163.8	2725.3	1.6718	6.9919	3.00
3.50	138.9	1.0786	0.5243	583.95	2546.9	584.33	2148.1	2732.4	1.7275	6.9405	3.50
4.00	143.6	1.0836	0.4625	604.31	2553.6	604.74	2133.8	2738.6	1.7766	6.8959	4.00
4.50	147.9	1.0882	0.4140	622.25	2557.6	623.25	2120.7	2743.9	1.8207	6.8565	4.50
5.00	151.9	1.0926	0.3749	639.68	2561.2	640.23	2108.5	2748.7	1.8607	6.8212	5.00
6.00	158.9	1.1006	0.3157	669.90	2567.4	670.56	2086.3	2756.8	1.9312	6.7600	6.00
7.00	165.0	1.1080	0.2729	696.44	2572.5	697.22	2066.3	2763.5	1.9922	6.7080	7.00
8.00	170.4	1.1148	0.2404	720.22	2576.8	721.11	2048.0	2769.1	2.0462	6.6628	8.00
9.00	175.4	1.1212	0.2150	741.83	2580.5	742.83	2031.1	2773.9	2.0946	6.6226	9.00
10.0	179.9	1.1273	0.1944	761.68	2583.6	762.81	2015.3	2778.1	2.1387	6.5863	10.0
15.0	198.3	1.1539	0.1318	843.16	2594.5	844.84	1947.3	2792.2	2.3150	6.4448	15.0
20.0	212.4	1.1767	0.09963	906.44	2600.3	908.79	1890.7	2799.5	2.4474	6.3409	20.0
25.0	224.0	1.1973	0.07998	959.11	2603.1	962.11	1841.0	2803.1	2.5547	6.2575	25.0
30.0	233.9	1.2165	0.06668	1004.8	2604.1	1008.4	1795.7	2804.2	2.6457	6.1869	30.0
35.0	242.6	1.2347	0.05707	1045.4	2603.7	1049.8	1753.7	2803.4	2.7253	6.1253	35.0
40.0	250.4	1.2522	0.04978	1082.3	2602.3	1087.3	1714.1	2801.4	2.7964	6.0701	40.0
45.0	257.5	1.2692	0.04406	1116.2	2600.1	1121.9	1676.4	2798.3	2.8610	6.0199	45.0
50.0	264.0	1.2859	0.03944	1147.8	2597.1	1154.2	1640.1	2794.3	2.9202	5.9734	50.0
60.0	275.6	1.3187	0.03244	1205.4	2589.7	1213.4	1571.0	2784.3	3.0267	5.8892	60.0
70.0	285.9	1.3513	0.02737	1257.6	2580.5	1267.0	1505.1	2772.1	3.1211	5.8133	70.0
80.0	295.1	1.3842	0.02352	1305.6	2569.8	1316.6	1441.3	2758.0	3.2068	5.7432	80.0
90.0	303.4	1.4178	0.02048	1350.5	2557.8	1363.3	1378.9	2742.1	3.2858	5.6772	90.0
100.	311.1	1.4524	0.01803	1393.0	2544.4	1407.6	1317.1	2724.7	3.3596	5.6141	100.
110.	318.2	1.4886	0.01599	1433.7	2529.8	1450.1	1255.5	2705.6	3.4295	5.5527	110.
120.	324.8	1.5267	0.01426	1473.0	2513.7	1491.3	1193.6	2684.9	3.4962	5.4924	120.
130.	330.9	1.5671	0.01278	1511.1	2496.1	1531.5	1130.7	2662.2	3.5606	5.4323	130.
140.	336.8	1.6107	0.01149	1548.6	2476.8	1571.1	1066.5	2637.6	3.6232	5.3717	140.
150.	342.2	1.6581	0.01034	1585.6	2455.5	1610.5	1000.0	2610.5	3.6848	5.3098	150.
160.	347.4	1.7107	0.009306	1622.7	2431.7	1650.1	930.6	2580.6	3.7461	5.2455	160.
170.	352.4	1.7702	0.008364	1660.2	2405.0	1690.3	856.9	2547.2	3.8079	5.1777	170.
180.	357.1	1.8397	0.007489	1698.9	2374.3	1732.0	777.1	2509.1	3.8715	5.1044	180.
190.	361.5	1.9243	0.006657	1739.9	2338.1	1776.5	688.0	2464.5	3.9388	5.0228	190.
200.	365.8	2.036	0.005834	1785.6	2293.0	1826.3	583.4	2409.7	4.0139	4.9269	200.
220.9	374.1	3.155	0.003155	2029.6	2029.6	2099.3	0	2099.3	4.4298	4.4298	220.9

Source: Adapted from M.J. Moran and H.N. Shapiro, *Fundamentals of Engineering Thermodynamics*, 3rd. ed., Wiley, New York, 1995, as extracted from J.H. Keenan, F.G. Keyes, P.G. Hill, and J.G. Moore, *Steam Tables*, Wiley, New York, 1969.

TABLE A.6 Properties of Superheated Steam

Symbols and Units:

T = temperature, °C

T_{sat} = Saturation temperature, °C

v = Specific volume, m³/kg

u = internal energy, kJ/kg

h = enthalpy, kJ/kg

S = entropy, kJ/kg·K

p = pressure, bar and μPa

T °C	v m³/kg	u kJ/kg	h kJ/kg	s kJ/kg · K	v m³/kg	u kJ/kg	h kJ/kg	s kJ/kg · K
	$p = 0.06$ bar $= 0.006$ MPa ($T_{sat} = 36.16$°C)				$p = 0.35$ bar $= 0.035$ MPa ($T_{sat} = 72.69$°C)			
Sat.	23.739	2425.0	2567.4	8.3304	4.526	2473.0	2631.4	7.7158
80	27.132	2487.3	2650.1	8.5804	4.625	2483.7	2645.6	7.7564
120	30.219	2544.7	2726.0	8.7840	5.163	2542.4	2723.1	7.9644
160	33.302	2602.7	2802.5	8.9693	5.696	2601.2	2800.6	8.1519
200	36.383	2661.4	2879.7	9.1398	6.228	2660.4	2878.4	8.3237
240	39.462	2721.0	2957.8	9.2982	6.758	2720.3	2956.8	8.4828
280	42.540	2781.5	3036.8	9.4464	7.287	2780.9	3036.0	8.6314
320	45.618	2843.0	3116.7	9.5859	7.815	2842.5	3116.1	8.7712
360	48.696	2905.5	3197.7	9.7180	8.344	2905.1	3197.1	8.9034
400	51.774	2969.0	3279.6	9.8435	8.872	2968.6	3279.2	9.0291
440	54.851	3033.5	3362.6	9.9633	9.400	3033.2	3362.2	9.1490
500	59.467	3132.3	3489.1	10.1336	10.192	3132.1	3488.8	9.3194

T °C	v m³/kg	u kJ/kg	h kJ/kg	s kJ/kg · K	v m³/kg	u kJ/kg	h kJ/kg	s kJ/kg · K
	$p = 0.70$ bar $= 0.07$ MPa ($T_{sat} = 89.95$°C)				$p = 1.0$ bar $= 0.10$ MPa ($T_{sat} = 99.63$°C)			
Sat.	2.365	2494.5	2660.0	7.4797	1.694	2506.1	2675.5	7.3594
100	2.434	2509.7	2680.0	7.5341	1.696	2506.7	2676.2	7.3614
120	2.571	2539.7	2719.6	7.6375	1.793	2537.3	2716.6	7.4668
160	2.841	2599.4	2798.2	7.8279	1.984	2597.8	2796.2	7.6597
200	3.108	2659.1	2876.7	8.0012	2.172	2658.1	2875.3	7.8343
240	3.374	2719.3	2955.5	8.1611	2.359	2718.5	2954.5	7.9949
280	3.640	2780.2	3035.0	8.3162	2.546	2779.6	3034.2	8.1445
320	3.905	2842.0	3115.3	8.4504	2.732	2841.5	3114.6	8.2849
360	4.170	2904.6	3196.5	8.5828	2.917	2904.2	3195.9	8.4175
400	4.434	2968.2	3278.6	8.7086	3.103	2967.9	3278.2	8.5435
440	4.698	3032.9	3361.8	8.8286	3.288	3032.6	3361.4	8.6636
500	5.095	3131.8	3488.5	8.9991	3.565	3131.6	3488.1	8.8342

T °C	v m³/kg	u kJ/kg	h kJ/kg	s kJ/kg · K	v m³/kg	u kJ/kg	h kJ/kg	s kJ/kg · K
	$p = 1.5$ bars $= 0.15$ MPa ($T_{sat} = 111.37$°C)				$p = 3.0$ bars $= 0.30$ MPa ($T_{sat} = 133.55$°C)			
Sat.	1.159	2519.7	2693.6	7.2233	0.606	2543.6	2725.3	6.9919
120	1.188	2533.3	2711.4	7.2693				
160	1.317	2595.2	2792.8	7.4665	0.651	2587.1	2782.3	7.1276
200	1.444	2656.2	2872.9	7.6433	0.716	2650.7	2865.5	7.3115
240	1.570	2717.2	2952.7	7.8052	0.781	2713.1	2947.3	7.4774
280	1.695	2778.6	3032.8	7.9555	0.844	2775.4	3028.6	7.6299
320	1.819	2840.6	3113.5	8.0964	0.907	2838.1	3110.1	7.7722
360	1.943	2903.5	3195.0	8.2293	0.969	2901.4	3192.2	7.9061
400	2.067	2967.3	3277.4	8.3555	1.032	2965.6	3275.0	8.0330
440	2.191	3032.1	3360.7	8.4757	1.094	3030.6	3358.7	8.1538
500	2.376	3131.2	3487.6	8.6466	1.187	3130.0	3486.0	8.3251
600	2.685	3301.7	3704.3	8.9101	1.341	3300.8	3703.2	8.5892

TABLE A.6 (continued) Properties of Superheated Steam

Symbols and Units:

T = temperature, °C

T_{sat} = Saturation temperature, °C

v = Specific volume, m³/kg

u = internal energy, kJ/kg

h = enthalpy, kJ/kg

S = entropy, kJ/kg·K

p = pressure, bar and µPa

T °C	v m³/kg	u kJ/kg	h kJ/kg	s kJ/kg · K	v m³/kg	u kJ/kg	h kJ/kg	s kJ/kg . k
	p = 5.0 bars = 0.50 MPa (T_{sat} = 151.86°C)				p = 7.0 bars = 0.70 MPa (T_{sat} = 164.97°C)			
Sat.	0.3749	2561.2	2748.7	6.8213	0.2729	2572.5	2763.5	6.7080
180	0.4045	2609.7	2812.0	6.9656	0.2847	2599.8	2799.1	6.7880
200	0.4249	2642.9	2855.4	7.0592	0.2999	2634.8	2844.8	6.8865
240	0.4646	2707.6	2939.9	7.2307	0.3292	2701.8	2932.2	7.0641
280	0.5034	2771.2	3022.9	7.3865	0.3574	2766.9	3017.1	7.2233
320	0.5416	2834.7	3105.6	7.5308	0.3852	2831.3	3100.9	7.3697
360	0.5796	2898.7	3188.4	7.6660	0.4126	2895.8	3184.7	7.5063
400	0.6173	2963.2	3271.9	7.7938	0.4397	2960.9	3268.7	7.6350
440	0.6548	3028.6	3356.0	7.9152	0.4667	3026.6	3353.3	7.7571
500	0.7109	3128.4	3483.9	8.0873	0.5070	3126.8	3481.7	7.9299
600	0.8041	3299.6	3701.7	8.3522	0.5738	3298.5	3700.2	8.1956
700	0.8969	3477.5	3925.9	8.5952	0.6403	3476.6	3924.8	8.4391

T °C	v m³/kg	u kJ/kg	h kJ/kg	s kJ/kg · K	v m³/kg	u kJ/kg	h kJ/kg	s kJ/kg . k
	p = 10.0 bars = 1.0 MPa (T_{sat} = 179.91°C)				p = 15.0 bars = 1.5 MPa (T_{sat} = 198.32°C)			
Sat.	0.1944	2583.6	2778.1	6.5865	0.1318	2594.5	2792.2	6.4448
200	0.2060	2621.9	2827.9	6.6940	0.1325	2598.1	2796.8	6.4546
240	0.2275	2692.9	2920.4	6.8817	0.1483	2676.9	2899.3	6.6628
280	0.2480	2760.2	3008.2	7.0465	0.1627	2748.6	2992.7	6.8381
320	0.2678	2826.1	3093.9	7.1962	0.1765	2817.1	3081.9	6.9938
360	0.2873	2891.6	3178.9	7.3349	0.1899	2884.4	3169.2	7.1363
400	0.3066	2957.3	3263.9	7.4651	0.2030	2951.3	3255.8	7.2690
440	0.3257	3023.6	3349.3	7.5883	0.2160	3018.5	3342.5	7.3940
500	0.3541	3124.4	3478.5	7.7622	0.2352	3120.3	3473.1	7.5698
540	0.3729	3192.6	3565.6	7.8720	0.2478	3189.1	3560.9	7.6805
600	0.4011	3296.8	3697.9	8.0290	0.2668	3293.9	3694.0	7.8385
640	0.4198	3367.4	3787.2	8.1290	0.2793	3364.8	3783.8	7.9391

T °C	v m³/kg	u kJ/kg	h kJ/kg	s kJ/kg · K	v m³/kg	u kJ/kg	h kJ/kg	s kJ/kg . k
	p = 20.0 bars = 2.0 MPa (T_{sat} = 212.42°C)				p = 30.0 bars = 3.0 MPa (T_{sat} = 233.90°C)			
Sat.	0.0996	2600.3	2799.5	6.3409	0.0667	2604.1	2804.2	6.1869
240	0.1085	2659.6	2876.5	6.4952	0.0682	2619.7	2824.3	6.2265
280	0.1200	2736.4	2976.4	6.6828	0.0771	2709.9	2941.3	6.4462
320	0.1308	2807.9	3069.5	6.8452	0.0850	2788.4	3043.4	6.6245
360	0.1411	2877.0	3159.3	6.9917	0.0923	2861.7	3138.7	6.7801
400	0.1512	2945.2	3247.6	7.1271	0.0994	2932.8	3230.9	6.9212
440	0.1611	3013.4	3335.5	7.2540	0.1062	3002.9	3321.5	7.0520
500	0.1757	3116.2	3467.6	7.4317	0.1162	3108.0	3456.5	7.2338
540	0.1853	3185.6	3556.1	7.5434	0.1227	3178.4	3546.6	7.3474
600	0.1996	3290.9	3690.1	7.7024	0.1324	3285.0	3682.3	7.5085
640	0.2091	3362.2	3780.4	7.8035	0.1388	3357.0	3773.5	7.6106
700	0.2232	3470.9	3917.4	7.9487	0.1484	3466.5	3911.7	7.7571

TABLE A.6 (continued)   Properties of Superheated Steam

Symbols and Units:

T = temperature, °C

T_{sat} = Saturation temperature, °C

v = Specific volume, m³/kg

u = internal energy, kJ/kg

h = enthalpy, kJ/kg

S = entropy, kJ/kg·K

p = pressure, bar and µPa

T °C	v m³/kg	u kJ/kg	h kJ/kg	s kJ/kg·K	v m³/kg	u kJ/kg	h kJ/kg	s kJ/kg·K
	p = 40 bars = 4.0 MPa (T_{sat} = 250.4°C)				p = 60 bars = 6.0 MPa (T_{sat} = 275.64°C)			
Sat.	0.04978	2602.3	2801.4	6.0701	0.03244	2589.7	2784.3	5.8892
280	0.05546	2680.0	2901.8	6.2568	0.03317	2605.2	2804.2	5.9252
320	0.06199	2767.4	3015.4	6.4553	0.03876	2720.0	2952.6	6.1846
360	0.06788	2845.7	3117.2	6.6215	0.04331	2811.2	3071.1	6.3782
400	0.07341	2919.9	3213.6	6.7690	0.04739	2892.9	3177.2	6.5408
440	0.07872	2992.2	3307.1	6.9041	0.05122	2970.0	3277.3	6.6853
500	0.08643	3099.5	3445.3	7.0901	0.05665	3082.2	3422.2	6.8803
540	0.09145	3171.1	3536.9	7.2056	0.06015	3156.1	3517.0	6.9999
600	0.09885	3279.1	3674.4	7.3688	0.06525	3266.9	3658.4	7.1677
640	0.1037	3351.8	3766.6	7.4720	0.06859	3341.0	3752.6	7.2731
700	0.1110	3462.1	3905.9	7.6198	0.07352	3453.1	3894.1	7.4234
740	0.1157	3536.6	3999.6	7.7141	0.07677	3528.3	3989.2	7.5190

T °C	v m³/kg	u kJ/kg	h kJ/kg	s kJ/kg·K	v m³/kg	u kJ/kg	h kJ/kg	s kJ/kg·K
	p = 80 bars = 8.0 MPa (T_{sat} = 295.06°C)				p = 100 bars = 10.0 MPa (T_{sat} = 311.06°C)			
Sat.	0.02352	2569.8	2758.0	5.7432	0.01803	2544.4	2724.7	5.6141
320	0.02682	2662.7	2877.2	5.9489	0.01925	2588.8	2781.3	5.7103
360	0.03089	2772.7	3019.8	6.1819	0.02331	2729.1	2962.1	6.0060
400	0.03432	2863.8	3138.3	6.3634	0.02641	2832.4	3096.5	6.2120
440	0.03742	2946.7	3246.1	6.5190	0.02911	2922.1	3213.2	6.3805
480	0.04034	3025.7	3348.4	6.6586	0.03160	3005.4	3321.4	6.5282
520	0.04313	3102.7	3447.7	6.7871	0.03394	3085.6	3425.1	6.6622
560	0.04582	3178.7	3545.3	6.9072	0.03619	3164.1	3526.0	6.7864
600	0.04845	3254.4	3642.0	7.0206	0.03837	3241.7	3625.3	6.9029
640	0.05102	3330.1	3738.3	7.1283	0.04048	3318.9	3723.7	7.0131
700	0.05481	3443.9	3882.4	7.2812	0.04358	3434.7	3870.5	7.1687
740	0.05729	3520.4	3978.7	7.3782	0.04560	3512.1	3968.1	7.2670

T °C	v m³/kg	u kJ/kg	h kJ/kg	s kJ/kg·K	v m³/kg	u kJ/kg	h kJ/kg	s kJ/kg·K
	p = 120 bars = 12.0 MPa (T_{sat} = 324.75°C)				p = 140 bars = 14.0 MPa (T_{sat} = 336.75°C)			
Sat.	0.01426	2513.7	2684.9	5.4924	0.01149	2476.8	2637.6	5.3717
360	0.01811	2678.4	2895.7	5.8361	0.01422	2617.4	2816.5	5.6602
400	0.02108	2798.3	3051.3	6.0747	0.01722	2760.9	3001.9	5.9448
440	0.02355	2896.1	3178.7	6.2586	0.01954	2868.6	3142.2	6.1474
480	0.02576	2984.4	3293.5	6.4154	0.02157	2962.5	3264.5	6.3143
520	0.02781	3068.0	3401.8	6.5555	0.02343	3049.8	3377.8	6.4610
560	0.02977	3149.0	3506.2	6.6840	0.02517	3133.6	3486.0	6.5941
600	0.03164	3228.7	3608.3	6.8037	0.02683	3215.4	3591.1	6.7172
640	0.03345	3307.5	3709.0	6.9164	0.02843	3296.0	3694.1	6.8326
700	0.03610	3425.2	3858.4	7.0749	0.03075	3415.7	3846.2	6.9939
740	0.03781	3503.7	3957.4	7.1746	0.03225	3495.2	3946.7	7.0952

TABLE A.7 Chemical, Physical, and Thermal Properties of Gases: Gases and Vapors, Including Fuels and Refrigerants, English and Metric Units

	Acetylene (Ethyne)	*Air [mixture]*	*Ammonia, anhyd.*	*Argon*
Common name(s)				
Chemical formula	C_2H_2		NH_3	*Ar*
Refrigerant number	—	729	717	740
CHEMICAL AND PHYSICAL PROPERTIES				
Molecular weight	26.04	28.966	17.02	39.948
Specific gravity, air = 1	0.90	1.00	0.59	1.38
Specific volume, ft^3/lb	14.9	13.5	23.0	9.80
Specific volume, m^3/kg	0.93	0.842	1.43	0.622
Density of liquid (at atm bp), lb/ft^3	43.0	54.6	42.6	87.0
Density of liquid (at atm bp), kg/m^3	693.	879.	686.	1 400.
Vapor pressure at 25 deg C, psia			145.4	
Vapor pressure at 25 deg C, MN/m^2			1.00	
Viscosity (abs), lbm/ft·sec	6.72×10^{-6}	12.1×10^{-6}	6.72×10^{-6}	13.4×10^{-6}
Viscosity (abs), centipoises"	0.01	0.018	0.010	0.02
Sound velocity in gas, m/sec	343	346	415	322
THERMAL AND THERMO-DYNAMIC PROPERTIES				
Specific heat, c_p, Btu/lb·deg F or cal/g·deg C	0.40	0.240 3	0.52	0.125
Specific heat, c_p, J/kg·K	1 674.	1 005.	2 175.	523.
Specific heat ratio, c_p/c_v	1.25	1.40	1.3	1.67
Gas constant R, ft-lb/lb·deg R	59.3	53.3	90.8	38.7
Gas constant R, J/kg·deg C	319	286.8	488.	208.
Thermal conductivity, Btu/hr·ft·deg F	0.014	0.015 1	0.015	0.010 2
Thermal conductivity, W/m·deg C	0.024	0.026	0.026	0.017 2
Boiling point (sat 14.7 psia), deg F	− 103	− 320	− 28.	− 303.
Boiling point (sat 760 mm), deg C	− 75	− 195	− 33.3	− 186
Latent heat of evap (at bp), Btu/lb	264	88.2	589.3	70.
Latent heat of evap (at bp), J/kg	614 000	205 000.	1 373 000	163 000
Freezing (melting) point, deg F (1 atm)	− 116	− 357.2	− 107.9	− 308.5
Freezing (melting) point, deg C (1 atm)	− 82.2	- 216.2	− 77.7	-- 189.2
Latent heat of fusion, Btu/lb	23.	10.0	143.0	
Latent heat of fusion, J/kg	53 500	23 200	332 300	
Critical temperature, deg F	97.1	− 220.5	271.4	− 187.6
Critical temperature, deg C	36.2	− 140.3	132.5	− 122
Critical pressure, psia	907.	550.	1 650.	707.
Critical pressure, MN/m^2	6.25	3.8	11.4	4.87
Critical volume, ft^3/lb		0.050	0.068	0.029 9
Critical volume, m^3/kg		0.003	0.004 24	0.001 86
Flammable (yes or no)	Yes	No	No	No
Heat of combustion, Btu/ft^3	1 450			
Heat of combustion, Btu/lb	21 600			
Heat of combustion, kJ/kg	50 200	—	—	—

"For N·sec/m^2 divide by 1 000.

Note: The properties of pure gases are given at 25°C (77°F, 298 K) and atmospheric pressure (except as stated).

TABLE A.7 (continued)   Chemical, Physical, and Thermal Properties of Gases: Gases and Vapors, Including Fuels and Refrigerants, English and Metric Units

Common name(s)	Butadiene	n-Butane	Isobutane (2-Methyl propane)	1-Butene (Butylene)
Chemical formula	C_4H_6	C_4H_{10}	C_4H_{10}	C_4H_8
Refrigerant number		600	600a	–
CHEMICAL AND PHYSICAL PROPERTIES				
Molecular weight	54.09	58.12	58.12	56.108
Specific gravity, air = 1	1.87	2.07	2.07	1.94
Specific volume, ft³/lb	7.1	6.5	6.5	6.7
Specific volume, m³/kg	0.44	0.405	0.418	0.42
Density of liquid (at atm bp), lb/ft³		37.5	37.2	
Density of liquid (at atm bp), kg/m³		604.	599.	
Vapor pressure at 25 deg C. psia		35.4	50.4	
Vapor pressure at 25 deg C. MN/m²		0.024 4	0.347	
Viscosity (abs), lbm/ft·sec		4.8×10^{-6}		
Viscosity (abs), centipoises[a]		0.007		
Sound velocity in gas, m/sec	226	216	216	222
THERMAL AND THERMO-DYNAMIC PROPERTIES				
Specific heat, c_p, Btu/lb·deg F or cal/g·deg C	0.341	0.39	0.39	0.36
Specific heat, c_p, J/kg·K	1 427.	1 675.	1 630.	1 505.
Specific heat ratio, c_p/c_v	1.12	1.096	1.10	1.112
Gas constant R, ft-lb/lb·deg F	28.55	26.56	26.56	27.52
Gas constant R, J/kg·deg C	154.	143.	143.	148.
Thermal conductivity, Btu/hr·ft·deg F		0.01	0.01	
Thermal conductivity, W/m·deg C		0.017	0.017	
Boiling point (sat 14.7 psia), deg F	24.1	31.2	10.8	20.6
Boiling point (sat 760 mm), deg C	−4.5	−0.4	−11.8	−6.3
Latent heat of evap (at bp), Btu/lb		165.6	157.5	167.9
Latent heat of evap (at bp), J/kg		386 000	366 000	391 000
Freezing (melting) point, deg F (1 atm)	−164.	−217.	−229	−301.6
Freezing (melting) point, deg C (1 atm)	−109.	−138	−145	−185.3
Latent heat of fusion, Btu/lb		19.2		16.4
Latent heat of fusion, J/kg		44 700		38 100
Critical temperature, deg F		306	273.	291.
Critical temperature, deg C	171.	152.	134.	144.
Critical pressure, psia	652.	550.	537.	621.
Critical pressure, MN/m²		3.8	3.7	4.28
Critical volume, ft³/lb		0.070		0.068
Critical volume, m³/kg		0.004 3		0.004 2
Flammable (yes or no)	Yes	Yes	Yes	Yes
Heat of combustion, Btu/ft³	2 950	3 300	3 300	3 150
Heat of combustion, Btu/lb	20 900	21 400	21 400	21 000
Heat of combustion, kJ/kg	48 600	49 700	49 700	48 800

[a]For N·sec/m² divide by 1 000.

TABLE A.7 (continued)   Chemical, Physical, and Thermal Properties of Gases: Gases and Vapors, Including Fuels and Refrigerants, English and Metric Units

	cis-2-Butene C_4H_8	trans-2-Butene C_4H_8	Isobutene C_4H_8	Carbon dioxide CO_2
Common name(s)	cis-2-Butene	trans-2-Butene	Isobutene	Carbon dioxide
Chemical formula	C_4H_8	C_4H_8	C_4H_8	CO_2
Refrigerant number	-	--	—	744
CHEMICAL AND PHYSICAL PROPERTIES				
Molecular weight	56.108	56.108	56.108	44.01
Specific gravity, air = 1	1.94	1.94	1.94	1.52
Specific volume, ft^3/lb	6.7	6.7	6.7	8.8
Specific volume, m^3/kg	0.42	0.42	0.42	0.55
Density of liquid (at atm bp), lb/ft^3				—
Density of liquid (at atm bp), kg/m^3				—
Vapor pressure at 25 deg C, psia				931.
Vapor pressure at 25 deg C, MN/m^2				6.42
Viscosity (abs), lbm/ft·sec				9.4×10^{-6}
Viscosity (abs), centipoises[a]				0.014
Sound velocity in gas, m/sec	223.	221.	221.	270.
THERMAL AND THERMO-DYNAMIC PROPERTIES				
Specific heat, c_p, Btu/lb·deg F or cal/g·deg C	0.327	0.365	0.37	0.205
Specific heat, c_p, J/kg·K	1 368.	1 527.	1 548.	876.
Specific heat ratio, c_p/c_v	1.121	1.107	1.10	1.30
Gas constant R, ft-lb/lb·deg F				35.1
Gas constant R, J/kg·deg C				189.
Thermal conductivity, Btu/hr·ft·deg F				0.01
Thermal conductivity, W/m·deg C				0.017
Boiling point (sat 14.7 psia), deg F	38.6	33.6	19.2	-109.4[b]
Boiling point (sat 760 mm), deg C	3.7	0.9	-7.1	-78.5
Latent heat of evap (at bp), Btu/lb	178.9	174.4	169.	246.
Latent heat of evap (at bp), J/kg	416 000.	406 000.	393 000.	572 000.
Freezing (melting) point, deg F (1 atm)	$-218.$	$-158.$		
Freezing (melting) point, deg C (1 atm)	-138.9	-105.5		
Latent heat of fusion, Btu/lb	31.2	41.6	25.3	—
Latent heat of fusion, J/kg	72 600.	96 800.	58 800.	—
Critical temperature, deg F				88.
Critical temperature, deg C	160.	155.		31.
Critical pressure, psia	595.	610.		1 072.
Critical pressure, MN/m^2	4.10	4.20		7.4
Critical volume, ft^3/lb				
Critical volume, m^3/kg				
Flammable (yes or no)	Yes	Yes	Yes	No
Heat of combustion, Btu/ft^3	3 150.	3 150.	3 150.	—
Heat of combustion, Btu/lb	21 000.	21 000.	21 000.	—
Heat of combustion, kJ/kg	48 800.	48 800.	48 800.	—

[a] For N·sec/m^2 divide by 1 000.
[b] Sublimes.

TABLE A.7 (continued) **Chemical, Physical, and Thermal Properties of Gases: Gases and Vapors, Including Fuels and Refrigerants, English and Metric Units**

	Carbon monoxide	Chlorine	Deuterium	Ethane
Common name(s)				
Chemical formula	CO	Cl_2	D_2	C_2H_6
Refrigerant number	—	—	—	170
CHEMICAL AND PHYSICAL PROPERTIES				
Molecular weight	28.011	70.906	2.014	30.070
Specific gravity, air = 1	0.967	2.45	0.070	1.04
Specific volume, ft³/lb	14.0	5.52	194.5	13.025
Specific volume, m³/kg	0.874	0.344	12.12	0.815
Density of liquid (at atm bp), lb/ft³		97.3		28.
Density of liquid (at atm bp), kg/m³		1 559.		449.
Vapor pressure at 25 deg C, psia			0.756	
Vapor pressure at 25 deg C, MN/m²			0.005 2	
Viscosity (abs), lbm/ft·sec	12.1×10^{-6}	9.4×10^{-6}	8.75×10^{-6}	$64. \times 10^{-6}$
Viscosity (abs), centipoises*	0.018	0.014	0.013	0.095
Sound velocity in gas, m/sec	352.	215.	930.	316.
THERMAL AND THERMO-DYNAMIC PROPERTIES				
Specific heat, c_p, Btu/lb·deg F or cal/g·deg C	0.25	0.114	1.73	0.41
Specific heat, c_p, J/kg·K	1 046.	477.	7 238.	1 715.
Specific heat ratio, c_p/c_v	1.40	1.35	1.40	1.20
Gas constant R, ft-lb/lb·deg F	55.2	21.8	384.	51.4
Gas constant R, J/kg·deg C	297.	117.	2 066.	276.
Thermal conductivity, Btu/hr·ft·deg F	0.014	0.005	0.081	0.010
Thermal conductivity, W/m·deg C	0.024	0.008 7	0.140	0.017
Boiling point (sat 14.7 psia), deg F	−312.7	−29.2		−127.
Boiling point (sat 760 mm), deg C	−191.5	−34.		−88.3
Latent heat of evap (at bp), Btu/lb	92.8	123.7		210.
Latent heat of evap (at bp), J/kg	216 000.	288 000.		488 000.
Freezing (melting) point, deg F (1 atm)	−337.	−150.		−278.
Freezing (melting) point, deg C (1 atm)	−205.	−101.		−172.2
Latent heat of fusion, Btu/lb	12.8	41.0		41.
Latent heat of fusion, J/kg		95 400.		95 300.
Critical temperature, deg F	−220.	291.	−390.6	90.1
Critical temperature, deg C	−140.	144.	−234.8	32.2
Critical pressure, psia	507.	1 120.	241.	709.
Critical pressure, MN/m²	3.49	7.72	1.66	4.89
Critical volume, ft³/lb	0.053	0.028	0.239	0.076
Critical volume, m³/kg	0.003 3	0.001 75	0.014 9	0.004 7
Flammable (yes or no)	Yes	No		Yes
Heat of combustion, Btu/ft³	310.	—		
Heat of combustion, Btu/lb	4 340.	—		22 300.
Heat of combustion, kJ/kg	10 100.	—		51 800.

*For N·sec/m² divide by 1 000.

TABLE A.7 (continued)   Chemical, Physical, and Thermal Properties of Gases: Gases and Vapors, Including Fuels and Refrigerants, English and Metric Units

Common name(s)	Ethyl chloride	Ethylene (Ethene)	Fluorine
Chemical formula	C_2H_5Cl	C_2H_4	F_2
Refrigerant number	160	1 150	
CHEMICAL AND PHYSICAL PROPERTIES			
Molecular weight	64.515	28.054	37.996
Specific gravity, air = 1	2.23	0.969	1.31
Specific volume, ft³/lb	6.07	13.9	10.31
Specific volume, m³/kg	0.378	0.87	0.706
Density of liquid (at atm bp), lb/ft³	56.5	35.5	
Density of liquid (at atm bp), kg/m³	905.	569.	
Vapor pressure at 25 deg C, psia			
Vapor pressure at 25 deg C, MN/m²			
Viscosity (abs), lbm/ft·sec		6.72×10^{-6}	16.1×10^{-6}
Viscosity (abs), centipoises[a]		0.010	0.024
Sound velocity in gas, m/sec	204.	331.	290.
THERMAL AND THERMO- DYNAMIC PROPERTIES			
Specific heat, c_p, Btu/lb·deg F or cal/g·deg C	0.27	0.37	0.198
Specific heat, c_p, J/kg·K	1 130.	1 548.	828.
Specific heat ratio, c_p/c_v	1.13	1.24	1.35
Gas constant R, ft-lb/lb·deg F	24.0	55.1	40.7
Gas constant R, J/kg·deg C	129.	296.	219.
Thermal conductivity, Btu/hr·ft·deg F		0.010	0.016
Thermal conductivity, W/m·deg C		0.017	0.028
Boiling point (sat 14.7 psia), deg F	54.	− 155.	− 306.4
Boiling point (sat 760 mm), deg C	12.2	− 103.8	− 188.
Latent heat of evap (at bp), Btu/lb	166.	208.	74.
Latent heat of evap (at bp), J/kg	386 000.	484 000.	172 000.
Freezing (melting) point, deg F (1 atm)	− 218.	− 272.	− 364.
Freezing (melting) point, deg C (1 atm)	− 138.9	− 169.	− 220.
Latent heat of fusion, Btu/lb	29.3	51.5	11.
Latent heat of fusion, J/kg	68 100.	120 000.	25 600.
Critical temperature, deg F	368.6	49.	− 200
Critical temperature, deg C	187.	9.5	− 129.
Critical pressure, psia	764.	741.	810.
Critical pressure, MN/m²	5.27	5.11	5.58
Critical volume, ft³/lb	0.049	0.073	
Critical volume, m³/kg	0.003 06	0.004 6	
Flammable (yes or no)	No	Yes	
Heat of combustion, Btu/ft³	—	1 480.	
Heat of combustion, Btu/lb	—	20 600.	
Heat of combustion, kJ/kg	—	47 800.	

[a] For N·sec/m² divide by 1 000.

TABLE A.7 (continued)   Chemical, Physical, and Thermal Properties of Gases: Gases and Vapors, Including Fuels and Refrigerants, English and Metric Units

Common name(s)	Fluorocarbons			
Chemical formula	CCl_3F	CCl_2F_2	$CClF_3$	$CBrF_3$
Refrigerant number	11	12	13	13B1
CHEMICAL AND PHYSICAL PROPERTIES				
Molecular weight	137.37	120.91	104.46	148.91
Specific gravity, air = 1	4.74	4.17	3.61	5.14
Specific volume, ft³/lb	2.74	3.12	3.58	2.50
Specific volume, m³/kg	0.171	0.195	0.224	0.975
Density of liquid (at atm bp), lb/ft³	92.1	93.0	95.0	124.4
Density of liquid (at atm bp), kg/m³	1 475.	1 490.	1 522.	1 993.
Vapor pressure at 25 deg C, psia		94.51	516.	234.8
Vapor pressure at 25 deg C, MN/m²		0.652	3.56	1.619
Viscosity (abs), lbm/ft·sec	7.39×10^{-6}	8.74×10^{-6}		
Viscosity (abs), centipoises^a	0.011	0.013		
Sound velocity in gas, m/sec				
THERMAL AND THERMO- DYNAMIC PROPERTIES				
Specific heat, c_p, Btu/lb·deg F or cal/g·deg C	0.14	0.146	0.154	
Specific heat, c_p, J/kg·K	586.	611.	644.	
Specific heat ratio, c_p/c_v	1.14	1.14	1.145	
Gas constant R, ft-lb/lb·deg F				
Gas constant R, J/kg·deg C				
Thermal conductivity, Btu/hr·ft·deg F	0.005	0.006		
Thermal conductivity, W/m·deg C	0.008 7	0.010 4		
Boiling point (sat 14.7 psia), deg F	74.9	−21.8	−114.6	−72.
Boiling point (sat 760 mm), deg C	23.8	−29.9	−81.4	−57.8
Latent heat of evap (at bp), Btu/lb	77.5	71.1	63.0	51.1
Latent heat of evap (at bp), J/kg	180 000.	165 000.	147 000.	119 000.
Freezing (melting) point, deg F (1 atm)	−168.	−252.	−294.	−270.
Freezing (melting) point, deg C (1 atm)	−111.	−157.8	−181.1	−167.8
Latent heat of fusion, Btu/lb				
Latent heat of fusion, J/kg				
Critical temperature, deg F	388.4	233.	83.9	152.
Critical temperature, deg C	198.	111.7	28.8	66.7
Critical pressure, psia	635.	582.	559.	573.
Critical pressure, MN/m²	4.38	4.01	3.85	3.95
Critical volume, ft³/lb	0.028 9	0.287	0.027 7	0.021 5
Critical volume, m³/kg	0.001 80	0.018	0.001 73	0.001 34
Flammable (yes or no)	No	No	No	No
Heat of combustion, Btu/ft³	—	—	··	—
Heat of combustion, Btu/lb	—	—	—	—
Heat of combustion, kJ/kg	—	—	—	—

^{a}For N·sec/m² divide by 1 000.

TABLE A.7 (continued) Chemical, Physical, and Thermal Properties of Gases: Gases and Vapors, Including Fuels and Refrigerants, English and Metric Units

Common name(s)	Fluorocarbons			
Chemical formula	CF_4	$CHCl_2F$	$CHClF_2$	$C_2Cl_2F_4$
Refrigerant number	14	21	22	114
CHEMICAL AND PHYSICAL PROPERTIES				
Molecular weight	88.00	102.92	86.468	170.92
Specific gravity, air = 1	3.04	3.55	2.99	5.90
Specific volume, ft^3/lb	4.34	3.7	4.35	2.6
Specific volume, m^3/kg	0.271	0.231	0.271	0.162
Density of liquid (at atm bp), lb/ft^3	102.0	87.7	88.2	94.8
Density of liquid (at atm bp), kg/m^3	1 634.	1 405.	1 413.	1 519.
Vapor pressure at 25 deg C, psia		26.4	151.4	30.9
Vapor pressure at 25 deg C, MN/m^2		0.182	1.044	0.213
Viscosity (abs), lbm/ft·sec		8.06×10^{-6}	8.74×10^{-6}	8.06×10^{-6}
Viscosity (abs), centipoises[a]		0.012	0.013	0.012
Sound velocity in gas, m/sec				
THERMAL AND THERMO-DYNAMIC PROPERTIES				
Specific heat, c_p, Btu/lb·deg F or cal/g·deg C		0.139	0.157	0.158
Specific heat, c_p, J/kg·K		582.	657.	661.
Specific heat ratio, c_p/c_v		1.18	1.185	1.09
Gas constant R, ft-lb/lb·deg F				
Gas constant R, J/kg·deg C				
Thermal conductivity, Btu/hr·ft·deg F			0.007	0.006
Thermal conductivity, W/m·deg C			0.012	0.010
Boiling point (sat 14.7 psia), deg F	−198.2	48.1	−41.3	38.4
Boiling point (sat 760 mm), deg C	−127.9	9.0	−40.7	3.55
Latent heat of evap (at bp), Btu/lb	58.5	104.1	100.4	58.4
Latent heat of evap (at bp), J/kg	136 000.	242 000.	234 000.	136 000.
Freezing (melting) point, deg F (1 atm)	−299.	−211.	−256.	−137.
Freezing (melting) point, deg C (1 atm)	−183.8	−135.	−160.	−93.8
Latent heat of fusion, Btu/lb	2.53			
Latent heat of fusion, J/kg	5 880.			
Critical temperature, deg F	−49.9	353.3	204.8	294.
Critical temperature, deg C	−45.5	178.5	96.5	
Critical pressure, psia	610.	750.	715.	475.
Critical pressure, MN/m^2	4.21	5.17	4.93	3.28
Critical volume, ft^3/lb	0.025	0.030 7	0.030 5	0.027 5
Critical volume, m^3/kg	0.001 6	0.001 91	0.001 90	0.001 71
Flammable (yes or no)	No	No	No	No
Heat of combustion, Btu/ft^3	—	—	—	—
Heat of combustion, Btu/lb	—	—	—	—
Heat of combustion, kJ/kg	—	—	—	—

[a]For N·sec/m² divide by 1 000.

TABLE A.7 (continued)   Chemical, Physical, and Thermal Properties of Gases: Gases and Vapors, Including Fuels and Refrigerants, English and Metric Units

Common name(s)	Fluorocarbons			Helium
Chemical formula	C_2ClF_5	$C_2H_3ClF_2$	$C_2H_4F_2$	He
Refrigerant number	115	142b	152a	704
CHEMICAL AND PHYSICAL PROPERTIES				
Molecular weight	154.47	100.50	66.05	4.002 6
Specific gravity, air = 1	5.33	3.47	2.28	0.138
Specific volume, ft³/lb	2.44	3.7	5.9	97.86
Specific volume, m³/kg	0.152	0.231	0.368	6.11
Density of liquid (at atm bp), lb/ft³	96.5	74.6	62.8	7.80
Density of liquid (at atm bp), kg/m³	1 546.	1 195.	1 006.	125.
Vapor pressure at 25 deg C, psia	132.1	49.1	86.8	
Vapor pressure at 25 deg C, MN/m²	0.911	0.338 5	0.596	
Viscosity (abs), lbm/ft·sec				13.4×10^{-6}
Viscosity (abs), centipoises[a]				0.02
Sound velocity in gas, m/sec				1 015.
THERMAL AND THERMO-DYNAMIC PROPERTIES				
Specific heat, c_p, Btu/lb·deg F or cal/g·deg C	0.161			1.24
Specific heat, c_p, J/kg·K	674.			5 188.
Specific heat ratio, c_p/c_v	1.091			1.66
Gas constant R, ft-lb/lb·deg F				386.
Gas constant R, J/kg·deg C				2 077.
Thermal conductivity, Btu/hr·ft·deg F				0.086
Thermal conductivity, W/m·deg C				0.149
Boiling point (sat 14.7 psia), deg F	−38.0	14.	−13.	−452.
Boiling point (sat 760 mm), deg C	−38.9	−10.0	−25.0	4.22 K
Latent heat of evap (at bp), Btu/lb	53.4	92.5	137.1	10.0
Latent heat of evap (at bp), J/kg	124 000.	215 000.	319 000.	23 300.
Freezing (melting) point, deg F (1 atm)	−149.			[b]
Freezing (melting) point, deg C (1 atm)	−100.6			−
Latent heat of fusion, Btu/lb				−
Latent heat of fusion, J/kg				−
Critical temperature, deg F	176.		387.	−450.3
Critical temperature, deg C				5.2 K
Critical pressure, psia	457.6			33.22
Critical pressure, MN/m²	3.155			
Critical volume, ft³/lb	0.026 1			0.231
Critical volume, m³/kg	0.001 63			0.014 4
Flammable (yes or no)	No	No	No	No
Heat of combustion, Btu/ft³	−	−	−	−
Heat of combustion, Btu/lb	−	−	−	−
Heat of combustion, kJ/kg	−	−	−	−

[a] For N·sec/m² divide by 1 000.
[b] Helium cannot be solidified at atmospheric pressure.

TABLE A.7 (continued)   Chemical, Physical, and Thermal Properties of Gases: Gases and Vapors, Including Fuels and Refrigerants, English and Metric Units

Common name(s)	*Hydrogen*	*Hydrogen chloride*	*Hydrogen sulfide*	*Krypton*
Chemical formula	H_2	*HCl*	H_2S	*Kr*
Refrigerant number	702	—	—	—
CHEMICAL AND PHYSICAL PROPERTIES				
Molecular weight	2.016	36.461	34.076	83.80
Specific gravity, air = 1	0.070	1.26	1.18	2.89
Specific volume, ft³/lb	194.	10.74	11.5	4.67
Specific volume, m³/kg	12.1	0.670	0.093·0	0.291
Density of liquid (at atm bp), lb/ft³	4.43	74.4	62.	150.6
Density of liquid (at atm bp), kg/m³	71.0	1 192.	993.	2 413.
Vapor pressure at 25 deg C, psia				
Vapor pressure at 25 deg C, MN/m²				
Viscosity (abs), lbm/ft·sec	6.05×10^{-6}	10.1×10^{-6}	8.74×10^{-6}	16.8×10^{-6}
Viscosity (abs), centipoises[a]	0.009	0.015	0.013	0.025
Sound velocity in gas, m/sec	1 315.	310.	302.	223.
THERMAL AND THERMO-DYNAMIC PROPERTIES				
Specific heat, c_p, Btu/lb·deg F or cal/g·deg C	3.42	0.194	0.23	0.059
Specific heat, c_p, J/kg·K	14 310.	812.	962.	247.
Specific heat ratio, c_p/c_v	1.405	1.39	1.33	1.68
Gas constant R, ft-lb/lb·deg F	767.	42.4	45.3	18.4
Gas constant R, J/kg·deg C	4 126.	228.	244.	99.0
Thermal conductivity, Btu/hr·ft·deg F	0.105	0.008	0.008	0.005 4
Thermal conductivity, W/m·deg C	0.018 2	0.014	0.014	0.009 3
Boiling point (sat 14.7 psia), deg F	− 423.	− 121.	− 76.	− 244.
Boiling point (sat 760 mm), deg C	20.4 K	− 85.	− 60.	− 153.
Latent heat of evap (at bp), Btu/lb	192.	190.5	234.	46.4
Latent heat of evap (at bp), J/kg	447 000.	443 000.	544 000.	108 000.
Freezing (melting) point, deg F (1 atm)	− 434.6	− 169.6	− 119.2	− 272.
Freezing (melting) point, deg C (1 atm)	− 259.1	− 112.	− 84.	− 169.
Latent heat of fusion, Btu/lb	25.0	23.4	30.2	4.7
Latent heat of fusion, J/kg	58 000.	54 400.	70 200.	10 900.
Critical temperature, deg F	− 399.8	124.	213.	
Critical temperature, deg C	− 240.0	51.2	100.4	− 63.8
Critical pressure, psia	189.	1 201.	1 309.	800.
Critical pressure, MN/m²	1.30	8.28	9.02	5.52
Critical volume, ft³/lb	0.53	0.038	0.046	0.017 7
Critical volume, m³/kg	0.033	0.002 4	0.002 9	0.001 1
Flammable (yes or no)	Yes	No	Yes	No
Heat of combustion, Btu/ft³	320.	—	700.	—
Heat of combustion, Btu/lb	62 050.	—	8 000.	—
Heat of combustion, kJ/kg	144 000.	—	18 600.	—

[a]For N·sec/m² divide by 1 000.

TABLE A.7 (continued) Chemical, Physical, and Thermal Properties of Gases: Gases and Vapors, Including Fuels and Refrigerants, English and Metric Units

Common name(s)	Methane	Methyl chloride	Neon	Nitric oxide
Chemical formula	CH_4	CH_3Cl	Ne	NO
Refrigerant number	50	40	720	—
CHEMICAL AND PHYSICAL PROPERTIES				
Molecular weight	16.044	50.488	20.179	30.006
Specific gravity, air = 1	0.554	1.74	0.697	1.04
Specific volume, ft³/lb	24.2	7.4	19.41	13.05
Specific volume, m³/kg	1.51	0.462	1.211	0.814
Density of liquid (at atm bp), lb/ft³	26.3	62.7	75.35	
Density of liquid (at atm bp), kg/m³	421.	1 004.	1 207.	
Vapor pressure at 25 deg C, psia		82.2		
Vapor pressure at 25 deg C, MN/m²		0.567		
Viscosity (abs), lbm/ft·sec	7.39×10^{-6}	7.39×10^{-6}	21.5×10^{-6}	12.8×10^{-6}
Viscosity (abs), centipoises*	0.011	0.011	0.032	0.019
Sound velocity in gas, m/sec	446.	251.	454.	341.
THERMAL AND THERMO-DYNAMIC PROPERTIES				
Specific heat, c_p, Btu/lb·deg F or cal/g·deg C	0.54	0.20	0.246	0.235
Specific heat, c_p, J/kg·K	2 260.	837.	1 030.	983.
Specific heat ratio, c_p/c_v	1.31	1.28	1.64	1.40
Gas constant R, ft-lb/lb·deg F	96.	30.6	76.6	51.5
Gas constant R, J/kg·deg C	518.	165.	412.	277.
Thermal conductivity, Btu/hr·ft·deg F	0.02	0.006	0.028	0.015
Thermal conductivity, W/m·deg C	0.035	0.010	0.048	0.026
Boiling point (sat 14.7 psia), deg F	−259.	−10.7	−410.9	−240.
Boiling point (sat 760 mm), deg C	−434.2	−23.7	−246.	−151.5
Latent heat of evap (at bp), Btu/lb	219.2	184.1	37.	
Latent heat of evap (at bp), J/kg	510 000.	428 000.	86 100.	
Freezing (melting) point, deg F (1 atm)	−296.6	−144.	−415.6	−258.
Freezing (melting) point, deg C (1 atm)	−182.6	−97.8	−248.7	−161.
Latent heat of fusion, Btu/lb	14.	56.	6.8	32.9
Latent heat of fusion, J/kg	32 600.	130 000.	15 800.	76 500.
Critical temperature, deg F	−116.	289.4	−379.8	−136.
Critical temperature, deg C	−82.3	143.	−228.8	−93.3
Critical pressure, psia	673.	968.	396.	945.
Critical pressure, MN/m²	4.64	6.67	2.73	6.52
Critical volume, ft³/lb	0.099	0.043	0.033	0.033 2
Critical volume, m³/kg	0.006 2	0.002 7	0.002 0	0.002 07
Flammable (yes or no)	Yes	Yes	No	No
Heat of combustion, Btu/ft³	985.		—	—
Heat of combustion, Btu/lb	2 290.		—	—
Heat of combustion, kJ/kg			—	—

*For N·sec/m² divide by 1 000.

TABLE A.7 (continued)   Chemical, Physical, and Thermal Properties of Gases: Gases and Vapors, Including Fuels and Refrigerants, English and Metric Units

Common name(s)	Nitrogen	Nitrous oxide	Oxygen	Ozone
Chemical formula	N_2	N_2O	O_2	O_3
Refrigerant number	728	744A	732	—
CHEMICAL AND PHYSICAL PROPERTIES				
Molecular weight	28.013 4	44.012	31.998 8	47.998
Specific gravity, air = 1	0.967	1.52	1.105	1.66
Specific volume, ft^3/lb	13.98	8.90	12.24	8.16
Specific volume, m^3/kg	0.872	0.555	0.764	0.509
Density of liquid (at atm bp), lb/ft^3	50.46	76.6	71.27	
Density of liquid (at atm bp), kg/m^3	808.4	1 227.	1 142.	
Vapor pressure at 25 deg C, psia				
Vapor pressure at 25 deg C, MN/m^2				
Viscosity (abs), lbm/ft·sec	12.1×10^{-6}	10.1×10^{-6}	13.4×10^{-6}	8.74×10^{-6}
Viscosity (abs), centipoises[a]	0.018	0.015	0.020	0.013
Sound velocity in gas, m/sec	353.	268.	329.	
THERMAL AND THERMO-DYNAMIC PROPERTIES				
Specific heat, c_p, Btu/lb·deg F or cal/g·deg C	0.249	0.21	0.220	0.196
Specific heat, c_p, J/kg·K	1 040.	879.	920.	820.
Specific heat ratio, c_p/c_v	1.40	1 31	1.40	
Gas constant R, ft-lb/lb·deg F	55.2	35.1	48.3	32.2
Gas constant R, J/kg·deg C	297.	189.	260.	173.
Thermal conductivity, Btu/hr·ft·deg F	0.015	0.010	0.015	0.019
Thermal conductivity, W/m·deg C	0.026	0.017	0.026	0.033
Boiling point (sat 14.7 psia), deg F	− 320.4	− 127.3	− 297.3	− 170.
Boiling point (sat 760 mm), deg C	− 195.8	− 88.5	− 182.97	− 112.
Latent heat of evap (at bp), Btu/lb	85.5	161.8	91.7	
Latent heat of evap (at bp), J/kg	199 000.	376 000.	213 000.	
Freezing (melting) point, deg F (1 atm)	− 346.	− 131.5	− 361.1	− 315.5
Freezing (melting) point, deg C (1 atm)	− 210.	− 90.8	− 218.4	− 193.
Latent heat of fusion, Btu/lb	11.1	63.9	5.9	97.2
Latent heat of fusion, J/kg	25 800.	149 000.	13 700.	226 000.
Critical temperature, deg F	− 232.6	97.7	− 181.5	16.
Critical temperature, deg C	− 147.	36.5	− 118.6	− 9.
Critical pressure, psia	493.	1 052.	726.	800.
Critical pressure, MN/m^2	3.40	7.25	5.01	5.52
Critical volume, ft^3/lb	0.051	0.036	0.040	0.029 8
Critical volume, m^3/kg	0.003 18	0.002 2	0.002 5	0.001 86
Flammable (yes or no)	No	No	No	No
Heat of combustion, Btu/ft^3	---	—	—	—
Heat of combustion, Btu/lb	—	—	—	—
Heat of combustion, kJ/kg	—	—	--	—

[a] For N·sec/m² divide by 1 000.

TABLE A.7 (continued)   Chemical, Physical, and Thermal Properties of Gases: Gases and Vapors, Including Fuels and Refrigerants, English and Metric Units

Common name(s)	Propane	Propylene (Propene)	Sulfur dioxide	Xenon
Chemical formula	C_3H_8	C_3H_6	SO_2	Xe
Refrigerant number	290	1 270	764	—
CHEMICAL AND PHYSICAL PROPERTIES				
Molecular weight	44.097	42.08	64.06	131.30
Specific gravity, air = 1	1.52	1.45	2.21	4.53
Specific volume, ft³/lb	8.84	9.3	6.11	2.98
Specific volume, m³/kg	0.552	0.58		
Density of liquid (at atm bp), lb/ft³	36.2	37.5	42.8	190.8
Density of liquid (at atm bp), kg/m³	580.	601.	585.	3 060.
Vapor pressure at 25 deg C, psia	135.7	166.4	56.6	
Vapor pressure at 25 deg C, MN/m²	0.936	1.147	0.390	
Viscosity (abs), lbm/ft·sec	53.8×10^{-6}	57.1×10^{-6}	8.74×10^{-6}	15.5×10^{-6}
Viscosity (abs), centipoises[a]	0.080	0.085	0.013	0.023
Sound velocity in gas, m/sec	253.	261.	220.	177.
THERMAL AND THERMO-DYNAMIC PROPERTIES				
Specific heat, c_p, Btu/lb·deg F or cal/g·deg C	0.39	0.36	0.11	0.115
Specific heat, c_p, J/kg·K	1 630.	1 506.	460.	481.
Specific heat ratio, c_p/c_v	1.2	1.16	1.29	1.67
Gas constant R, ft-lb/lb·deg F	35.0	36.7	24.1	11.8
Gas constant R, J/kg·deg C	188.	197.	130.	63.5
Thermal conductivity, Btu/hr·ft·deg F	0.010	0.010	0.006	0.003
Thermal conductivity, W/m·deg C	0.017	0.017	0.010	0.005 2
Boiling point (sat 14.7 psia), deg F	− 44.	− 54.	14.0	− 162.5
Boiling point (sat 760 mm), deg C	− 42.2	− 48.3	− 10.	− 108.
Latent heat of evap (at bp), Btu/lb	184.	188.2	155.5	41.4
Latent heat of evap (at bp), J/kg	428 000.	438 000.	362 000.	96 000.
Freezing (melting) point, deg F (1 atm)	− 309.8	− 301.	− 104.	− 220.
Freezing (melting) point, deg C (1 atm)	− 189.9	− 185.	− 75.5	− 140.
Latent heat of fusion, Btu/lb	19.1		58.0	10.
Latent heat of fusion, J/kg	44 400.		135 000.	23 300.
Critical temperature, deg F	205.	197.	315.5	61.9
Critical temperature, deg C	96.	91.7	157.6	16.6
Critical pressure, psia	618.	668.	1 141.	852.
Critical pressure, MN/m²	4.26	4.61	7.87	5.87
Critical volume, ft³/lb	0.073	0.069	0.03	0.014 5
Critical volume, m³/kg	0.004 5	0.004 3	0.001 9	0.000 90
Flammable (yes or no)	Yes	Yes	No	No
Heat of combustion, Btu/ft³	2 450.	2 310.	—	—
Heat of combustion, Btu/lb	21 660.	21 500.	—	—
Heat of combustion, kJ/kg	50 340.	50 000.	—	—

[a]For N·sec/m² divide by 1 000.

TABLE A.8 Ideal Gas Properties of Air

Part a. SI Units

$T(\text{K})$, h and $u(\text{kJ/kg})$, $s^\circ(\text{kJ/kg} \cdot \text{K})$

T	h	p_r	u	v_r	s°	T	h	p_r	u	v_r	s°
200	199.97	0.3363	142.56	1707.	1.29559	450	451.80	5.775	322.62	223.6	2.11161
210	209.97	0.3987	149.69	1512.	1.34444	460	462.02	6.245	329.97	211.4	2.13407
220	219.97	0.4690	156.82	1346.	1.39105	470	472.24	6.742	337.32	200.1	2.15604
230	230.02	0.5477	164.00	1205.	1.43557	480	482.49	7.268	344.70	189.5	2.17760
240	240.02	0.6355	171.13	1084.	1.47824	490	492.74	7.824	352.08	179.7	2.19876
250	250.05	0.7329	178.28	979.	1.51917	500	503.02	8.411	359.49	170.6	2.21952
260	260.09	0.8405	185.45	887.8	1.55848	510	513.32	9.031	366.92	162.1	2.23993
270	270.11	0.9590	192.60	808.0	1.59634	520	523.63	9.684	374.36	154.1	2.25997
280	280.13	1.0889	199.75	738.0	1.63279	530	533.98	10.37	381.84	146.7	2.27967
285	285.14	1.1584	203.33	706.1	1.65055	540	544.35	11.10	389.34	139.7	2.29906
290	290.16	1.2311	206.91	676.1	1.66802	550	554.74	11.86	396.86	133.1	2.31809
295	295.17	1.3068	210.49	647.9	1.68515	560	565.17	12.66	404.42	127.0	2.33685
300	300.19	1.3860	214.07	621.2	1.70203	570	575.59	13.50	411.97	121.2	2.35531
305	305.22	1.4686	217.67	596.0	1.71865	580	586.04	14.38	419.55	115.7	2.37348
310	310.24	1.5546	221.25	572.3	1.73498	590	596.52	15.31	427.15	110.6	2.39140
315	315.27	1.6442	224.85	549.8	1.75106	600	607.02	16.28	434.78	105.8	2.40902
320	320.29	1.7375	228.42	528.6	1.76690	610	617.53	17.30	442.42	101.2	2.42644
325	325.31	1.8345	232.02	508.4	1.78249	620	628.07	18.36	450.09	96.92	2.44356
330	330.34	1.9352	235.61	489.4	1.79783	630	638.63	19.84	457.78	92.84	2.46048
340	340.42	2.149	242.82	454.1	1.82790	640	649.22	20.64	465.50	88.99	2.47716
350	350.49	2.379	250.02	422.2	1.85708	650	659.84	21.86	473.25	85.34	2.49364
360	360.58	2.626	257.24	393.4	1.88543	660	670.47	23.13	481.01	81.89	2.50985
370	370.67	2.892	264.46	367.2	1.91313	670	681.14	24.46	488.81	78.61	2.52589
380	380.77	3.176	271.69	343.4	1.94001	680	691.82	25.85	496.62	75.50	2.54175
390	390.88	3.481	278.93	321.5	1.96633	690	702.52	27.29	504.45	72.56	2.55731
400	400.98	3.806	286.16	301.6	1.99194	700	713.27	28.80	512.33	69.76	2.57277
410	411.12	4.153	293.43	283.3	2.01699	710	724.04	30.38	520.23	67.07	2.58810
420	421.26	4.522	300.69	266.6	2.04142	720	734.82	32.02	528.14	64.53	2.60319
430	431.43	4.915	307.99	251.1	2.06533	730	745.62	33.72	536.07	62.13	2.61803
440	441.61	5.332	315.30	236.8	2.08870	740	756.44	35.50	544.02	59.82	2.63280

TABLE A.8 (continued) **Ideal Gas Properties of Air**

$T(K)$, h and $u(kJ/kg)$, $s^o(kJ/kg \cdot K)$

T	h	p_r	u	v_r	s^o	T	h	p_r	u	v_r	s^o
750	767.29	37.35	551.99	57.63	2.64737	1300	1395.97	330.9	1022.82	11.275	3.27345
760	778.18	39.27	560.01	55.54	2.66176	1320	1419.76	352.5	1040.88	10.747	3.29160
770	789.11	41.31	568.07	53.39	2.67595	1340	1443.60	375.3	1058.94	10.247	3.30959
780	800.03	43.35	576.12	51.64	2.69013	1360	1467.49	399.1	1077.10	9.780	3.32724
790	810.99	45.55	584.21	49.86	2.70400	1380	1491.44	424.2	1095.26	9.337	3.34474
800	821.95	47.75	592.30	48.08	2.71787	1400	1515.42	450.5	1113.52	8.919	3.36200
820	843.98	52.59	608.59	44.84	2.74504	1420	1539.44	478.0	1131.77	8.526	3.37901
840	866.08	57.60	624.95	41.85	2.77170	1440	1563.51	506.9	1150.13	8.153	3.39586
860	888.27	63.09	641.40	39.12	2.79783	1460	1587.63	537.1	1168.49	7.801	3.41247
880	910.56	68.98	657.95	36.61	2.82344	1480	1611.79	568.8	1186.95	7.468	3.42892
900	932.93	75.29	674.58	34.31	2.84856	1500	1635.97	601.9	1205.41	7.152	3.44516
920	955.38	82.05	691.28	32.18	2.87324	1520	1660.23	636.5	1223.87	6.854	3.46120
940	977.92	89.28	708.08	30.22	2.89748	1540	1684.51	672.8	1242.43	6.569	3.47712
960	1000.55	97.00	725.02	28.40	2.92128	1560	1708.82	710.5	1260.99	6.301	3.49276
980	1023.25	105.2	741.98	26.73	2.94468	1580	1733.17	750.0	1279.65	6.046	3.50829
1000	1046.04	114.0	758.94	25.17	2.96770	1600	1757.57	791.2	1298.30	5.804	3.52364
1020	1068.89	123.4	776.10	23.72	2.99034	1620	1782.00	834.1	1316.96	5.574	3.53879
1040	1091.85	133.3	793.36	22.39	3.01260	1640	1806.46	878.9	1335.72	5.355	3.55381
1060	1114.86	143.9	810.62	21.14	3.03449	1660	1830.96	925.6	1354.48	5.147	3.56867
1080	1137.89	155.2	827.88	19.98	3.05608	1680	1855.50	974.2	1373.24	4.949	3.58335
1100	1161.07	167.1	845.33	18.896	3.07732	1700	1880.1	1025	1392.7	4.761	3.5979
1120	1184.28	179.7	862.79	17.886	3.09825	1750	1941.6	1161	1439.8	4.328	3.6336
1140	1207.57	193.1	880.35	16.946	3.11883	1800	2003.3	1310	1487.2	3.944	3.6684
1160	1230.92	207.2	897.91	16.064	3.13916	1850	2065.3	1475	1534.9	3.601	3.7023
1180	1254.34	222.2	915.57	15.241	3.15916	1900	2127.4	1655	1582.6	3.295	3.7354
1200	1277.79	238.0	933.33	14.470	3.17888	1950	2189.7	1852	1630.6	3:022	3.7677
1220	1301.31	254.7	951.09	13.747	3.19834	2000	2252.1	2068	1678.7	2.776	3.7994
1240	1324.93	272.3	968.95	13.069	3.21751	2050	2314.6	2303	1726.8	2.555	3.8303
1260	1348.55	290.8	986.90	12.435	3.23638	2100	2377.4	2559	1775.3	2.356	3.8605
1280	1372.24	310.4	1004.76	11.835	3.25510	2150	2440.3	2837	1823.8	2.175	3.8901
						2200	2503.2	3138	1872.4	2.012	3.9191
						2250	2566.4	3464	1921.3	1.864	3.9474

TABLE A.8 (continued) Ideal Gas Properties of Air

Part b. English Units

$T(°R)$, h and u (Btu/lb), $s°$ (Btu/lb · °R)

T	h	p_r	u	v_r	$s°$	T	h	p_r	u	v_r	$s°$
360	85.97	0.3363	61.29	396.6	0.50369	940	226.11	9.834	161.68	35.41	0.73509
380	90.75	0.4061	64.70	346.6	0.51663	960	231.06	10.61	165.26	33.52	0.74030
400	95.53	0.4858	68.11	305.0	0.52890	980	236.02	11.43	168.83	31.76	0.74540
420	100.32	0.5760	71.52	270.1	0.54058	1000	240.98	12.30	172.43	30.12	0.75042
440	105.11	0.6776	74.93	240.6	0.55172	1040	250.95	14.18	179.66	27.17	0.76019
460	109.90	0.7913	78.36	215.33	0.56235	1080	260.97	16.28	186.93	24.58	0.76964
480	114.69	0.9182	81.77	193.65	0.57255	1120	271.03	18.60	194.25	22.30	0.77880
500	119.48	1.0590	85.20	174.90	0.58233	1160	281.14	21.18	201.63	20.29	0.78767
520	124.27	1.2147	88.62	158.58	0.59172	1200	291.30	24.01	209.05	18.51	0.79628
537	128.34	1.3593	91.53	146.34	0.59945	1240	301.52	27.13	216.53	16.93	0.80466
540	129.06	1.3860	92.04	144.32	0.60078	1280	311.79	30.55	224.05	15.52	0.81280
560	133.86	1.5742	95.47	131.78	0.60950	1320	322.11	34.31	231.63	14.25	0.82075
580	138.66	1.7800	98.90	120.70	0.61793	1360	332.48	38.41	239.25	13.12	0.82848
600	143.47	2.005	102.34	110.88	0.62607	1400	342.90	42.88	246.93	12.10	0.83604
620	148.28	2.249	105.78	102.12	0.63395	1440	353.37	47.75	254.66	11.17	0.84341
640	153.09	2.514	109.21	94.30	0.64159	1480	363.89	53.04	262.44	10.34	0.85062
660	157.92	2.801	112.67	87.27	0.64902	1520	374.47	58.78	270.26	9.578	0.85767
680	162.73	3.111	116.12	80.96	0.65621	1560	385.08	65.00	278.13	8.890	0.86456
700	167.56	3.446	119.58	75.25	0.66321	1600	395.74	71.73	286.06	8.263	0.87130
720	172.39	3.806	123.04	70.07	0.67002	1650	409.13	80.89	296.03	7.556	0.87954
740	177.23	4.193	126.51	65.38	0.67665	1700	422.59	90.95	306.06	6.924	0.88758
760	182.08	4.607	129.99	61.10	0.68312	1750	436.12	101.98	316.16	6.357	0.89542
780	186.94	5.051	133.47	57.20	0.68942	1800	449.71	114.0	326.32	5.847	0.90308
800	191.81	5.526	136.97	53.63	0.69558	1850	463.37	127.2	336.55	5.388	0.91056
820	196.69	6.033	140.47	50.35	0.70160	1900	477.09	141.5	346.85	4.974	0.91788
840	201.56	6.573	143.98	47.34	0.70747	1950	490.88	157.1	357.20	4.598	0.92504
860	206.46	7.149	147.50	44.57	0.71323	2000	504.71	174.0	367.61	4.258	0.93205
880	211.35	7.761	151.02	42.01	0.71886	2050	518.61	192.3	378.08	3.949	0.93891
900	216.26	8.411	154.57	39.64	0.72438	2100	532.55	212.1	388.60	3.667	0.94564
920	221.18	9.102	158.12	37.44	0.72979	2150	546.54	233.5	399.17	3.410	0.95222

TABLE A.8 (continued) Ideal Gas Properties of Air

$T(^\circ R)$, h and u (Btu/lb), s° (Btu/lb · $^\circ R$)

T	h	p_r	u	v_r	s°	T	h	p_r	u	v_r	s°
2200	560.59	256.6	409.78	3.176	0.95868	3700	998.11	2330	744.48	.5882	1.10991
2250	574.69	281.4	420.46	2.961	0.96501	3750	1013.1	2471	756.04	.5621	1.11393
2300	588.82	308.1	431.16	2.765	0.97123	3800	1028.1	2618	767.60	.5376	1.11791
2350	603.00	336.8	441.91	2.585	0.97732	3850	1043.1	2773	779.19	.5143	1.12183
2400	617.22	367.6	452.70	2.419	0.98331	3900	1058.1	2934	790.80	.4923	1.12571
2450	631.48	400.5	463.54	2.266	0.98919	3950	1073.2	3103	802.43	.4715	1.12955
2500	645.78	435.7	474.40	2.125	0.99497	4000	1088.3	3280	814.06	.4518	1.13334
2550	660.12	473.3	485.31	1.996	1.00064	4050	1103.4	3464	825.72	.4331	1.13709
2600	674.49	513.5	496.26	1.876	1.00623	4100	1118.5	3656	837.40	.4154	1.14079
2650	688.90	556.3	507.25	1.765	1.01172	4150	1133.6	3858	849.09	.3985	1.14446
2700	703.35	601.9	518.26	1.662	1.01712	4200	1148.7	4067	860.81	.3826	1.14809
2750	717.83	650.4	529.31	1.566	1.02244	4300	1179.0	4513	884.28	.3529	1.15522
2800	732.33	702.0	540.40	1.478	1.02767	4400	1209.4	4997	907.81	.3262	1.16221
2850	746.88	756.7	551.52	1.395	1.03282	4500	1239.9	5521	931.39	.3019	1.16905
2900	761.45	814.8	562.66	1.318	1.03788	4600	1270.4	6089	955.04	.2799	1.17575
2950	776.05	876.4	573.84	1.247	1.04288	4700	1300.9	6701	978.73	.2598	1.18232
3000	790.68	941.4	585.04	1.180	1.04779	4800	1331.5	7362	1002.5	.2415	1.18876
3050	805.34	1011	596.28	1.118	1.05264	4900	1362.2	8073	1026.3	.2248	1.19508
3100	820.03	1083	607.53	1.060	1.05741	5000	1392.9	8837	1050.1	.2096	1.20129
3150	834.75	1161	618.82	1.006	1.06212	5100	1423.6	9658	1074.0	.1956	1.20738
3200	849.48	1242	630.12	.9546	1.06676	5200	1454.4	10539	1098.0	.1828	1.21336
3250	864.24	1328	641.46	.9069	1.07134	5300	1485.3	11481	1122.0	.1710	1.21923
3300	879.02	1418	652.81	.8621	1.07585						
3350	893.83	1513	664.20	.8202	1.08031						
3400	908.66	1613	675.60	.7807	1.08470						
3450	923.52	1719	687.04	.7436	1.08904						
3500	938.40	1829	698.48	.7087	1.09332						
3550	953.30	1946	709.95	.6759	1.09755						
3600	968.21	2068	721.44	.6449	1.10172						
3650	983.15	2196	732.95	.6157	1.10584						

Source: Adapted from M.J. Moran and H.N. Shapiro, *Fundamentals of Engineering Thermodynamics*, 3rd. ed., Wiley, New York, 1995, as based on J.H. Keenan and J. Kaye, *Gas Tables*, Wiley, New York, 1945.

TABLE A.9 Equations for Gas Properties

Gas	Molar Mass M kg/kmol	Gas Constant R kJ/kg·K	c_p kJ/kg·K	c_v kJ/kg·K	k	Temperature Range	a	$b \times 10^3$ K⁻¹	$c \times 10^6$ K⁻²	$d \times 10^{10}$ K⁻³	$e \times 10^{13}$ K⁻⁴	p_c MPa	T_c K	a kPa·m⁶·K^0.5 kmol²	b m³/kmol
Acetylene, C_2H_2	26.04	0.319	1.69	1.37	1.232	300–1000K	0.8021	23.51	-35.95	286.1	-87.64	6.14	308	8030	0.0362
						1000–3000K	3.825	6.767	-3.014	6.931	-0.6469				
Air	28.97	0.287	1.01	0.718	1.400	300–1000K	3.721	-1.874	4.719	-34.45	8.531	3.77	132	1580	0.0253
						1000–3000K	2.786	1.925	-0.9465	2.321	-0.2229				
Argon, Ar	39.95	0.208	0.520	0.312	1.667	1000–3000K	2.50	0	0	0	0	4.90	151	1680	0.0222
Butane, C_4H_{10}	58.12	0.143	1.67	1.53	1.094	300–1500K	0.4756	44.65	-22.04	42.07	0	3.80	425	29000	0.0806
Carbon Dioxide CO_2	44.01	0.189	0.844	0.655	1.289	300–1000K	2.227	9.992	-9.802	53.97	-12.81	7.38	304	6450	0.0297
						1000–3000K	3.247	5.847	-3.412	9.469	-1.009				
Carbon Monoxide CO	28.01	0.297	1.04	0.744	1.399	300–1000K	3.776	-2.093	4.880	-32.71	6.984	3.50	133	1720	0.0274
						1000–3000K	2.654	2.226	-1.146	2.851	-0.2762				
Ethane, C_2H_6	30.07	0.276	1.75	1.48	1.187	300–1500K	0.8293	20.75	-7.704	8.756	0	4.88	306	9860	0.0450
Ethylene, C_2H_4	28.05	0.296	1.53	1.23	1.240	300–1000K	1.575	10.19	11.25	-199.1	81.98	5.03	282	7860	0.0404
						1000–3000K	0.2530	18.67	-9.978	26.03	-2.668				
Helium, He	4.003	2.08	5.19	3.12	1.667	1000–3000K	2.50	0	0	0	0	0.228	5.20	8.00	0.0165
Hydrogen, H_2	2.016	4.12	14.3	10.2	1.405	300–1000K	2.892	3.884	-8.850	86.94	-29.88	1.31	33.2	143	0.0182
						1000–3000K	3.717	-0.9220	1.221	-4.328	0.5202				
Hydrogen, H	1.008	8.25	20.6	12.4	1.667	300–1000K	2.496	0.02977	-0.07655	0.8238	-0.3158				
						1000–3000K	2.567	-0.1509	0.1219	-0.4184	0.05182				
Hydroxyl, OH	17.01	0.489	1.76	1.27	1.384	300–1000K	3.874	-1.349	1.670	-5.670	0.6189				
						1000–3000K	3.229	0.2014	0.4357	-2.043	0.2696				
Methane, CH_4	16.04	0.518	2.22	1.70	1.304	300–1000K	4.503	-8.965	37.38	-364.9	122.2	4.60	191	3210	0.0298
						1000–3000K	-0.6992	15.31	-7.695	18.96	-1.849				
Neon, Ne	20.18	0.412	1.03	0.618	1.667	1000–3000K	2.50	0	0	0	0	2.65	44.4	146	0.0120
Nitric Oxide, NO	30.01	0.277	0.995	0.718	1.386	300–1000K	4.120	-4.225	10.77	-97.64	31.85	6.48	180	1980	0.0200
						1000–3000K	2.730	2.372	-1.338	3.604	-0.3743				
Nitrogen, N_2	28.01	0.297	1.04	0.743	1.400	300–1000K	3.725	-1.562	3.208	-15.54	1.154	3.39	126	1550	0.0267
						1000–3000K	2.469	2.467	-1.312	3.401	-0.3454				
Nitrogen, N	14.01	0.594	1.48	0.890	1.667	300–1000K	2.496	0.02977	-0.07655	0.8238	-0.3158				
						1000–3000K	2.483	0.03033	-0.01517	0.001879	0.009657				
Oxygen, O_2	32.00	0.260	0.919	0.659	1.395	300–1000K	3.837	-3.420	10.99	-109.6	37.47	5.04	155	1740	0.0221
						1000–3000K	3.156	1.809	-1.052	3.190	-0.3629				
Oxygen, O	16.00	0.520	1.37	0.850	1.612	300–1000K	3.020	-2.176	3.793	-30.62	9.402				
						1000–3000K	2.662	-0.3051	0.2250	-0.7447	0.09383				
Propane, C_3H_8	44.10	0.189	1.67	1.48	1.127	300–1500K	-0.4861	36.63	-18.91	38.14	0	4.26	370	18300	0.0626
Water, H_2O	18.02	0.462	1.86	1.40	1.329	300–1000K	4.132	-1.559	5.315	-42.09	12.84	22.1	647	14300	0.0211
						1000–3000K	2.798	2.693	-0.5392	-0.01783	0.09027				

Source: Adapted from J.B. Jones and R.E. Dugan, *Engineering Thermodynamics*, Prentice-Hall, Englewood Cliffs, NJ 1996 from various sources: *JANAF Thermochemical Tables*, 3rd ed., published by the American Chemical Society and the American Institute of Physics for the National Bureau of Standards, 1986. Data for butane, ethane, and propane from K.A. Kobe and E.G. Long, "Thermochemistry for the Petrochemical Industry, Part II — Paraffinic Hydrocarbons, C_1–C_{16}," *Petroleum Refiner,* Vol. 28, No. 2, 1949, pp. 113–116.

Appendix B. Properties of Liquids

TABLE B.1 Properties of Liquid Water*

Symbols and Units:

ρ = density, lbm/ft³. For g/cm³ multiply by 0.016018. For kg/m³ multiply by 16.018.

c_p = specific heat, Btu/lbm·deg R = cal/g·K. For J/kg·K multiply by 4186.8

μ = viscosity. For lbf·sec/ft² = slugs/sec·ft, multiply by 10^{-7}. For lbm·sec·ft multiply by 10^{-7} and by 32.174. For g/sec·cm (poises) multiply by 10^{-7} and by 478.80. For N·sec/m² multiply by 10^{-7} and by 478.880.

k = thermal conductivity, Btu/hr·ft·deg R. For W/m·K multiply by 1.7307.

Temp, °F	At 1 atm or 14.7 psia				At 1,000 psia				At 10,000 psia			
	ρ	c_p	μ	k	ρ	c_p	μ	k	ρ	c_p	μ	$k\dagger$
32	62.42	1.007	366	0.3286	62.62	0.999	365	0.3319	64.5	0.937	357	0.3508
40	62.42	1.004	323	0.334	62.62	0.997	323	0.337	64.5	0.945	315	0.356
50	62.42	1.002	272	0.3392	62.62	0.995	272	0.3425	64.5	0.951	267	0.3610
60	62.38	1.000	235	0.345	62.58	0.994	235	0.348	64.1	0.956	233	0.366
70	62.31	0.999	204	0.350	62.50	0.994	204	0.353	64.1	0.960	203	0.371
80	62.23	0.998	177	0.354	62.42	0.994	177	0.358	64.1	0.962	176	0.376
90	62.11	0.998	160	0.359	62.31	0.994	160	0.362	63.7	0.964	159	0.380
100	62.00	0.998	142	0.3633	62.19	0.994	142	0.3666	63.7	0.965	142	0.3841
110	61.88	0.999	126	0.367	62.03	0.994	126	0.371	63.7	0.966	126	0.388
120	61.73	0.999	114	0.371	61.88	0.995	114	0.374	63.3	0.967	114	0.391
130	61.54	0.999	105	0.374	61.73	0.995	105	0.378	63.3	0.968	105	0.395
140	61.39	0.999	96	0.378	61.58	0.996	96	0.381	63.3	0.969	98	0.398
150	61.20	1.000	89	0.3806	61.39	0.996	89	0.3837	63.0	0.970	91	0.4003
160	61.01	1.001	83	0.383	61.20	0.997	83	0.386	62.9	0.971	85	0.403
170	60.79	1.002	77	0.386	60.98	0.998	77	0.389	62.5	0.972	79	0.405
180	60.57	1.003	72	0.388	60.75	0.999	72	0.391	62.5	0.973	74	0.407
190	60.35	1.004	68	0.390	60.53	1.001	68	0.393	62.1	0.974	70	0.409
200	60.10	1.005	62.5	0.3916	60.31	1.002	62.9	0.3944	62.1	0.975	65.4	0.4106
250		boiling point 212°F			59.03	1.001	47.8	0.3994	60.6	0.981	50.6	0.4158
300					57.54	1.024	38.4	0.3993	59.5	0.988	41.3	0.4164
350					55.83	1.044	32.1	0.3944	58.1	0.999	35.1	0.4132
400					53.91	1.072	27.6	0.3849	56.5	1.011	30.6	0.4064
500					49.11	1.181	21.6	0.3508	52.9	1.051	24.8	0.3836
600						boiling point 544.58°F			48.3	1.118	21.0	0.3493

†At 7,500 psia

*From: "1967 ASME Steam Tables," American Society of Mechanical Engineers, Tables 9, 10, and 11 and Figures 6, 7, 8, and 9.

The ASME compilation is a 330-page book of tables and charts, including a 2-1/2x3-1/2-ft Mollier chart. All values have been computed in accordance with the 1967 specifications of the International Formulation Committee (IFC) and are in conformity with the 1963 International Skeleton Tables. This standardization of tables began in 1921 and was extended through the International Conferences in London (1929), Berlin (1930), Washington (1934), Philadelphia (1954), London (1956), New York (1963) and Glasgow (1966). Based on these world-wide standard data, the 1967 ASME volume represents detailed computer output in both tabular and graphic form. Included are density and volume, enthalpy, entropy, specific heat, viscosity, thermal conductivity, Prandtl number, isentropic exponent, choking velocity, p-v product, etc., over the entire range (to 1500 psia 1500°F). English units are used, but all conversion factors are given.

TABLE B.2 Physical and Thermal Properties of Common Liquids

Part a. SI Units

(At 1.0 Atm Pressure (0.101 325 MN/m²), 300 K, except as noted.)

Common name	Density, kg/m³	Specific heat, kJ/kg·K	Viscosity, N·s/m²	Thermal conductivity, W/m·K	Freezing point, K	Latent heat of fusion, kJ/kg	Boiling point, K	Latent heat of evapora-tion, kJ/kg	Coefficient of cubical expansion per K
Acetic acid	1 049	2.18	.001 155	0.171	290	181	391	402	0.001 1
Acetone	784.6	2.15	.000 316	0.161	179.0	98.3	329	518	0.001 5
Alcohol, ethyl	785.1	2.44	.001 095	0.171	158.6	108	351.46	846	0.001 1
Alcohol, methyl	786.5	2.54	.000 56	0.202	175.5	98.8	337.8	1 100	0.001 4
Alcohol, propyl	800.0	2.37	.001 92	0.161	146	86.5	371	779	
Ammonia (aqua)	823.5	4.38		0.353					
Benzene	873.8	1.73	.000 601	0.144	278.68	126	353.3	390	0.001 3
Bromine		.473	.000 95		245.84	66.7	331.6	193	0.001 2
Carbon disulfide	1 261	.992	.000 36	0.161	161.2	57.6	319.40	351	0.001 3
Carbon tetrachloride	1 584	.866	.000 91	0.104	250.35	174	349.6	194	0.001 3
Castor oil	956.1	1.97	.650	0.180	263.2				
Chloroform	1 465	1.05	.000 53	0.118	209.6	77.0	334.4	247	0.001 3
Decane	726.3	2.21	.000 859	0.147	243.5	201	447.2	263	
Dodecane	754.6	2.21	.001 374	0.140	247.18	216	489.4	256	
Ether	713.5	2.21	.000 223	0.130	157	96.2	307.7	372	0.001 6
Ethylene glycol	1 097	2.36	.016 2	0.258	260.2	181	470	800	
Fluorine refrigerant R-11	1 476	.870ᵃ	.000 42	0.093ᵃ	162		297.0	180ᵇ	
Fluorine refrigerant R-12	1 311	.971ᵃ		0.071ᵃ	115	34.4	243.4	165ᵇ	
Fluorine refrigerant R-22	1 194	1.26ᵃ		0.086ᵃ	113	183	232.4	232ᵇ	
Glycerine	1 259	2.62	.950	0.287	264.8	200	563.4	974	0.000 54
Heptane	679.5	2.24	.000 376	0.128	182.54	140	371.5	318	
Hexane	654.8	2.26	.000 297	0.124	178.0	152	341.84	365	
Iodine		2.15			386.6	62.2	457.5	164	
Kerosene	820.1	2.09	.001 64	0.145				251	
Linseed oil	929.1	1.84	.033 1		253		560		
Mercury		.139	.001 53		234.3	11.6	630	295	0.000 18
Octane	698.6	2.15	.000 51	0.131	216.4	181	398	298	0.000 72
Phenol	1 072	1.43	.008 0	0.190	316.2	121	455		0.000 90
Propane	493.5	2.41ᵃ	.000 11		85.5	79.9	231.08	428ᵇ	
Propylene	514.4	2.85	.000 09		87.9	71.4	225.45	342	
Propylene glycol	965.3	2.50	.042		213		460	914	
Sea water	1 025	3.76–4.10			270.6				
Toluene	862.3	1.72	.000 550	0.133	178	71.8	383.6	363	
Turpentine	868.2	1.78	.001 375	0.121	214		433	293	0.000 99
Water	997.1	4.18	.000 89	0.609	273	333	373	2 260	0.000 20

ᵃAt 297 K, liquid.
ᵇAt .101 325 meganewtons, saturation temperature.

TABLE B.2 (continued) Physical and Thermal Properties of Common Liquids

Part b. English Units

(At 1.0 Atm Pressure 77°F (25°C), except as noted.)

For viscosity in N·s/m² (=kg m·s), multiply values in centipoises by 0.001. For surface tension in N/m, multiply values in dyne/cm by 0.001.

Common name	Density, $\frac{lb}{ft^3}$	Specific gravity	Viscosity $lb_m/ft\ sec \times 10^4$	cp	Sound velocity, $\frac{meters}{sec}$	Dielectric constant	Refractive index
Acetic acid	65.493	1.049	7.76	1.155	1584⁵⁰	6.15	1.37
Acetone	48.98	.787	2.12	0.316	1174	20.7	1.36
Alcohol, ethyl	49.01	.787	7.36	1.095	1144	24.3	1.36
Alcohol, methyl	49.10	.789	3.76	0.56	1103	32.6	1.33
Alcohol, propyl	49.94	.802	12.9	1.92	1205	20.1	1.38
Ammonia (aqua)	51.411	.826	—	—	—	16.9	—
Benzene	54.55	.876	4.04	0.601	1298	2.2	1.50
Bromine	—	—	6.38	0.95	—	3.20	—
Carbon disulfide	78.72	1.265	2.42	0.36	1149	2.64	1.63
Carbon tetrachloride	98.91	1.59	6.11	0.91	924	2.23	1.46
Castor oil	59.69	0.960	—	650	1474	4.7	—
Chloroform	91.44	1.47	3.56	0.53	995	4.8	1.44
Decane	45.34	.728	5.77	0.859	—	2.0	1.41
Dodecane	47.11	—	9.23	1.374	—	—	1.41
Ether	44.54	0.715	1.50	0.223	985	4.3	1.35
Ethylene glycol	68.47	1.100	109	16.2	1644	37.7	1.43
Fluorine refrigerant R–11	92.14	1.480	2.82	0.42	—	2.0	1.37
Fluorine refrigerant R–12	81.84	1.315	—	—	—	2.0	1.29
Fluorine refrigerant R–22	74.53	1.197	—	—	—	2.0	1.26
Glycerine	78.62	1.263	6380	950	1909	40	1.47
Heptane	42.42	.681	2.53	0.376	1138	1.92	1.38
Hexane	40.88	.657	2.00	0.297	1203	—	1.37
Iodine	—	—	—	—	—	11	—
Kerosene	51.2	0.823	11.0	1.64	1320	—	—
Linseed oil	58.0	0.93	222	33.1	—	3.3	—
Mercury	—	13.633	10.3	1.53	1450	—	—
Octane	43.61	.701	3.43	0.51	1171	—	1.40
Phenol	66.94	1.071	54	8.0	1274¹⁰⁰	9.8	—
Propane	30.81	.495	0.74	0.11	—	1.27	1.34
Propylene	32.11	.516	0.60	0.09	—	—	1.36
Propylene glycol	60.26	.968	—	42	—	—	1.43
Sea water	64.0	1.03	—	—	1535	—	—
Toluene	53.83	0.865	3.70	0.550	1275³⁰	2.4	1.49
Turpentine	54.2	0.87	9.24	1.375	1240	—	1.47
Water	62.247	1.00	6.0	0.89	1498	78.54ᵃ	1.33

ᵃThe dielectric constant of water near the freezing point is 87.8; it decreases with increase in temperature to about 55.6 near the boiling point.

Appendix C. Properties of Solids

TABLE C.1 Properties of Common Solids*

Material	Specific gravity	Specific heat		Thermal conductivity	
		$\dfrac{Btu}{lbm \cdot deg\ R}$	$\dfrac{kJ}{kg \cdot K}$	$\dfrac{Btu}{hr \cdot ft \cdot deg\ F}$	$\dfrac{W}{m \cdot K}$
Asbestos cement board	1.4	0.2	.837	0.35	0.607
Asbestos millboard	1.0	0.2	.837	0.08	0.14
Asphalt	1.1	0.4	1.67		
Beeswax	0.95	0.82	3.43		
Brick, common	1.75	0.22	.920	0.42	0.71
Brick, hard	2.0	0.24	1.00	0.75	1.3
Chalk	2.0	0.215	.900	0.48	0.84
Charcoal, wood	0.4	0.24	1.00	0.05	0.088
Coal, anthracite	1.5	0.3	1.26		
Coal, bituminous	1.2	0.33	1.38		
Concrete, light	1.4	0.23	.962	0.25	0.42
Concrete, stone	2.2	0.18	.753	1.0	1.7
Corkboard	0.2	0.45	1.88	0.025	0.04
Earth, dry	1.4	0.3	1.26	0.85	1.5
Fiberboard, light	0.24	0.6	2.51	0.035	0.058
Fiber hardboard	1.1	0.5	2.09	0.12	0.2
Firebrick	2.1	0.25	1.05	0.8	1.4
Glass, window	2.5	0.2	.837	0.55	0.96
Gypsum board	0.8	0.26	1.09	0.1	0.17
Hairfelt	0.1	0.5	2.09	0.03	0.050
Ice (32°)	0.9	0.5	2.09	1.25	2.2
Leather, dry	0.9	0.36	1.51	0.09	0.2
Limestone	2.5	0.217	.908	1.1	1.9
Magnesia (85%)	0.25	0.2	.837	0.04	0.071
Marble	2.6	0.21	.879	1.5	2.6
Mica	2.7	0.12	.502	0.4	0.71
Mineral wool blanket	0.1	0.2	.837	0.025	0.04
Paper	0.9	0.33	1.38	0.07	0.1
Paraffin wax	0.9	0.69	2.89	0.15	0.2
Plaster, light	0.7	0.24	1.00	0.15	0.2
Plaster, sand	1.8	0.22	.920	0.42	0.71
Plastics, foamed	0.2	0.3	1.26	0.02	0.03
Plastics, solid	1.2	0.4	1.67	0.11	0.19
Porcelain	2.5	0.22	.920	0.9	1.5
Sandstone	2.3	0.22	.920	1.0	1.7
Sawdust	0.15	0.21	.879	0.05	0.08
Silica aerogel	0.11	0.2	.837	0.015	0.02
Vermiculite	0.13	0.2	.837	0.035	0.058
Wood, balsa	0.16	0.7	2.93	0.03	0.050
Wood, oak	0.7	0.5	2.09	0.10	0.17
Wood, white pine	0.5	0.6	2.51	0.07	0.12
Wool, felt	0.3	0.33	1.38	0.04	0.071
Wool, loose	0.1	0.3	1.26	0.02	0.3

*Compiled from several sources.

TABLE C.2 Density of Various Solids:* Approximate Density of Solids at Ordinary Atmospheric Temperature

Substance	Grams per cu cm	Pounds per cu ft	Substance	Grams per cu cm	Pounds per cu ft	Substance	Grams per cu cm	Pounds per cu ft
Agate	2.5–2.7	156–168	Glass			Tallow		
Alabaster			Common	2.4–2.8	150–175	Beef	0.94	59
Carbonate	2.69–2.78	168–173	Flint	2.9–5.9	180–370	Mutton	0.94	59
Sulfate	2.26–2.32	141–145	Glue	1.27	79	Tar	1.02	66
Albite	2.62–2.65	163–165	Granite	2.64–2.76	165–172	Topaz	3.5–3.6	219–223
Amber	1.06–1.11	66–69	Graphite†	2.30–2.72	144–170	Tourmaline	3.0–3.2	190–200
Amphiboles	2.9–3.2	180–200	Gum arabic	1.3–1.4	81–87	Wax, sealing	1.8	112
Anorthite	2.74–2.76	171–172	Gypsum	2.31–2.33	144–145	Wood (seasoned)		
Asbestos	2.0–2.8	125–175	Hematite	4.9–5.3	306–330	Alder	0.42–0.68	26–42
Asbestos slate	1.8	112	Hornblende	3.0	187	Apple	0.66–0.84	41–52
Asphalt	1.1–1.5	69–94	Ice	0.917	57.2	Ash	0.65–0.85	40–53
Basalt	2.4–3.1	150–190	Ivory	1.83–1.92	114–120	Balsa	0.11–0.14	7–9
Beeswax	0.96–0.97	60–61	Leather, dry	0.86	54	Bamboo	0.31–0.40	19–25
Beryl	2.69–2.7	168–169	Lime, slaked	1.3–1.4	81–87	Basswood	0.32–0.59	20–37
Biotite	2.7–3.1	170–190	Limestone	2.68–2.76	167–171	Beech	0.70–0.90	32–56
Bone	1.7–2.0	106–125	Linoleum	1.18	74	Birch	0.51–0.77	32–48
Brick	1.4–2.2	87–137	Magnetite	4.9–5.2	306–324	Blue gum	1.00	62
Butter	0.86–0.87	53–54	Malachite	3.7–4.1	231–256	Box	0.95–1.16	59–72
Calamine	4.1–4.5	255–280	Marble	2.6–2.84	160–177	Butternut	0.38	24
Calcspar	2.6–2.8	162–175	Meerschaum	0.99–1.28	62–80	Cedar	0.49–0.57	30–35
Camphor	0.99	62	Mica	2.6–3.2	165–200	Cherry	0.70–0.90	43–56
Caoutchouc	0.92–0.99	57–62	Muscovite	2.76–3.00	172–187	Dogwood	0.76	47
Cardboard	0.69	43	Ochre	3.5	218	Ebony	1.11–1.33	69–83
Celluloid	1.4	87	Opal	2.2	137	Elm	0.54–0.60	34–37
Cement, set	2.7–3.0	170–190	Paper	0.7–1.15	44–72	Hickory	0.60–0.93	37–58
Chalk	1.9–2.8	118–175	Paraffin	0.87–0.91	54–57	Holly	0.76	47
Charcoal			Peat blocks	0.84	52	Juniper	0.56	35
Oak	0.57	35	Pitch	1.07	67	Larch	0.50–0.56	31–35
Pine	0.28–0.44	18–28	Porcelain	2.3–2.5	143–156	Lignum vitae	1.17–1.33	73–83
Cinnabar	8.12	507	Porphyry	2.6–2.9	162–181	Locust	0.67–0.71	42–44
Clay	1.8–2.6	112–162	Pressed wood			Logwood	0.91	57
Coal			pulp board	0.19	12	Mahogany		
Anthracite	1.4–1.8	87–112	Pyrite	4.95–5.1	309–318	Honduras	0.66	41
Bituminous	1.2–1.5	75–94	Quartz	2.65	165	Spanish	0.85	53
Cocoa butter	0.89–0.91	56–57	Resin	1.07	67	Maple	0.62–0.75	39–47
Coke	1.0–1.7	62–105	Rock salt	2.18	136	Oak	0.60–0.90	37–56
Copal	1.04–1.14	65–71	Rubber, hard	1.19	74	Pear	0.61–0.73	38–45
Cork	0.22–0.26	14–16	Rubber, soft			Pine		
Cork linoleum	0.54	34	Commercial	1.1	69	Pitch	0.83–0.85	52–53
Corundum	3.9–4.0	245–250	Pure gum	0.91–0.93	57–58	White	0.35–0.50	22–31
Diamond	3.01–3.52	188–220	Sandstone	2.14–2.36	134–147	Yellow	0.37–0.60	23–37
Dolomite	2.84	177	Serpentine	2.50–2.65	156–165	Plum	0.66–0.78	41–49
Ebonite	1.15	72	Silica			Poplar	0.35–0.5	22–31
Emery	4.0	250	Fused trans-			Satinwood	0.95	59
Epidote	3.25–3.50	203–218	parent	2.21	138	Spruce	0.48–0.70	30–44
Feldspar	2.55–2.75	159–172	Translucent	2.07	129	Sycamore	0.40–0.60	24–37
Flint	2.63	164	Slag	2.0–3.9	125–240	Teak		
Fluorite	3.18	198	Slate	2.6–3.3	162–205	Indian	0.66–0.88	41–55
Galena	7.3–7.6	460–470	Soapstone	2.6–2.8	162–175	African	0.98	61
Gamboge	1.2	75	Spermaceti	0.95	59	Walnut	0.64–0.70	40–43
Garnet	3.15–4.3	197–268	Starch	1.53	95	Water gum	1.00	62
Gas carbon	1.88	117	Sugar	1.59	99	Willow	0.40–0.60	24–37
Gelatin	1.27	79	Talc	2.7–2.8	168–174			

†Some values reported as low as 1.6
*Based largely on: "Smithsonian Physical Tables", 9th rev. ed., W.E. Forsythe, Ed., The Smithsonian Institution, 1956, p. 292.

Note: In the case of substances with voids, such as paper or leather, the bulk density is indicated rather than the density of the solid portion. For density in kg/m³, multiply values in g/cm³ by 1,000.

TABLE C.3 Specific Stiffness of Metals, Alloys, and Certain Non-Metallics*

Specific stiffness is usually expressed as the modulus of elasticity (in tension) per unit weight-density, i.e., E/ρ, in units of pounds and inches. While the stiffness of similar alloys varies considerably, there are definite ranges and groups to be recognized. Since the specific stiffness of steel is about 100 million, the values in the following table are also approximately the percentage stiffness, referred to steel.

Material	Specific stiffness, millions
Beryllium	650
Silicon carbide	600
Alumina ceramics	400
Mica	350
Titanium carbide cermet	250
Alumina cermet	200
Molybdenum and alloys; silica glass	130
Titanium and alloys; cobalt superalloys; soda-lime glass	110
Carbon and low-alloy steel; wrought iron	105
Stainless steel; nodular cast iron; magnesium and alloys; aluminum and alloys	100
Nickel and alloys; malleable iron	95
Iron silicon alloys (cast); iridium; vanadium	90
Monel alloys; tungsten	80
Gray cast iron; columbium alloys	70
Aluminum bronze; beryllium copper	65
Nickel silver; cupronickel; zirconium	55
Yellow brass; nickel cast iron; bronze; Muntz metal; antimony	50
Copper; red brass; tantalum	45
Silver and alloys; pewter; platinum and alloys; white gold	30
Tin; thorium	25
Gold	20
Tin-lead alloy	10
Lead	5

*Compiled from several sources.

TABLE C.4 Thermal Properties of Pure Metals—Metric Units

Metal	Melting point, °C	Boiling point, °C	AT ATMOSPHERIC PRESSURE							LIQUID METAL		
				At 100°K		At 25°C (77°F)			Specific heat (liquid) at 2000°K, cal/g °C**	Vapor pressure		
			Latent heat of fusion, cal/g**	Thermal conductivity, watts/cm °C	Specific heat, cal/g °C**	Specific heat, cal/g °C**	Coeff. of linear expansion, ($\times 10^6$) (°C)$^{-1}$	Thermal conductivity, watts/cm °C		10^{-3} atm	10^{-6} atm	10^{-9} atm
										Boiling point temperatures, °K		
Aluminum	660.	2441.	95	3.00*	.115	0.215	25	2.37	.26	1,782	1,333	1,063
Antimony	630.	1440.	38.5	—	.040	.050	9	.185	.062	1,007	741	612
Beryllium	1285.	2475.	324.	—	.049	.436	12	2.18	.78	1,793	1,347	1,085
Bismuth	271.4	1660.	12.4	—	.026	.030	13	.084	.036	1,155	851	677
Cadmium	321.	767.	13.2	1.03	.047	.055	30	.93	.063	655	486	388
Chromium	1860.	2670.	79	1.58	.046	.110	6	.91	.224	1,992	1,530	1,247
Cobalt	1495.	2925.	66	—	.057	.092	12	.69	.164	2,167	1,652	1,345
Copper	1084.	2575.	49	4.83*	.061	.031	16.6	3.98	.118	1,862	1,391	1,120
Gold	1063.	2800.	15	3.45*	.026	.031	14.2	3.15	.0355	2,023	1,510	1,211
Iridium	2450.	4390.	33	—	.022	.031	6	1.47	.0434	3,253	2,515	2,062
Iron	1536.	2870.	65	1.32*	.052	.108	12	.803	.197	2,093	1,594	1,297
Lead	327.5	1750.	5.5	0.396	.028	.031	29	.346	.033	1,230	889	698
Magnesium	650.	1090.	88.0	1.69	.016	.243	25	1.59	.32	857	638	509
Manganese	1244.	2060.	64	—	.064	.114	22	—	.20	1,495	1,131	913
Mercury	−38.86	356.55	2.7	—		.033	—	.0839	—	393	287	227
Molybdenum	2620.	4651.	69	1.79	.029	.060	5	1.4	.089	3,344	2,558	2,079
Nickel	1453.	2800.	71	1.58	.033	.106	13	.899	.175	2,156	1,646	1,343
Niobium (Columbium)	2470.	4740.	68	—	.055	—	7	.52	—	—	—	—
Osmium	3025.	4225.	34	0.552	.045	.064	5	.61	.083	3,523	2,721	2,232
Platinum	1770.	3825.	24	—	.024	.031	9	.73	.039	2,817	2,155	1,757
Plutonium	640.	3230.	3	0.79*	.019	.032	54	.08	.043	2,200	1,596	1,252
Potassium	63.3	760.	14.5	—	.150	.180	83	.99	.041	606	430	335
Rhodium	1965.	3700.	50	—	—	.058	8	1.50	—	—	—	—
Selenium	217.	700.	16	—	—	.077	37	.005	—	—	—	—
Silicon	1411.	3280.	430	—	—	.17	3	.835	.092	2,340	1,749	1,427
Silver	961.	2212.	26.5	4.50*	.045	.057	19	4.27	.217	1,582	1,179	952
Sodium	97.83	884.	27	—	.234	.293	70	1.34	.068	701	504	394
Tantalum	2980.	5365.	41	0.592	.026	.034	6.5	.54	.040	3,959	3,052	2,495
Thorium	1750.	4800.	17	—	.024	.03	12	.41	.047	3,251	2,407	1,919
Tin	232.	2600.	14.1	0.85	.039	.054	20	.64	.058	1,857	1,366	1,080
Titanium	1670.	3290.	100	—	.072	.125	8.5	.2	.188	2,405	1,827	1,484
Tungsten	3400.	5550.	46	0.312	.021	.032	4.5	1.78	.040	4,139	3,228	2,656
Uranium	1132.	4140.	12	2.35*	.022	.028	13.4	.25	.048	2,861	2,128	1,699
Vanadium	1900.	3400.	98	—	.061	.116	8	.60	.207	2,525	1,948	1,591
Zinc	419.5	910.	27	1.32	.063	.093	35	1.15	—	752	559	449

* Temperatures of maximum thermal conductivity (conductivity values in watts/cm °C): Aluminum 13°K, cond. = 71.5; copper 10°K, cond. = 196; gold 10°K, cond. = 28.2; iron 20°K, cond. = 9.97; platinum 8°K, cond. = 12.9; silver 7°K, cond. = 193; tungsten 8°K, cond. = 85.3.

** To convert to SI units note that 1 cal = 4.186 J.

TABLE C.5 Mechanical Properties of Metals and Alloys:* Typical Composition, Properties, and Uses of Common Materials

For MN/m² multiply strength in thousands of psi by 6.895.

FERROUS ALLOYS

Ferrous alloys comprise the largest volume of metal alloys used in engineering. The actual range of mechanical properties in any particular grade of alloy steel depends on the particular history and heat treatment. The steels listed in this table are intended to give some idea of the range of properties readily obtainable. Many hundreds of steels are available. Cost is frequently an important criterion in the choice of material; in general the greater the percentage of alloying elements present in the alloy, the greater will be the cost.

No.	Material	Nominal composition	Form and condition	Typical mechanical properties				Comments
				Yield strength (0.2% offset), 1000 lb/sq in.	Tensile strength, 1000 lb/sq in.	Elongation, in 2 in., %	Hardness, Brinell	
1	*IRON* Ingot iron (Included for comparison)	Fe 99.9	Hot-rolled	29	45	26	90	
			Annealed	19	38	45	67	
	PLAIN CARBON STEELS							
2	AISI-SAE 1020	C 0.20 Mn 0.45 Si 0.25 Fe bal.	Hot-rolled	30	55	25	111	Bolts, crankshafts, gears, connecting rods; easily weldable
			Hardened (water-quenched, 1000°F-tempered)	62	90	25	179	
3	AISI 1025	C 0.25 Mn 0.45 Fe bal.	Bar stock Hot-rolled	32	58	25	116	
			Cold-drawn	54	64	15	126	
4	AISI-SAE 1035	C 0.35 Mn 0.75 Fe bal.	Hot-rolled	39	72	18	143	Medium-strength, engineering steel
			Cold-rolled	67	80	12	163	
5	AISI-SAE 1045	C 0.45 Mn 0.75 Fe bal.	Bar stock Annealed	73	80	12	170	
			Hot-rolled	45	82	16	163	
			Cold-drawn	77	91	12	179	
6	AISI-SAE 1078	C 0.78 Mn 0.45 Fe bal.	Bar stock Hot-rolled: spheroidized	55	100	12	207	
			Annealed	72	94	10	192	
7	AISI-SAE 1095	C 0.95 Mn 0.40 Fe bal.						
8	AISI-SAE 1120	C 0.2 Mn 0.8 S 0.1	Cold-drawn	58	69	—	137	Free-cutting, leaded, resulphurized steel; high-speed, automatic machining
9	*ALLOY STEELS* ASTM A202/56	C 0.17 Mn 1.2 Cr 0.5 Si 0.75	Stress-relieved	45	75	18	—	Low alloy; boilers, pressure vessels

TABLE C.5 (continued) Mechanical Properties of Metals and Alloys:* Typical Composition, Properties, and Uses of Common Materials

No.	Material	Nominal composition	Form and condition	Typical mechanical properties				Comments
				Yield strength (0.2% offset), 1000 lb/sq in.	Tensile strength, 1000 lb/sq in.	Elongation, in 2 in., %	Hardness, Brinell	
10	AISI 4140	C 0.40 Si 0.3; Cr 1.0 Mo 0.2; Mn 0.9	Fully-tempered Optimum properties	95 132	108 150	22 18	240 —	High strength; gears, shafts
11	12% Manganese steel	12% Mn C	Tempered 600°F Rolled and heat-treated stock	200	220	10	—	Machine tool parts; wear, abrasion-resistant
12	VASCO 300	Ni 18.5 Ti 0.6; Co 9.0 C 0.03; Mo 4.8	Solution treatment 1500°F; aged 900°F	44 110	160 150	40 18	170 —	Very high strength, maraging, good machining properties in annealed state
13	T1 (AISI)	W 18.0 V 1.0; Cr 4.0 C 0.7	Quenched; tempered				R(c)	High speed tool steel, cutting tools, punches, etc.
14	M2 (AISI)	W 6.5 Mo 5.0; Cr 4.0 C 0.85; V 2.0	Quenched; tempered				65–66	M-grade, cheaper, tougher
15	Stainless steel type 304	Ni 9.0 C 0.08; Cr 19.0 max	Annealed; cold-rolled	35 to 160	85 to 185	60 8	160 to 400	General purpose, weldable; nonmagnetic austenitic steel
16	Stainless steel type 316	Cr 18.0 C 0.10; Ni 11.0 max; Mo 2.5 Fe bal.	Annealed	30 to 120	90 to 150	50 8	165 275	For severe corrosive media, under stress; nonmagnetic austenitic steel
17	Stainless steel type 431	Cr 16.0 Si 1.0; Ni 2.0 C 0.20; Mn 1.0 Fe bal.	Annealed Heat-treated	85 150	120 195	25 20	250 400	Heat-treated stainless steel, with good mechanical strength; magnetic
18	Stainless steel 17–4 PH	Cr 17.0 Co 0.35; Ni 4.0 C 0.07; Cu 4.0 Fe bal.	Annealed	110	150	10	363	Precipitation hardening; heat-resisting type; retains strength up to approx. 600°F

TABLE C.5 (continued) Mechanical Properties of Metals and Alloys:* Typical Composition, Properties, and Uses of Common Materials

No.	Material	Nominal composition	Form and condition	Yield strength (0.2% offset), 1000 lb/sq in.	Tensile strength, 1000 lb/sq in.	Elongation, in 2 in., %	Hardness, Brinell	Comments
					Typical mechanical properties			

CAST IRONS AND CAST STEELS

These alloys are used where large and/or intricate-shaped articles are required or where over-all dimensional tolerances are not critical. Thus the article can be produced with the fabrication and machining costs held to a minimum. Except for a few heat-treatable cast steels, this class of alloys does not demonstrate high-strength qualities.

No.	Material	Nominal composition	Form and condition	Yield strength (0.2% offset), 1000 lb/sq in.	Tensile strength, 1000 lb/sq in.	Elongation, in 2 in., %	Hardness, Brinell	Comments
	CAST IRONS							
19	Cast gray iron ASTM A48–48, Class 25	C 3.4 Mn 0.5 Si 1.8	Cast (as cast)	—	25 min	0.5 max	180	Engine blocks, fly-wheels, gears, machine-tool bases
20	White	C 3.4 Mn 0.6 Si 0.7	Cast	—	25	0	450	
21	Malleable iron ASTM A47	C 2.5 Mn 0.55 max Si 1.0	Cast (annealed)	33	52	12	130	Automotives, axle bearings, track wheels, crankshafts
22	Ductile or nodular iron (Mg-containing) ASTM A339 ASTM A395	C 3.4 Mn 0.40 Ni 1% Si 2.5 P 0.1 max Mg 0.06 Fe bal.	Cast (as cast) Cast (quenched, tempered)	53 68	70 90	18 7	170 235	Heavy-duty machines, gears, cams, crankshafts
23	Ni-hard type 2	C 2.7 Mn 0.5 Cr 2.0 Si 0.6 Ni 4.5 Fe bal.	Sand-cast Chill-cast (tempered)	— —	55 75	— —	550 625	Strength, with heat- and corrosion-resistance
24	Ni-resist type 2	C 3.0 Mn 1.0 Cr 2.5 Si 2.0 Ni 20.0 Fe bal.	Cast (as cast)	—	27	2	140	
	CAST STEELS							
25	ASTM A27–62 (60–30)	C 0.3 Si 0.8 Cr 0.4 Mn 0.6 Ni 0.5 Mo 0.2		30	60	24	—	Low alloy, medium strength, general application
26	ASTM A148–60 (105–85)			85	105	17	—	High strength; structural application

TABLE C.5 (continued) **Mechanical Properties of Metals and Alloys:** Typical Composition, Properties, and Uses of Common Materials

No.	Material	Nominal composition	Form and condition	Typical mechanical properties				Comments
				Yield strength (0.2% offset), 1000 lb/sq in.	Tensile strength, 1000 lb/sq in.	Elongation, in 2 in., %	Hardness, Brinell	
27	Cast 12 Cr alloy (CA–15)	C 0.15 max; Si 1.50 max; Ni 1.00 max; Mn 1.00 max; Cr 11.5–14; Fe bal.	Air-cooled from 1800°F; tempered at 600°F; Air-cooled from 1800°F; tempered at 1400°F	150; 75	200; 100	7; 30	390; 185	Stainless, corrosion-resistant to mildly corrosive alkalis and acids
28	Cast 29–9 alloy (CE–30) ASTM A296 63T	C 0.30 max; Si 2.00 max; Ni 8–11; Mn 1.50 max; Cr 26–30; Fe bal.	As cast	60	95	15	170	Greater corrosion resistance, especially for oxidizing condition
29	Cast 28–7 alloy (HD) ASTM A297–63T	C 0.50 max; Si 2.00 max; Ni 4–7; Mn 1.50 max; Cr 26–30; Fe bal.	As cast	48	85	16	190	Heat-resistant

SUPER ALLOYS

The advent of engineering applications requiring high temperature and high strength, as in jet engines and rocket motors, has lead to the development of a range of alloys collectively called super alloys. These alloys require excellent resistance to oxidation together with strength at high temperatures, typically 1800°F in existing engines. These alloys are continually being modified to develop better specific properties, and therefore entries in this group of alloys should be considered "fluid". Both wrought and casting-type alloys are represented. As the high temperature properties of cast materials improve, these alloys become more attractive, since great dimensional precision is now attainable in investment castings.

NICKEL BASE

No.	Material	Nominal composition	Form and condition	Yield strength (0.2% offset), 1000 lb/sq in.	Tensile strength, 1000 lb/sq in.	Elongation, in 2 in., %	Hardness, Brinell	Comments
30	Hastelloy X	Co 1.5 max; Cr 22.0; W 0.6; C 0.20 max (cast); Fe 18.5; Mo 9.0; C 0.15 max (wrought); Ni bal.	Wrought sheet; Mill-annealed; As investment cast	52; –; 46.5	113.2; 67; –	43; 17; –	194; 172; –	
31	Hastelloy C	Cr 16.0; W 4.0; Mo 17.0; Fe 6.0; C 0.15 max; Ni bal.	Sand-cast (annealed); Rolled (annealed); Investment cast	50; 71; 50	78; 130; 80	5; 45; 10	199; 204; 215	

TABLE C.5 (continued) Mechanical Properties of Metals and Alloys:* Typical Composition, Properties, and Uses of Common Materials

No.	Material	Nominal composition		Form and condition	Yield strength (0.2% offset), 1000 lb/sq in.	Tensile strength, 1000 lb/sq in.	Elongation, in 2 in., %	Hardness, Brinell	Comments
	NICKEL BASE (Cont.)								
32	Inconel 713C	Ni (+Co) bal. Mo 4.5 Al 6.0	Cr 13.0 Cb 2.0 Ti 0.6	Investment cast	102	120	6	—	
33	In 100	C 18.0 Mo 3.0 Al 55.0 V 1.0	Cr 10.0 Ti 4.7 Co 15.0	Cast					
34	Taz 8	C 125.0 Mo 4.0 W 4.0 Ta 8.0	Cr 6.0 Al 6.0 Zr 1.0 V 2.5	Cast					
35	Nimonic 90	Ni (+Co) 57.00 Mn 0.50 S 0.007 Cu 0.05 Al 1.65 Co 16.90	C 0.05 Fe 0.45 Si 0.20 Cr 20.55 Ti 2.60	Annealed; wrought	90	155	—	260	General elevated temperature applications
36	Inconel X	Ni (+Co) 72.85 Mn 0.65 S 0.007 Cu 0.05 Al 0.75 Cb (+Ta) 0.85	C 0.04 Fe 6.80 Si 0.30 Cr 15.0 Ti 2.50	Annealed Annealed; age-hardened	50 115	115 175	50 25	150 300	
37	Waspaloy	C 0.08 Mo 4.3 Co 13.5	Cr 19.5 Ti 3.0	Cold-rolled	270	275	8	Rc 51	
38	Rene 41	C 0.09 Mo 10.0 Al 1.5	Cr 19.0 Ti 3.1 Co 11.0	Wrought	100	145	—	—	

TABLE C.5 (continued) Mechanical Properties of Metals and Alloys:* Typical Composition, Properties, and Uses of Common Materials

No.	Material	Nominal composition	Form and condition	Typical mechanical properties				Comments
				Yield strength (0.2% offset), 1000 lb/sq in.	Tensile strength, 1000 lb/sq in.	Elongation, in 2 in., %	Hardness, Brinell	
39	Udimet 700	C 0.08 Cr 15.0 / Mo 5.0 Ti 3.5 / Al 4.3 Co 18.5	Cold-rolled	280	285	6	Rc 53	
40	T.D. Nickel	Ni 97.5 ThO$_2$ 2.4	Extended and cold-worked	85	100	13	—	High temperature; jet engine parts
41	COBALT BASE Haynes Stellite alloy 25 (L605)	C 0.15 max Cr 20.0 / Ni 10.0 W 15.0 / Mn 1.5 Co bal.	Wrought sheet; mill annealed	63	140	60	244	Wrought products
42	Haynes Stellite alloy 21 AMS 5385 (cast)	C 0.25 Mo 5.5 / Ni 2.5 Co bal. / Cr 28.5	As investment cast	82	103	8	313 max	For castings

ALUMINUM ALLOYS

Although the strength of aluminum alloys is in general less than that attainable in ferrous alloys or copper-base alloys, their major advantage lies in their high strength-to-weight ratio due to the low density of aluminum. Aluminum alloys have good corrosion resistance for most applications except in alkaline solutions.

No.	Material	Nominal composition	Form and condition	Yield strength	Tensile strength	Elongation	Hardness	Comments
43	3003 ASTM B221	Cu 0.12 Al bal. / Mn 1.2	Annealed-O / Cold-rolled-H14 / Cold-rolled-H18	6 / 21 / 27	16 / 22 / 29	40 / 16 / 10	28 / 40 / 55	Good formability, weldable, medium strength; chemical equipment
44	2017 ASTM B221	Mn 0.5 Mg 0.5 / Cu 4.0 Al bal.	Annealed-O / Heat-treated-T4	10 / 40	26 / 62	22 / 22	45 / 105	High strength; structural parts, aircraft, heavy forgings
45	2024 ASTM B211	Cu 4.5 Mg 1.5 / Mn 0.6 Al bal.	Heat-treated-T4	47	68	19	120	
46	5052 ASTM B211	Cr 0.25 Al bal. / Mg 2.5	Annealed-O / Cold-rolled and stabilized-H34	13 / 31	28 / 38	30 / 14	47 / 68	Medium strength, good fatigue properties; street-light standards
47	ASTM B209		Cold-rolled and stabilized-H38	37	42	8	77	

TABLE C.5 (continued)   Mechanical Properties of Metals and Alloys:* Typical Composition, Properties, and Uses of Common Materials

No.	Material	Nominal composition	Form and condition	Yield strength (0.2% offset), 1000 lb/sq in.	Tensile strength, 1000 lb/sq in.	Elongation, in 2 in., %	Hardness, Brinell	Comments
					Typical mechanical properties			
48	7075 ASTM B211	Cu 1.6, Cr 0.3, Zn 5.6, Mg 2.5, Al bal.	Annealed-O / Heat-treated and artificially aged-T6	15 / 73	33 / 83	17 / 11	60 / 150	High strength, good corrosion resistance
49	380 ASTM SC84B	Si 9.0, Cu 3.5, Al bal.	Die-cast	24	48	3	—	General purpose die casting
50	195 ASTM C4A	Si 0.8, Cu 4.5, Al bal.	Sand-cast; heat-treated-T4 / Sand-cast; heat-treated and artificially aged-T6	16 / 24	32 / 36	8.5 / 5	60 / 75	Structural elements, aircraft, and machines
51	214 ASTM G4A	Mg 3.8, Al bal.	Sand-cast-F	12	25	9	50	Chemical equipment, marine hardware, architectural
52	220 ASTM G10A	Mg 10.0, Al bal.	Sand-cast; heat-treated-T4	26	48	16	75	Strength with shock resistance; aircraft

COPPER ALLOYS

Because of their corrosion resistance and the fact that copper alloys have been used for many thousands of years, the number of copper alloys available is second only to the ferrous alloys. In general copper alloys do not have the high-strength qualities of the ferrous alloys, while their density is comparable. The cost per strength-weight ratio is high; however, they have the advantage of ease of joining by soldering, which is not shared by other metals that have reasonable corrosion resistance.

No.	Material	Nominal composition	Form and condition	Yield strength (0.2% offset), 1000 lb/sq in.	Tensile strength, 1000 lb/sq in.	Elongation, in 2 in., %	Hardness, Brinell	Comments
53	Copper ASTM B152, B124, B133, ASTM B1, B2, B3	Cu 99.9 plus	Annealed / Cold-drawn / Cold-rolled	10 / 40 / 40	32 / 45 / 46	45 / 15 / 5	42 / 90 / 100	Bus-bars, switches, architectural, roofing, screens
54	Gilding metal ASTM B36	Cu 95.0, Zn 5.0	Cold-rolled	50	56	5	114	Coinage, ammunition
55	Cartridge 70–30 brass ASTM B14, ASTM B19, ASTM B36, ASTM B134, ASTM B135	Cu 70.0, Zn 30.0	Cold-rolled	63	76	8	155	Good cold-working properties; radiator covers, hardware, electrical
56	Phosphor bronze 10% ASTM B103, ASTM B139, ASTM B159	Cu 90.0, Sn 10.0, P 0.25	Spring temper	—	122	4	241	Good spring qualities, high-fatigue strength

TABLE C.5 (continued) Mechanical Properties of Metals and Alloys:* Typical Composition, Properties, and Uses of Common Materials

No.	Material	Nominal composition	Form and condition	Typical mechanical properties				Comments
				Yield strength (0.2% offset), 1000 lb/sq in.	Tensile strength, 1000 lb/sq in.	Elongation, in 2 in., %	Hardness, Brinell	
57	Yellow brass (high brass) ASTM B36 ASTM B134 ASTM B135	Cu 65.0 Zn 35.0	Annealed Cold-drawn Cold-rolled (HT)	18 55 60	48 70 74	60 15 10	55 115 180	Good corrosion resistance; plumbing, architectural
58	Manganese bronze ASTM B138	Cu 58.5 Zn 39.2 Fe 1.0 Sn 1.0 Mn 0.3	Annealed Cold-drawn	30 50	60 80	30 20	95 180	Forgings
59	Naval brass ASTM B21	Cu 60.0 Zn 39.25 Sn 0.75	Annealed Cold-drawn	22 40	56 65	40 35	90 150	Condensor tubing; high resistance to salt-water corrosion
60	Muntz metal ASTM B111	Cu 60.0 Zn 40.0	Annealed	20	54	45	80	Condensor tubes; valve stress
61	Aluminum bronze ASTM B169, alloy A ASTM B124 ASTM B150	Cu 92.0 Al 8.0	Annealed Hard	25 65	70 105	60 7	80 210	
62	Beryllium copper 25 ASTM B194 ASTM B197 ASTM B196	Be 1.9 Co or Ni 0.25 Cu bal.	Annealed, solution-treated Cold-rolled Cold-rolled	32 104 70	70 110 190	45 5 3	B60 (Rockwell) B81 C40	Bellows, fuse clips, electrical relay parts, valves, pumps
63	Free-cutting brass	Cu 62.0 Zn 35.5 Pb 2.5	Cold-drawn	44	70	18	B80 (Rockwell)	Screws, nuts, gears, keys
64	Nickel silver 18% Alloy A (wrought) ASTM B122, No. 2	Cu 65.0 Zn 17.0 Ni 18.0	Annealed Cold-rolled Cold-drawn wire	25 70 —	58 85 105	40 4 —	70 170 —	Hardware, optical goods, camera parts
65	Nickel silver 13% (cast) 10A ASTM B149, No. 10A	Ni 12.5 Pb 9.0 Sn 2.0 Cu bal. Zn 20.0	Cast	18	35	15	55	Ornamental castings, plumbing; good machining qualities
66	Cupronickel 10% ASTM B111 ASTM B171	Cu 88.35 Ni 10.0 Fe 1.25 Mn 0.4	Annealed Cold-drawn tube	22 57	44 60	45 15	— —	Condensor, salt-water piping

TABLE C.5 (continued) Mechanical Properties of Metals and Alloys:* Typical Composition, Properties, and Uses of Common Materials

No.	Material	Nominal composition	Form and condition	Yield strength (0.2% offset), 1000 lb/sq in.	Tensile strength, 1000 lb/sq in.	Elongation, in 2 in., %	Hardness, Brinell	Comments
67	Cupronickel	Cu 70.0 Ni 30.0	Wrought					Heat-exchanger process equipment, valves
68	Red brass (cast) ASTM B30, No. 4A	Cu 85.0 Zn 5.0 Pb 5.0 Sn 5.0	As-cast	17	35	25	60	
69	Silicon bronze ASTM B30, alloy 12A	Si 4.0 Fe 2.0 Zn 4.0 Al 1.0 Mn 1.0	Castings					Cheaper substitute for tin bronze
70	Tin bronze ASTM B30, alloy 1B	Sn 8% Zn 4.0	Castings					Bearings, high-pressure bushings, pump impellers
71	Navy bronze		Cast					

TIN AND LEAD-BASE ALLOYS

Major uses for these alloys are as "white" metal bearing alloys, extruded cable sheathing, and solders. Tin forms the basis of pewter used for culinary applications.

No.	Material	Nominal composition	Form and condition	Yield strength (0.2% offset), 1000 lb/sq in.	Tensile strength, 1000 lb/sq in.	Elongation, in 2 in., %	Hardness, Brinell	Comments
72	Lead-base Babbitt ASTM B23, alloy 19	Pb 85.0 Sn 5.0 Sb 10.0 As 0.6 Cu 0.5	Chill cast	—	10	5	19	Bearings, light loads and low speeds
73	Arsenical-lead Babbitt ASTM B23, alloy 15	Pb 83.0 Sn 1.0 Sb 16.0 As 1.1 Cu 0.6	Chill cast	—	10.3	2	20	Bearings, high loads and speeds, diesel engines, steel mills
74	Chemical lead	Pb 99.9 Cu 0.06 Bi 0.005 max	Rolled 95%	1.9	2.5	50	5	
75	Antimonial lead (hard lead)	Pb 94.0 Sb 6.0	Chill cast Rolled 95%	— —	6.8 4.1	22 47	(500 kg) 9	Good corrosion resistance and strength
76	Calcium lead	Pb 99.9 Ca 0.025 Cu 0.10	Extruded and aged	—	4.5	25	—	Cable sheathing, creep-resistant pipe
77	Tin Babbitt alloy ASTM B23-61, grade 1	Sb 4.5 Sn bal. Cu 4.5	Chill cast	—	9.3	2	17	General bearings and die casting
78	Tin die-casting alloy ASTM B102-52	Sb 13.0 Sn bal. Cu 5.0	Die-cast	—	10	1	29	Die-casting alloy

TABLE C.5 (continued) Mechanical Properties of Metals and Alloys:* Typical Composition, Properties, and Uses of Common Materials

No.	Material	Nominal composition	Form and condition	Yield strength (0.2% offset), 1000 lb/sq in.	Tensile strength, 1000 lb/sq in.	Elongation, in 2 in., %	Hardness, Brinell	Comments
				Typical mechanical properties				
79	Pewter	Sn 91.0 Cu 2.0 Sb 7.0	Rolled sheet, annealed	—	8.6	40	9.5	Ornamental and household items
80	Solder 50–50	Sn 50.0 Pb 50.0	Cast	4.8	6.1	60	14	General-purpose solder
81	Solder	Sn 20.0 Pb 80.0	Cast	3.6	5.8	16	11	Coating and joining, filling seams on automobile bodies

MAGNESIUM ALLOYS

Because of their low density these alloys are attractive for use where weight is at a premium. The major drawback to the use of these alloys is their ability to ignite in air (this can be a problem in machining); they are also costly. Magnesium alloys are used in both the wrought and die-cast forms, the latter being the most frequently used form.

No.	Material	Nominal composition	Form and condition	Yield strength (0.2% offset), 1000 lb/sq in.	Tensile strength, 1000 lb/sq in.	Elongation, in 2 in., %	Hardness, Brinell	Comments
82	Magnesium alloy AZ31B	Zn 1.0 Al 3.0 Mn 0.20 min Mg bal.	Rolled-plate (strain-hardened, then partially annealed)	24	37	18	—	Structural applications of medium strength
			Rolled-sheet (strain-hardened, then partially annealed)	32	42	15	73	
			Annealed	22	37	21	56	
			Extruded	28	38	14	—	
83	Magnesium alloy AZ80A	Zn 0.5 Al 8.5 Mn 0.15 min Mg bal.	Extruded	36	49	11	60	General extruded and forged products
			Extruded (age-hardened)	39	53	6	82	
			Forged (age-hardened)	34	50	6	72	
84	Magnesium alloy AZ92A	Zn 2.0 Al 9.0 Mn 0.10 min Mg bal.	Sand-cast (as cast)	14	24	6	50	Pressure-tight sand and permanent mold castings; high UTS and good yield strength
			Sand-cast (solution heat-treated)	14	40	12	55	
			Sand-cast (solution heat-treated and aged)	19	40	5	83	
			Sand-cast (age-hardened)	16	30	18	—	
			Sand-cast and tempered	22	40	3	81	
85	Magnesium alloy ZK60A	Zn 5.7 Zr 0.55 Mg bal.	Extruded	43	52	12	82	

TABLE C.5 (continued) Mechanical Properties of Metals and Alloys:* Typical Composition, Properties, and Uses of Common Materials

No.	Material	Nominal composition	Form and condition	Yield strength (0.2% offset), 1000 lb/sq in.	Tensile strength, 1000 lb/sq in.	Elongation, in 2 in., %	Hardness, Brinell	Comments
86	Magnesium alloy AZ91A and AZ91B	Zn 0.6, Al 9.0, Mn 0.13 min, Mg bal.	Die-cast (as cast)	22	33	3	67	General die-casting applications
			BERYLLIUM					
87	Beryllium		Hot-pressed / Cross-rolled	27 / 38 / 40 / 60	33 / 51 / 60 / 90	1–3 / 10–40	— / —	Windows, X-ray tubes / Moderator- and reflector-cladding nuclear reactors; heat-shield and structural-member missiles

NICKEL ALLOYS

Nickel and its alloys are expensive and used mainly either for their high-corrosion resistance in many environments or for high-temperature and strength applications. (See Super Alloys, above.)

No.	Material	Nominal composition	Form and condition	Yield strength (0.2% offset), 1000 lb/sq in.	Tensile strength, 1000 lb/sq in.	Elongation, in 2 in., %	Hardness, Brinell	Comments
88	Nickel (cast)	Ni (+Co) 95.6, Fe 0.5, Si 1.5, Cu 0.5, Mn 0.8, C 0.8	As cast	25	57	22	110	Good corrosion-resistance applications
89	K Monel	Ni (+Co) 65.25, Mn 0.60, S 0.005, Cu 29.60, Ti 0.45, C 0.15, Fe 1.00, Si 0.15, Al 2.75	Annealed / Annealed, age-hardened / Spring / Spring, age-hardened	45 / 100 / 140 / 160	100 / 155 / 150 / 185	40 / 25 / 5 / 10	155 / 270 / 300 / 335	High strength and corrosion resistance; aircraft parts, valve stems, pumps
90	A nickel ASTM B160 ASTM B161 ASTM B162	Ni (+Co) 99.40, Mn 0.25, S 0.005, Cu 0.05, C 0.06, Fe 0.15, Si 0.05	Annealed / Hot-rolled / Cold-drawn / Cold-rolled	20 / 25 / 70 / 95	70 / 75 / 95 / 105	40 / 40 / 25 / 5	100 / 110 / 170 / 210	Chemical industry for resistance to strong alkalis, plating nickel
91	Duranickel	Ni (+Co) 93.90, Mn 0.25, S 0.005, Cu 0.05, Ti 0.45, C 0.15, Fe 0.15, Si 0.55, Al 4.50	Annealed / Annealed, age-hardened / Spring / Spring, age-hardened	45 / 125 / — / —	100 / 170 / 175 / 205	40 / 25 / 5 / 10	160 / 330 / 320 / 370	High strength and corrosion resistance; pump rods, shafts, springs

TABLE C.5 (continued) Mechanical Properties of Metals and Alloys:* Typical Composition, Properties, and Uses of Common Materials

No.	Material	Nominal composition	Form and condition	Typical mechanical properties				Comments
				Yield strength (0.2% offset), 1000 lb/sq in.	Tensile strength, 1000 lb/sq in.	Elongation, in 2 in., %	Hardness, Brinell	
92	Cupronickel 55-45 (Constantan)	Cu 55.0 Ni 45.0	Annealed / Cold-drawn / Cold-rolled	30 / 50 / 65	60 / 65 / 85	45 / 30 / 20	— / — / —	Electrical-resistance wire; low temperature coefficient, high resistivity
93	Nichrome	Ni 80.0 Cr 20.0					—	Heating elements for furnaces
94	"S" Monel	Ni 60.0 Cu 29.0 / Fe 2.50 max Mn 1.5 max / Si 4.0 max Al 0.5 max	Sand-casting	80–115	110–145	2	270–350	High-strength casting alloy; good bearing properties for valve seats

TITANIUM ALLOYS

The main application for these alloys is in the aerospace industry. Because of the low density and high strength of titanium alloys, they present excellent strength-to-weight ratios.

No.	Material	Nominal composition	Form and condition	Yield strength (0.2% offset), 1000 lb/sq in.	Tensile strength, 1000 lb/sq in.	Elongation, in 2 in., %	Hardness, Brinell	Comments
95	Commercial titanium ASTM B265–58T	Ti 99.4	Annealed at 1100 to 1350°F (593 to 732°C)	70	80	20	—	Moderate strength, excellent fabricability; chemical industry pipes
96	Titanium alloy ASTM B265–58T–5 Ti-6 Al-4V		Water-quenched from 1750°F (954°C); aged at 1000°F (538°C) for 2 hr	160	170	13	—	High-temperature strength needed in gas-turbine compressor blades
97	Titanium alloy Ti-4 Al-4Mn		Water-quenched from 1450°F (788°C); aged at 900°F (482°C) for 8 hr	170	185	13	—	Aircraft forgings and compressor parts
98	Ti-Mn alloy ASTM B265–58T–7	Fe 0.5 Ti bal. Mn 7.0–8.0	Sheet	140	150	18	—	Good formability, moderate high-temperature strength; aircraft skin

ZINC ALLOYS

A major use for these alloys is for low-cost die-cast products, such as household fixtures, automotive parts, and trim.

No.	Material	Nominal composition	Form and condition	Yield strength (0.2% offset), 1000 lb/sq in.	Tensile strength, 1000 lb/sq in.	Elongation, in 2 in., %	Hardness, Brinell	Comments
99	Zinc ASTM B69	Cd 0.35 Zn bal. Pb 0.08	Hot-rolled	—	19.5	65	38	Battery cans, grommets, lithographer's sheet

TABLE C.5 (continued)   Mechanical Properties of Metals and Alloys:* Typical Composition, Properties, and Uses of Common Materials

No.	Material	Nominal composition	Form and condition	Yield strength (0.2% offset), 1000 lb/sq in.	Tensile strength, 1000 lb/sq in.	Elongation, in 2 in., %	Hardness, Brinell	Comments
				Typical mechanical properties				
100	Zilloy-15	Cu 1.00 Zn bal. Mg 0.010	Hot-rolled Cold-rolled	— —	29 36	20 25	61 80	Corrugated roofs, articles with maximum stiffness
101	Zilloy-40	Cu 1.00 Zn bal.	Hot-rolled Cold-rolled	— —	24 31	50 40	52 60	Weatherstrip, spun articles
102	Zamac-5 ASTM 25	Zn (99.99% pure remainder) Al 3.5–4.3 Cu 0.75–1.25 Mg 0.03–0.08	Die-cast	—	47.6	7	91	Die casting for automobile parts, padlocks; used also for die material

ZIRCONIUM ALLOYS

These alloys have good corrosion resistance but are easily oxidized at elevated temperatures in air. The major application is for use in nuclear reactors.

103	Zirconium, commercial	O₂ 0.07 C 0.15 Hf 1.90 Zr bal.	Annealed	40	65	27	B80 (Rockwell)	
104	Zircaloy 2	Hf 0.02 Ni 0.05 Fe 0.15 Other 0.25 Sn 1.46 Zr bal.	Annealed	50	75	22	B90 (Rockwell)	Nuclear power-reactor cores at elevated temperatures

*Compiled from various sources.

TABLE C.6 Miscellaneous Properties of Metals and Alloys

Part a. Pure Metals

At Room Temperature

Common name		PROPERTIES (TYPICAL ONLY)					
	Thermal conductivity, Btu/hr ft °F	Specific gravity	Coeff. of linear expansion, μ in./ in. °F	Electrical resistivity, microhm-cm	Poisson's ratio	Modulus of elasticity, millions of psi	Approximate melting point, °F
Aluminum	137	2.70	14	2.655	0.33	10.0	1220
Antimony	10.7	6.69	5	41.8		11.3	1170
Beryllium	126	1.85	6.7	4.0	0.024–.030	42	2345
Bismuth	4.9	9.75	7.2	115		4.6	521
Cadmium	54	8.65	17	7.4		8	610
Chromium	52	7.2	3.3	13		36	3380
Cobalt	40	8.9	6.7	9		30	2723
Copper	230	8.96	9.2	1.673	0.36	17	1983
Gold	182	19.32	7.9	2.35	0.42	10.8	1945
Iridium	85.0	22.42	3.3	5.3		75	4440
Iron	46.4	7.87	6.7	9.7		28.5	2797
Lead	20.0	11.35	16	20.6	0.40–.45	2.0	621
Magnesium	91.9	1.74	14	4.45	0.35	6.4	1200
Manganese		7.21–7.44	12	185		23	2271
Mercury	4.85	13.546		98.4			−38
Molybdenum	81	10.22	3.0	5.2	0.32	40	4750
Nickel	52.0	8.90	7.4	6.85	0.31	31	2647
Niobium (Columbium)	30	8.57	3.9	13		15	4473
Osmium	35	22.57	2.8	9		80	5477
Platinum	42	21.45	5	10.5	0.39	21.3	3220
Plutonium	4.6	19.84	30	141.4	0.15–.21	14	1180
Potassium	57.8	0.86	46	7.01			146
Rhodium	86.7	12.41	4.4	4.6		42	3569
Selenium	0.3	4.8	21	12.0		8.4	423
Silicon	48.3	2.33	2.8	1×10^5		16	2572
Silver	247	10.50	11	1.59	0.37	10.5	1760
Sodium	77.5	0.97	39	4.2			208
Tantalum	31	16.6	3.6	12.4	0.35	27	5400
Thorium	24	11.7	6.7	18	0.27	8.5	3180
Tin	37	7.31	11	11.0	0.33	6	450
Titanium	12	4.54	4.7	43	0.3	16	3040
Tungsten	103	19.3	2.5	5.65	0.28	50	6150
Uranium	14	18.8	7.4	30	0.21	24	2070
Vanadium	35	6.1	4.4	25		19	3450
Zinc	66.5	7	19	5.92	0.25	12	787

TABLE C.6 **Miscellaneous Properties of Metals and Alloys**

Part b. Commercial Metals and Alloys

Material No. (from Table 1-57)	Common name and classification	Thermal conductivity, Btu/hr ft °F	Specific gravity	Coeff. of linear expansion, μ in./in. °F	Electrical resistivity, microhm-cm	Modulus of elasticity, millions of psi	Approximate melting point, °F
1	Ingot iron (included for comparison)	42.	7.86	6.8	9.	30	2800
2	Plain carbon steel						
	AISI–SAE 1020	30.	7.86	6.7	10.	30	2760
15	Stainless steel type 304	10.	8.02	9.6	72.	28	2600
19	Cast gray iron						
	ASTM A48–48, Class 25	26.	7.2	6.7	67.	13	2150
21	Malleable iron						
	ASTM A47	—	7.32	6.6	30.	25	2250
22	Ductile cast iron						
	ASTM A339, A395	19	7.2	7.5	60.	25	2100
24	Ni-resist cast iron, type 2	23	7.3	9.6	170.	15.6	2250
29	Cast 28–7 alloy (HD)						
	ASTM A297–63T	1.5	7.6	9.2	41.	27	2700
31	Hastelloy C	5	3.94	6.3	139.	30	2350
36	Inconel X, annealed	9	8.25	6.7	122.	31	2550
41	Haynes Stellite alloy 25 (L605)	5.5	9.15	7.61	88.	34	2500
43	Aluminum alloy 3003, rolled						
	ASTM B221	90	2.73	12.9	4.	10	1200
44	Aluminum alloy 2017, annealed						
	ASTM B221	95	2.8	12.7	4.	10.5	1185
49	Aluminum alloy 380						
	ASTM SC84B	56	2.7	11.6	7.5	10.3	1050
53	Copper						
	ASTM B152, B124, B133,						
	B1, B2, B3	225	8.91	9.3	1.7	17	1980
57	Yellow brass (high brass)						
	ASTM B36, B134, B135	69	8.47	10.5	7.	15	1710
61	Aluminum bronze						
	ASTM B169, alloy A;						
	ASTM B124, B150	41	7.8	9.2	12.	17	1900
62	Beryllium copper 25						
	ASTM B194	7	8.25	9.3	—	19	1700
64	Nickel silver 18% alloy A (wrought)						
	ASTM B122, No. 2	19	8.8	9.0	29.	18	2030
67	Cupronickel 30%	17	8.95	8.5	35.	22	2240
68	Red brass (cast)						
	ASTM B30, No. 4A	42	8.7	10.	11.	13	1825
74	Chemical lead	20	11.35	16.4	21.	2	621
75	Antimonial lead (hard lead)	17	10.9	15.1	23.	3	554
80	Solder 50–50	26	8.89	13.1	15.	—	420
82	Magnesium alloy AZ31B	45	1.77	14.5	9.	6.5	1160
89	K Monel	11	8.47	7.4	58.	26	2430
90	Nickel						
	ASTM B160, B161, B162	35	8.89	6.6	10.	30	2625
92	Cupronickel 55–45 (Constantan)	13	8.9	8.1	49.	24	2300
95	Commercial titanium	10	5.	4.9	80.	16.5	3300
99	Zinc						
	ASTM B69	62	7.14	18	6.	—	785
103	Zirconium, commercial	10	6.5	2.9	41.	12	3350

The column headers span: *CLASSIFICATION AND DESIGNATION* (Material No., Common name and classification) and *PROPERTIES (TYPICAL ONLY)* (remaining columns).

*Compiled from several sources.

TABLE C.7 Composition and Melting Points of Binary Eutectic Alloys:* Binary Alloys and Solid Solutions of Metallic Components

This table represents most of the common binary combinations of metals. For many pairs no eutectic exists; for many others, the informations is uncertain or unavailable. In a fair number of cases, there is complete mutual solubility in all proportions; hence, there is a smooth temperature vs. composition curve, with no point of inflection from the melting point of one constituent to that of the other. For purposes of comparison, all values must be considered approximate in view of the experimental difficulties and the many sources of data.

Those pairs for which the liquidus curve exhibits more than one cusp are designated by a superscript *a*. In a few cases, the cusp selected for this table does not represent the lowest possible melting point for the binary mixture.

Constituents		*Composition*		*Melting point*		*Constituents*		*Composition*		*Melting point*	
A	*B*	*Mol % B*	*Weight % B*	*K*	*deg F*	*A*	*B*	*Mol % B*	*Weight % B*	*K*	*deg F*
Ag	Al	57	25	835	1 044	Au	Bi	86.8	85	514	466
Ag	As	24	18	813	1 004	Au	Cd	70	57.1	773	932
Ag	Ca^a	37	18	820	1 017	Au	Ce^a	86	81	793	968
Ag	Ce^a	80	84	798	977	Au	Ge	27	12	629	673
Ag	Cu	40	28	1 050	1 431	Au	La^a	83	78	834	1 042
Ag	Ge	25	18	924	1 204	Au	Mg	93	62	848	1 067
Ag	La^a	72	77	791	964	Au	Mn^a	32	12	1 233	1 760
Ag	Li	99	89	418	293	Au	Na	17	2.3	1 149	1 609
Ag	Mg^a	83	52	745	882	Au	Pb	84	85	488	419
Ag	Pb	95.3	97.5	577	579	Au	Sb	34.8	24.8	633	680
Ag	Pd	25.9	25.6	924	1 204	Au	Si	18.6	3.15	636	685
Ag	Sb	41	44	758	905	Au	Sn^a	29.3	19.9	551	532
Ag	Si	10.5	2.96	1 110	1 539	Au	Te	88	83	689	781
Ag	Sr^a	77	73	709	817	Au	Tl	72	73	404	268
Ag	Te	65	69	623	662	Au	U	14	16	1 128	1 571
Ag	Th	7.6	15	1 167	1 641	B	Hf	13	71	2 130	3 375
Ag	Zr	97	93	1 100	1 521	B	Ni	57	88	1 263	1 814
Al	Au^a	59,5	90.0	842	1 056	B	Ti	7	25	1 700	2 601
Al	Ca^a	65	73	818	1 013	B	Zr	88	98	1 920	2 997
Al	Cd	81	90	1 650	2 511	Ba	Mg	97	87	891	1 144
Al	Ce	69	92	928	1 211	Be	Ni	33	76	1 468	2 183
Al	Cu^a	17.3	33.0	821	1 018	Be	Pu	97	99	910	1 179
Al	Fe	32	49.34	1 426	2 107	Be	Si	33	61	1 363	1 994
Al	Ge	29	55	700	801	Be	Ti	75	94	1 300	2 061
Al	In	5	18	910	1 179	Be	Y	61	94	1 390	2 043
Al	Mg	70	67.0	710	819	Be	Zr	65	95	1 250	1 791
Al	Ni^a	76	87	1 658	2 525	Bi	Ca	88	58.5	1 059	1 447
Al	Pt^a	57	90	1 533	2 300	Bi	Cd	56	40	420	297
Al	Si	13	13	850	1 071	Bi	In^a	78	66	340	153
Al	Th	80	97	1 510	2 259	Bi	K	50	16	615	648
Al	Zn	88.7	95.0	655	720	Bi	Mg	85	40	820	1 017
As	Co	75	70	1 189	1 681	Bi	Na	22	3.0	500	441
As	Cu^a	81.6	78.0	958	1 265	Bi	Pb	44	44	397	255
As	Fe	75	69	1 103	2 017	Bi	Sn	57	43	415	288
As	In	13	18	1 004	1 348	Bi	Te	90	84	686	775
As	Mn	57	49	1 143	1 598	Bi	Tl^a	53	52	465	378
As	Ni^a	63	57	1 077	1 479	C	Cr	87	96	1 775	2 736
As	Sb	80	87	878	1 121	C	Hf	35	88	3 450	5 751
As	Sn^a	40	51	852	1 074	C	Mo	17	45	2 480	4 005
As	Zn^a	20	18	996	1 333	C	Nb	40	84	3 580	5 985

*Compiled from several sources.

TABLE C.7 (continued) Composition and Melting Points of Binary Eutectic Alloys:* Binary Alloys and Solid Solutions of Metallic Components

Constituents		Composition		Melting point		Constituents		Composition		Melting point	
A	B	Mol % B	Weight % B	K	deg F	A	B	Mol % B	Weight % B	K	deg F
C	Ti	36	69	3 050	5 031	Gd	Ni^a	32	15	943	1 238
C	V	84	96	1 900	2 961	Ge	Mg	38	17	953	1 256
C	W	59	96	2 980	4 905	Ge	Mn^a	48	41	970	1 287
Ca	Cu	51	62	833	1 040	Hf	Ta	24	24	1 300	1 881
Ca	Mg^a	32	22	718	833	In	Ni	30	17.97	1 143	1 598
Ca	Na	22	14	983	1 310	In	Sb	68	69	780	945
Ca	Ni	16	22	878	1 121	In	Sn	47	48	390	243
Ca	Sn	19	41	1 010	1 359	Ir	Mo	68	52	2 350	3 771
Cd	Cu	52	38	810	999	Ir	Nb	55	23	2 110	3 339
Cd	In	74	74	400	261	Ir	W	22	12	2 590	4 203
Cd	Pb	71	82	540	513	K	Na	32	21.67	260	− 8.6
Cd	Pu	40	59	1 170	1 647	K	Rb	70	84	307	93
Cd	Sb	7.4	8	563	554	K	Sb^a	68	84	680	765
Cd	Sn	68	69	450	351	K	Tl	84	96	440	333
Cd	Tl	73	83	475	396	La	Mg^a	38	9.7	970	1 287
Cd	Zn	27	18	540	513	La	Pb^a	11	15	1 049	1 429
Ce	Cu^a	28	15	688	779	La	Sn^a	10	9	993	1 328
Ce	Ru	33	26	923	1 202	La	Tl	16	22	913	1 184
Co	Gd	65	83	913	1 184	Mg	Ni	11	22.98	780	945
Co	Mo	27	38	1 610	2 439	Mg	Pr	4.9	23	858	1 085
Co	Nb	15	22	1 500	2 241	Mg	Pu	15	63	815	1 008
Co	Si^a	71	54	1 486	2 215	Mg	Sb^a	86	97	855	1 080
Co	Sn	21	35	1 380	2 025	Mg	Si	53	57	1 223	1 742
Co	Ti^a	22	19	1 430	2 115	Mg	Sr^a	70	89	699	799
Co	V	41	38	1 521	2 278	Mg	Th	7	42	855	1 080
Cr	Mo	14	23	1 973	3 092	Mg	Zn	30	53	615	648
Cr	Ni	46	47	1 610	2 439	Mn	Ni	40	42	1 300	1 881
Cr	Ta	13	34	1 950	3 051	Mn	Pd	26	41	1 398	2 057
Cr	Ti	86	85	950	1 251	Mn	Sb	82	91	843	1 058
Cr	V	33	32	2 050	3 231	Mn	Ti^a	9	7.9	1 460	2 169
Cs	K	50	23	235	− 36	Mn	U^a	75	93	988	1 319
Cs	Na	20.9	4.37	241	− 26	Mn	Y^a	65	75	1 163	1 634
Cs	Rb	50	39	282	48	Mo	Nb	66	65	2 570	4 167
Cu	Ge	34	37	913	1 184	Mo	Ni	64	52	1 590	2 403
Cu	Mg^a	85.5	69.3	758	905	Mo	Os	21	34	2 650	4 311
Cu	Mn	37	34	1 143	1 598	Mo	Pd	54	57	2 020	3 177
Cu	Pb	15	36	1 230	1 755	Mo	Re	48	64	2 780	4 545
Cu	Pr^a	69	83	745	882	Mo	Ru	41	42	2 200	3 501
Cu	Sb^a	63	76	800	981	Mo	Si^a	17	5.7	2 350	3 771
Cu	Si	30	16	1 075	1 476	Na	Rb	82.1	94.5	269	25
Cu	Te	69	82	617	207	Na	Sb	60	89	678	761
Cu	Ti^a	27	22	1 133	1 580	Na	Sn	37	75	718	833
Cu	Tl	14.5	35.3	1 357	1 983	Na	Te	55	87	592	606
Cu	U	8.2	25	1 213	1 724	Nb	Ni	58	47	1 450	2 151
Cu	Zr	9.4	13	1 253	1 796	Nb	Pt	54	71	1 970	3 087
Fe	Gd	69	86	1 123	1 562	Nb	Rh	45	31	1 770	2 727
Fe	Mo	21	31	1 725	2 646	Nb	Ru^a	64	49	2 050	3 231
Fe	Nb	12	18.49	1 643	2 498	Nb	Zr	77	77	2 010	3 159
Fe	Sb	88	94.10	1 021	1 378	Ni	Sb	22	36.90	1 375	2 016
Fe	Si^a	35	21	· 1 475	2 196	Ni	Sn	19	32.16	1 403	2 066
Fe	Sn	31	49	1 400	2 061	Ni	Th^a	35	68	1 303	1 886
Fe	Y	65	75	1 173	1 652	Ni	Ti^a	39	34	1 390	2 043
Fe	Zr^a	11	17	1 600	2 421	Ni	V	52	48	1 473	2 192
Ga	Mg^a	80	58	698	797	Ni	W	20.7	45	1 773	2 732
Ga	Ni	70	66	1 477	2 199	Ni	Zn	69	71	1 148	1 607

TABLE C.7 (continued) Composition and Melting Points of Binary Eutectic Alloys:* Binary Alloys and Solid Solutions of Metallic Components

Constituents		Composition		Melting point		Constituents		Composition		Melting point	
A	B	Mol % B	Weight % B	K	deg F	A	B	Mol % B	Weight % B	K	deg F
Pb	Pr	40	31	1 315	1 908	Si	Th[a]	88	98	1 710	2 619
Pb	Pt	5.3	5.0	563	554	Si	Ti[a]	86	91	1 600	2 421
Pb	Sb	18	11	520	477	Si	Zr[a]	9	24	1 570	2 367
Pb	Sn	73	61	460	369	Sn	Te	84	85	678	761
Pb	Te	85	78	680	765	Sn	Tl	31	44	440	333
Pb	Ti	92	74	998	1 337	Sn	Zn	16	9.5	465	378
Pd	Sb	89	90	868	1 103	Te	Tl	30	41	483	410
Pt	Sn	40	29	1 345	1 962	Th	Ti	40	12	1 463	2 174
Pu	Zn	73	42	1 100	1 521	Th	Zn[a]	49	21	1 220	1 737
Re	W	26	26	3 100	5 121	Ti	U	17	51	933	1 220
Sb	Tl	70	80	468	383	Ti	Y	6.8	12	1 593	2 408
Sb	Zn	33	21	780	945	Ti	Zr	50	66	790	963
Sb	Zr	82	77	1 700	2 601	U	Zr	70	47	879	1 123
Se	Sn	39	49	913	1 184						
Se	Tl	26	48	424	304						

REFERENCES

"Selected Values of Thermodynamic Properties of Metals and Alloys," R. Hultgren, R.L. Orr, P.D. Anderson, K.K. Kelley, John Wiley & Sons, inc., 1963; a supplement to this publication has been issued periodically by the University of California, 1964-1971.

"Constitution of Binary Alloys," 2nd ed., M. Hansen, McGraw-Hill Book Company, 1958.

"Metals Reference Book," 4th ed., C.J. Smithells, Vol. 2, Butterworth & co., London, 1967.

"Handbook of Binary Metallic Systems," 2 volumes; translated from Russian, Israel Program for Scientific Translations, JJerusalem. Available from Clearinghouse for Federal Scientific and Technical Information, Springfield, Virginia 22151.

See also *Trans. AIME, J. Inst. Metals,* and *Z. Metallkunde,* by indexes.

TABLE C.8 Melting Points of Mixtures of Metals**

Melting Points, °C

Metals	0%	10%	20%	30%	40%	50%	60%	70%	80%	90%	100%
Pb. Sn.	326	295	276	262	240	220	190	185	200	216	232
Bi.	322	290	...	...	179	145	126	168	205	...	268
Te.	322	710	790	880	917	760	600	480	410	425	446
Ag.	328	460	545	590	620	650	705	775	840	905	959
Na.	...	360	420	400	370	330	290	250	200	130	96
Cu.	326	870	920	925	945	950	955	985	1005	1020	1084
Sb.	326	250	275	330	395	440	490	525	560	600	632
Al. Sb.	650	750	840	925	945	950	970	1000	1040	1010	632
Cu.	650	630	600	580	540	580	610	755	930	1055	1084
Au.	655	675	740	800	855	915	970	1025	1055	675	1062
Ag.	650	625	615	600	590	580	575	570	650	750	954
Zn.	654	640	620	600	580	560	530	510	475	425	419
Fe.	653	860	1015	1110	1145	1145	1220	1315	1425	1500	1515
Sn.	650	645	635	625	620	605	590	570	560	540	232
Sb. Bi.	632	610	590	575	555	540	520	470	405	330	268
Ag.	630	595	570	545	520	500	505	545	680	850	959
Sn.	622	600	570	525	480	430	395	350	310	255	232
Zn.	632	555	510	540	570	565	540	525	510	470	419
Ni. Sn.	1455	1380	1290	1200	1235	1290	1305	1230	1060	800	232
Na. Bi.	96	425	520	590	645	690	720	730	715	570	268
Cd.	96	125	185	245	285	325	330	340	360	390	322
Cd. Ag.	322	420	520	610	700	760	805	895	940	954	
Zn.	322	280	270	295	313	327	340	355	370	390	419
Au. Cu.	1063	910	890	895	905	925	975	1000	1025	1060	1084
Ag.	1064	1062	1061	1058	1054	1049	1039	1025	1006	982	963
Pt.	1075	1125	1190	1250	1320	1380	1455	1530	1610	1685	1775
K. Na.	62	17.5	-10	-3.5	5	11	26	41	58	77	97.5
Hg.	...	...	...	...	...	90	110	135	162	265	...
Tl.	62.5	133	165	188	205	215	220	240	280	305	301
Cu. Ni.	1080	1180	1240	1290	1320	1355	1380	1410	1430	1440	1455
Ag.	1082	1035	990	945	910	870	830	788	814	875	960
Sn.	1084	1005	890	755	680	630	580	530	440	232	
Zn.	1084	1040	995	930	880	820	780	700	580	419	
Ag. Zn.	959	850	755	705	690	660	630	610	570	505	419
Sn.	959	870	750	630	550	495	450	420	375	300	232
Na. Hg.	96.5	90	80	70	60	45	22	55	95	215	...

*The data in this table are compiled from various sources—hence, the variations in the melting point of the metals as shown in this column.

**Based largely on: "Smithsonian Physical Tables," 9th rev. ed., W.E. Forsythe, Ed., The Smithsonian Institution, 1956.

TABLE C.9 Trade Names, Composition, and Manufacturers of Various Plastics

Trade name	Composition	Manufacturer	Trade name	Composition	Manufacturer
Abson	Acrylonitrile-butadiene, ABS polymers	B. F. Goodrich Chemical Co.	Forticel	Cellulose propionate sheet films, molding powders	Celanese Plastics Co.
Alathon	Polyethylene resins	E. I. du Pont de Nemours & Co., Inc.	Fortiflex	Polyethylene resins	Celanese Plastics Co.
Alkor	Furane resin cement	Atlas Minerals & Chemicals Div., The Electric Storage Battery Co.	Fosta-Tuf-Flex	Polystyrene, high-impact	Foster-Grant, Inc.
Amres	Phenolics, urea, and melamine resins	American Marietta Co., Pacific Resins & Chemicals, Inc.	Furnane	Furanes	Atlas Minerals & Chemicals Div., The Electric Storage Battery Co.
Araldite	Epoxy resins	CIBA Products Co., Div. CIBA Corp.	GenEpoxy	Epoxy resins for adhesives, coatings, etc.	General Mills, Inc., Chemical Div.
Atlac	Polyester resins	Atlas Chemical Industries, Inc.	Genetron	Fluorinated hydrocarbons, monomers, and polymers	Allied Chemical Corp., General Chemical Div.
Bakelite	Acrylics, epoxies, phenolics, polyethylenes, copolymers	Union Carbide Corp., Chemicals and Plastics Div.	Geon	Polyvinyl chloride materials	B. F. Goodrich Chemical Co.
Bavick-11	Methyl methacrylate and methylstyrene copolymer	J. T. Baker Chemical Co.	Grex	High-density polyethylenes	Allied Chemical Corp., Plastics Div.
Boltaflex	Supported and unsupported flexible vinyl sheeting	The General Tire & Rubber Co.	Halon	Fluorohalocarbon resins	Allied Chemical Corp.
Boltaron	Rigid polyvinyl chloride sheet	The General Tire & Rubber Co., Chemical & Plastics Div.	Hetron	Fire-retardant polyester resin	Hooker Chemical Corp., Durez Plastics Div.
Butacite	Polyvinyl butyral resins	E. I. du Pont de Nemours & Co., Inc.	Isothane	Polyurethane foam, ester, and ether	Bernel Foam Products Co., Inc.
Conolite	Polyester resins and laminates	Shellmar-Betner, Div. Continental Can Co. Woodall Industries, Inc., Conolite Div.	Kel-F	Chlorotrifluoroethylene, molding resins, and dispersions	3M Company
Corvel	Epoxies, vinyls Fusion-bond finishes	The Polymer Corp., Export-Polypenco Div.	Kralac	High-styrene resins, styrene-butadiene copolymers	Uniroyal Chemical, Div. of Uniroyal Inc.
Cumar	Paracoumarone-indene resins	Allied Chemical Corp., Plastics Div	Kralastic	ABS polymers, copolymers	Uniroyal Chemical, Div. of Uniroyal Inc.
Cycolac	ABS polymers, acrylonitrile-butadiene-styrene copolymers	Marbon Chemical Div., Borg-Warner Corporation	Kynar	Polyvinylidene fluoride	Pennsalt Chemical Corp.
			Lexan	Polycarbonate resin, film, and sheet	General Electric Company, Plastics Dept.
Dacovin	Polyvinyl chlorides	Diamond Shamrock Corp.	Lucite	Acrylic resin and syrup	E. I. du Pont de Nemours & Co., Inc.
Dapon	Diallyl phthalate resins	FMC Corp., Organic Div.	Lustran	ABS polymers	Monsanto Co.
Delrin	Acetal resin and pipe	E. I. du Pont de Nemours & Co., Inc.	Lustrex	Styrene molding and extrusion resins	Monsanto Co.
Dylan	Polyethylene	Sinclair-Koppers Co.	Lytron	Styrene molding and extrusion resins	Monsanto Co.
Dylene	Polystyrene	Sinclair-Koppers Co.	Madurit	Melamine resins and compounds	Cassella Farbwerke Mainkur, A.G.
Dylite	Expandable polystyrene	Sinclair-Koppers Co	Maraglas	Epoxy-casting resin	The Marblette Corporation, Div. of Allied Products
Epi-Rez	Epoxy resins	Celanese Coatings Co., Celanese Resin Div.	Marlex	Polyethylenes, polypropylenes, copolymers	Phillips Petroleum Co
Epolene	Low molecular-weight polyethylene resins	Eastman Chemical Products, Inc., Sub. Eastman Kodak Company	Marvinol	Vinyl chloride resins and compounds	Uniroyal Chemical, Div. of Uniroyal Inc.
Epoxical	Epoxy resins	United States Gypsum Co.	Merlon	Polycarbonate resins	Mobay Chemical Co.
Epon	Epoxy resins and curing agents	The Shell Chemical Company, Plastics and Resins Div.	Micarta	Melamines, phenolics, polyesters	Westinghouse Electric Co., Industrial Micarta Div.
Escon	Polypropylene resins	Enjay Chemical Co., Div. Humble Oil & Refining Company	Microthene	Polyethylenes, polyolefins	U.S. Industrial Chemicals Co.
Estane	Polyurethane materials	B. F. Goodrich Chemical Company	Multrathane	Urethane elastomers	Mobay Chemical Company
			Nopcofoam	Polyurethane plastics	Nopco Chemical Co., Plastics Div.
Fluorogreen	Teflon with glass and ceramic fibers, fluorocarbons	John L. Dore Co.	Novodur	Polyacrylonitrile-butadiene-styrene	Farbenfabriken Bayer, A. G.
Fluororay	Ceramic-filled fluorocarbons	Raybestos-Manhattan, Inc., Plastic Products Div.	Opalon	Vinyl chloride resins and compounds	Monsanto Co
Formica	Melamines	Formica Corp. of American Cyanamid	Paraplex	Polyester resins, acrylic-modified polyester resins	Rohm & Haas Company

TABLE C.9 (continued)   Trade Names, Composition, and Manufacturers of Various Plastics

Trade name	Composition	Manufacturer	Trade name	Composition	Manufacturer
Permelite	Melamines	Melamine Plastics, Inc., Div. of Fiberlite Corp.	Super Dylan	Polyethylene	Sinclair-Koppers Co.
Petrothene	Polyethylene resins, polypropylene resins	U.S. Industrial Chemicals Co.	Supreme	Polyethylenes	Johns-Manville Company
			Sylplast	Urea-formaldehyde compounds	FMC Corp., Organic Chemicals Div.
Piccoflex	Styrene-copolymer resins	Pennsylvania Industrial Chemical Corp.	Teflon	Fluorocarbon resins	E. I. du Pont de Nemours & Co., Inc.
Piccolastic	Styrene-polymer resins	Pennsylvania Industrial Chemical Corp.	Tenite	Cellulose acetate, cellulose-acetate-polyethylene, poly-propylenes, urethane elastomers, copolymers	Eastman Chemical Products, Inc., Sub. Eastman Kodak Co.
Plaskon	Nylons, melamines, phenolics, polyesters	Allied Chemical Corp.			
Pleogen	Alkyds, polyesters, copolymers	Mol-Rez Div., American Petrochemical Corp.			
Plexiglas	Acrylics	Rohm & Haas Company	Tetran	Fluorocarbons	Pennsalt Chemicals Corp.
Pliovic	Polyvinyl chlorides	The Goodyear Tire & Rubber Co., Chemical Div.	Texin	Urethane elastomers	Mobay Chemical Company
			Thioment	Polyisoprenes	Atlas Minerals & Chemicals Div., The Electric Storage Battery Co.
Plyophen	Phenolic resins	Reichhold Chemicals, Inc.			
Poly-Eth	Polyethylene resins	Gulf Oil Corp., U.S. Div. of Gulf Oil Corp.	Ultrapas	Melamine resins	Dynamit Nobel, A. G.
			Ultrathene	Ethylene-vinyl acetates	U.S. Industrial Chemicals Co.
Polylite	Polyester resins	Reichhold Chemicals, Inc.			
Polypenco	Acrylics, chlorinated polyethers, fluoro-carbons, nylons, polycarbonates	Polymer Corp.	Ultron	Polyvinyl chlorides	Monsanto Co.
			Vibrathane	Urethane elastomers	Uniroyal Chemical, Div. of Uniroyal Inc.
			Vibrin	Polyester resins	Uniroyal Chemical, Div. of Uniroyal Inc.
Resimene	Urea and melamine resins	Monsanto Co.	Vitel	Polyesters	The Goodyear Tire & Rubber Co., Chemical Div.
Resinox	Phenolic resins and compounds	Monsanto Co.			
Rhonite	Urea resins	Rohm & Haas Company	Viton	Synthetic rubbers	E. I. du Pont de Nemours & Co., Inc.
Roylar	Polyurethanes	Uniroyal Chemical, Div. of Uniroyal Inc.			
Ryertex	Laminated phenolics and rigid polyvinyl chloride extrusions	Joseph T. Ryerson & Son, Inc., Industrial Plastics and Bearings Sales Div.	Vitroplast	Polyester cements	Atlas Minerals & Chemicals Div., The Electric Storage Battery Co.
			Vyron	Polyvinyl chlorides	Industrial Vinyls, Inc.

TABLE C.10 Properties of Commercial Nylon Resins*

Property	Type 6/6	Type 6	Type 6/10	Type 11	Glass-reinforced Type 6/6, 40%	MoS₂-filled, 2½%	Direct polymerized, castable
Mechanical							
Tensile strength, psi	11,800	11,800	8200	8500	30,000	10,000 to 14,000	11,000 to 14,000
Elongation, %	60	200	240	120	1.9	5 to 150	10 to 50
Tensile yield stress, psi	11,800	11,800	8500		30,000		
Flexural modulus, psi	410.000	395,000	280,000	151,000	1,800,000	450,000	
Tensile modulus, psi	420,000	380,000	280,000	178,000		450,000 to 600,000	350,000 to 450,000
Hardness, Rockwell	118R	119R	111R	55A	75E–80E	110R–125R	112R–120R
Impact strength, tensile, ft-lb/sq in.	76		160			50–180	80–100
Impact strength, Izod, ft-lb/in. of notch	0.9	1.0	1.2	3.3	3.7**	0.6	0.9
Deformation under load, 2000 psi, 122°F, %	1.4	1.8	4.2	2.02†	0.4§	0.5 to 2 5	0.5 to 1
Thermal							
Heat-deflection temp, °F							
At 66 psi	360	365	300	154	509	400 to 490	400 to 425
At 264 psi	150	152	135	118	502	200 to 470	300 to 425
Coefficient of thermal expansion, per °F	4.5×10^{-5}	4.6×10^{-5}	5×10^{-5}	10×10^{-5}	0.9×10^{-5}	3.5×10^{-5}	5.0×10^{-5}
Coefficient of thermal conductivity, Btu in./hr ft³ °F	1.7	1.7	1.5				
Specific heat	0.3–0.5	0.4	0.3 0.5	0.58			
Brittleness temp, °F	− 112		− 166				
Electrical							
Dielectric strength, short time, v/mil	385	420	470	425	480	300 to 400	500 to 600‡
Dielectric constant							
At 60 hz	4.0	3.8	3.9		4.45		3.7
At 10³ hz	3.9	3.7	3.6	3.3	4.40		3.7
At 10⁶ hz	3 6	3 4	3.5		4.10		3.7
Power factor							
At 60 hz	0.014	0.010	0.04	0.03	0.009		0.02
At 10³ hz	0.02	0.016	0.04	0.03	0.011		0.02
At 10⁶ hz	0.04	0.020	0.03	0.02	0.018		0.02
Volume resistivity, ohm-cm	10^{14} to 10^{15}	3×10^{15}	10^{14} to 10^{15}	2×10^{13}	2.6×10^{15}	2.5×10^{13}	
General							
Water absorption, 24 hr, %	1.5	1.6	0.4	0.4	0.6	0.5 to 1.4	0.9
Specific gravity	1.13 to 1.15	1.13	1.07 to 1.09	1.04	1.52	1.14 to 1.18	1.15 to 1.17
Melting point, °F	482 to 500	420 to 435	405 to 430	367	480 to 490	496 ± 9	430 ± 10
Flammability	self ext	self ext	self ext	self ext	self ext	self ext	self ext
Chemical resistance to							
Strong acids	Poor	Poor	Poor	Poor	Poor	Poor	Poor
Strong bases	Good	Good	Good	Fair	Good	Good	Good
Hydrocarbons	Excellent	Excellent	Excellent	Good	Excellent	Excellent	Excellent
Chlorinated hydrocarbons	Good	Good	Good	Fair	Good	Good	Good
Aromatic alcohols	Good	Good	Good	Good	Good	Good	Good
Aliphatic alcohols	Good	Good	Good	Fair	Good	Good	Good

Notes:

Most nylon resins listed in this table are used for injection molding, and test values are determined from standard injection-molded specimens. In these cases, a single typical value is listed. Exceptions are MoS2-filled nylon and direct-polymerized (castable) nylon, which are sold principally in semifinished stock shapes. Ranges of values listed are based on tests on various forms and sizes produced under varying conditions.

Because single values apply only to standard molded specimens, and properties vary in finished parts of different sizes and forms produced by various processes, these values should be used for comparison and preliminary design considerations only. For final design purposes the manufacturer should be consulted for test experience with the form being considered. Limited values should not be used for specification purposes.

÷ 2000 psi, 73°F.

‡0.040-in. thick.

**$1/_2$x$1/_4$ in. bar

§4000 psi, 122°F.

*From: "Nylons," D.D. Carswell, *Machine Design*, 40(29):62, Dec. 12, 1968.

For Conversion factors see Table C.10.

TABLE C.11 Properties of Silicate Glasses*

Most of the commercially produced glass is for windows, bottles, and inexpensive containers; it is a soda-lime-silica glass of fairly uniform composition, similar to glass No. 0080 in the table below. The following tables on glasses deal largely with that one-tenth of the glass output for which special properties are required. All data are subject to normal manufacturing variations.

Silica glass is inherently high in viscosity and melting point. These are reduced by fluxes such as Na_2O, K2O, and B2O. Soda and potash glasses have a high expansion coefficient (column 7), while that of fused silica is very low. Because the borosilicate glasses are intermediate, and their thermal shock resistance is very high (e.g., Corning Code 7740 glasses), they are widely used for laboratory and kitchen glassware. Aluminosilicate glasses are hard, heat-resisting, and of high chemical durability. Glass hardness (indentation) correlates closely with elastic modulus (column 14). Lead oxide is laso used with flux, with a result of reduced softening point and high refractive index: hence, its uses for optical glass and art glass.

Sealing of glass with metal calls for close control of the coefficient of expansion (column 7).

EXPLANATION OF COLUMNS:

Column 5:

B—blown ware	P—pressed ware	S—plate glass
M—multiform	R—rolled sheet	T—tubing and rod
U—panels	LC—large castings	

Column 6:

[2]Since weathering is determined primarily by clouding, which changes transmission, a rating for the opal glasses is omitted.

[3]These borosilicate glasses may rate differently if subjected to excessive heat treatment.

Column 8:

Normal service: No breakage from excessive thermal shock is assumed.

Extreme limits: Glass will be very vulnerable to thermal shock. Recommendations in this range are based on mechanical stability considerations only. Tests should be made before adopting final designs. These data are approximate only.

Column 9:

Based on plunging sample into cold water after oven heating. Resistance of 100°C means no breakage if heated to 110°C and plunged into water at 10°C. Tempered samples have over twice the resistance of annealed glass.

these data are approximate only.

Column 10:

[4]These data are estimated.

Resistance in °C is the temperature differential between the two surfaces of a tube or a constrained plate that will cause a tensile stress of 1000 psi on the cooler surface.

Column 11:

Viscosity is given in poises. At the strain point the stresses are significantly reduced in a matter of hours, while at the annealing point there is adequate stress reduction in minutes.

Column 12:

Data show relative resistance to sandblasting.

Column 15:

Data at 25°C are extrapolated from high temperature readings and are approximate only.

---

*From: "Properties of Selected Commercial Glasses," Publications B-83, Corning Glass Works.

TABLE C.11 (continued)   Properties of Silicate Glasses*

1	2	3	4	5	6			7		8				9		
					Corrosion Resistance			Thermal Expansion 10⁻⁷ in /in /°C		Upper Working Temperatures (Mechanical Considerations Only)				Thermal Shock Res Plates 6″ × 6″		
										Annealed		Tempered		Annealed		
Glass Code†	Type	Color	Principal Use	Forms Usually Available	Weath-ering	Water	Acid	0-300°C 32-572°F	Room Temp.-Setting Point	Normal Service °C.	Extreme Limit °C.	Normal Service °C.	Extreme Limit °C.	⅛″ Thk. °C	¼″ Thk. °C	½″ Thk. °C
0010	Potash Soda Lead	Clear	Lamp Tubing	T	2	2	2	93	100	110	380			65	50	35
0080	Soda Lime	Clear	Lamp Bulbs	B M T	3	2	2	92	103	110	460	220	250	65	50	35
0120	Potash Soda Lead	Clear	Lamp Tubing	T M	2	2	2	89	98	110	380	—	—	65	50	35
1720	Aluminosilicate	Clear	Ignition Tube	B T	1	1	3	42	52	200	650	400	450	135	115	75
1723	Aluminosilicate	Clear	Electron Tube	B T	1	1	3	46	54	200	650	400	450	125	100	70
1990	Potash Soda Lead	Clear	Iron Sealing	-	3	3	4	124	136	100	310	—	—	45	35	25
2405	Borosilicate	Red	General	B P U			..	43	51	200	480	—	—	135	115	75
2475	Soda Zinc	Red	Neon Signs	T	3	2	2	93		110	440	—	..	65	50	35
3320	Borosilicate	Canary	Tungsten Sealing		¹1	¹1	²2	40	43	200	480	.	—	145	110	80
6720	Soda Zinc	Opal	General	P	²₋	1	2	80	92	110	480	220	275	70	60	40
6750	Soda Barium	Opal	Lighting Ware	B P R	²₋	2	2	88	—	110	420	220	220	65	50	35
6810	Soda Zinc	Opal	Lighting Ware	B P R	²	1	2	69	--	120	470	240	270	85	70	45
7040	Borosilicate	Clear	Kovar Sealing	B T	³3	³3	²4	48	54	200	430		—	—	—	—
7050	Borosilicate	Clear	Series Sealing	T	¹3	¹3	²4	46	51	200	440	235	235	125	100	70
7052	Borosilicate	Clear	Kovar Sealing	B M P T	²2	²2	²4	46	53	200	420	210	210	125	100	70
7056	Borosilicate	Clear	Kovar Sealing	B T P	2	2	4	51	57	200	460	—	—	—	—	.
7070	Borosilicate	Clear	Low Loss Electrical	B M P T	³2	³2	²2	32	39	230	430	230	230	180	150	100
7250	Borosilicate	Clear	Seal Beam Lamps	P	²1	²2	²2	36	38	230	460	260	260	160	130	90
7570	High Lead	Clear	Solder Sealing	—	1	1	4	84	92	100	300	—	—	—	—	—
7720	Borosilicate	Clear	Tungsten Sealing	B P T	²2	²2	²2	36	43	230	460	260	260	160	130	90
7740	Borosilicate	Clear	General	B P S T U	¹1	²1	²1	33	35	230	490	260	290	180	150	100
7760	Borosilicate	Clear	General	B P	2	2	2	34	37	230	450	250	250	160	130	90
7900¹	96″ Silica	Clear	High Temp.	B P T U M	1	1	1	8	7	800	1100	—	—	1250	1000	750
7913	96″ Silica	Clear	High Temp.	B P R S T	1	1	1	8	7	900	1200	—		.	—	—
7940	Fused Silica	Clear	Ultrasonic	U	1	1	1	5.5	7	900	1100			1250	1000	750
8160	Potash Soda Lead	Clear	Electron Tubes	P T	2	2	3	91	100	110	380		.	65	50	35
8161	Potash Lead	Clear	Electron Tubes	P T	2	1	4	90	97	110	390	--	—	—	—	—
8363	High Lead	Clear	Radiation Shielding	L C	3	1	4	104	112	100	200	..	-	—		—
8871	Potash Lead	Clear	Capacitors		2	1	4	102	113	125	300	—	—	55	45	35
9010	Potash Soda Barium	Grey	TV Bulbs	P	2	2	2	89	102	110	380					
9700	Borosilicate	Clear	u v Trans-mission	T U	¹1	¹1	²2	39	39	220	500	--	...	150	120	80
9741	Borosilicate	Clear	u v Trans-mission	B U T	³3	¹3	²4	39	49	200	390	.		150	120	80

† Corning Glass Works code numbers are used in this table.

TABLE C.11 (continued)   Properties of Silicate Glasses*

10	11				12	13	14			15			16			17	18
Thermal Stress Resistance °C.	Viscosity Data†				Impact Abrasion Resistance	Density grams per C.C	Young's Modulus		Poisson's Ratio	Log₁₀ of Volume Resistivity			Dielectric Properties at 1 Mc and 20°C			Refractive Index Sod. D Line (.5893 Microns)	Glass Code
	Strain Point °C.	Annealing Point °C.	Softening Point °C.	Working Point °C.			$(10^6$lb/sq. in)	$(10^4$kg/cm²)		25°C. 77°F	250°C. 482°F	350°C. 662°F	Power Factor	Dielectric Const.	Loss Factor		
19	395	435	625	985	0.8	2.86	8.9	0.63	.21	17.+	8.9	7.0	.16"„	6.7	1."„	1.539	0010
17	470	510	695	1005	1.2	2.47	10.0	0.70	.24	12.4	6.4	5.1	.9	7.2	6.5	1.512	0080
20	395	435	630	980	0.8	3.05	8.6	0.60	.22	17.+	10.1	8.0	.12	6.7	.8	1.560	0120
28	670	715	915	1190	2.0	2.52	12.7	0.89	0.25	.	11.4	9.5	.38	7.2	2.7	1.530	1720
25	670	710	910	1175	2.0	2.64	12.5	0.88	0.25	.	13.5	11.3	.16"„	6.3	1.0"„	1.547	1723
14	330	360	500	755	..	3.47	8.4	0.59	.25		10.1	7.7	.04	8.3	33		1990
'37	500	530	770	1085	—	2.50	9.9	0.70	0.21	..	..	..			.	1.507	2405
'17	440	480	690	1040	.	2.59	10.0	0.70	–	—	7.8	6.2		...	.	1.511	2475
'40	500	540	780	1155		2.27	9.4	0.66	0.19	—	8.6	7.1	.30	4.9	1.5	1.481	3320
19	510	550	775	1010	—	2.58	10.2	0.72	.21	—	..	.	—	...	—	1.507	6720
'18	445	485	670	1040		2.59	—	..			.	—		.	—	1.513	6750
'23	490	530	770	1010	..	2.65	—	—	—	.	.	—		.	...	1.508	6810
37	450	490	700	1080	..	2.24	8.6	0.60	.23	—	9.6	7.8	.20	4.8	1.0	1.480	7040
39	460	500	705	1025	...	2.24	8.7	0.61	.22	16.	8.8	7.2	.33	4.9	1.6	1.479	7050
41	435	480	710	1115	—	2.28	8.2	0.58	.22	17.	9.2	7.4	.26	4.9	1.3	1.484	7052
34	470	510	720	1045	.	2.29	9.2	0.65	.21	..	10.2	8.3	.27	5.7	1.5	1.487	7056
66	455	495	...	1070	4.1	2.13	7.4	0.52	.22	17.+	11.2	9.1	.06	4.1	.25	1.469	7070
48	490	540	780	1190	3.2	2.24	9.2	0.65	.20	15.	8.2	6.7	.27	4.7	1.3	1.475	7250
21	340	365	440	560	—	5.42	8.0	0.56	.28	..	10.6	8.7	.22	15.	3.3	—	7570
49	485	525	755	1140	3.2	2.35	9.1	0.64	.20	16.	8.8	7.2	.27	4.7	1.3	1.487	7720
53	515	565	820	1245	3.1	2.23	9.1	0.64	.20	15.	8.1	6.6	.50	4.6	2.6	1.474	7740
52	480	525	780	1210	—	2.23	9.1	0.64	—	17.	9.4	7.7	.18	4.5	.79	1.473	7760
202	820	910	1500	—	3.5	2.18	10.0	0.70	.19	17.	9.7	8.1	.05	3.8	.19	1.458	7900'
211	820	910	1500	—	3.5	2.18	9.6	0.67	.19	..	9.7	8.1	.04	3.8	0.15	1.458	7913
290	990	1050	1580	–	3.6	2.20	10.5	0.74	.16	—	11.8	10.2	.001	3.8	.0038	1.459	7940
'18	395	435	630	975	—	2.98	..	—	—	—	10.6	8.4	.09	7.0	.63	1.553	8160
22	400	435	600	860	...	4.00	7.8	0.55	.24	—	12.0	9.9	.06	8.3	0.50	1.659	8161
19	300	315	380	460	.	6.22	7.4	0.52	.27	—	9.2	7.5	.19	17.0	3.2	1.97	8363
17	350	385	525	785	..	3.84	8.4	0.59	.26	–	11.1	8.8	.05	8.4	.42	-	8871
18	405	445	650	1010		2.64	9.8	0.69	.21	—	8.9	7.0	.17	6.3	1.1	1.507	9010
45	520	565	805	1200	.	2.26	9.6	0.67	.20	15.	8.0	6.5	..	—	—	1.478	9700
55	410	450	705		..	2.16	7.2	0.51	.23	17.+	9.4	7.6	–	—	..	1.468	9741

†Viscosities at these four temperatures are approximately as follows: $10^{14.5}$ poises at the strain point, 10^{13} poises at the annealing point, $10^{7.8}$ poises at the softening point, at 10^4 poises at the working point.

TABLE C.12 Properties of Window Glass*: Transmittance of Sheet and Plate Glass

Type or tint	Nominal thickness, in.	Weight, lb/ft²	Transmittance	
			Total visible daylight, %	Direct 90° solar energy, %
Sheet	$\frac{1}{16}$	0.81	91	89
Sheet	$\frac{5}{64}$	1.00	91	88
Sheet	$\frac{3}{32}$	1.22	90	87
Sheet	$\frac{1}{8}$	1.64	90	86
Sheet	$\frac{3}{16}$	2.47	89	84
Sheet	$\frac{7}{32}$	2.85	89	82
Plate or float	$\frac{1}{8}$	1.64	90	86
Plate or float	$\frac{1}{4}$	3.28	88	79
Plate or float	$\frac{5}{16}$	4.09	88	77
Plate or float	$\frac{3}{8}$	4.91	87	74
Plate or float	$\frac{1}{2}$	6.55	86	70
Plate or float	$\frac{5}{8}$	8.18	85	65
Plate or float	$\frac{3}{4}$	9.83	83	60
Plate or float	$\frac{7}{8}$	11.45	81	55
Plate or float	1	13.13	79	49
Gray[a]	$\frac{1}{4}$	3.28	43	46
Bronze[a]	$\frac{1}{4}$	3.28	49	45
Green[a]	$\frac{1}{4}$	3.28	75	46
Double[b]	$\frac{1}{4}$ each	6.56	78	—

Note: Many types of glass are available, including tempered heat-strengthened glass, laminated shatter-proof glass, conductive-coated glass, and reflective-coated glass. Several double-pane combinations are offered.

Direct 90° transmittance of solar ultraviolet radiation through non-tinted window glass is about 85 percent az high as the values for toal solar energy transmittance. Ultraviolet transmittance of gray or bronze glass is lower.

Infrared transmittance is considerably lower than visual transmittance. This is significant in view of the large percentage of infrared radiation from most sources.

Approximate shading coefficients, ASHRAE, 1/4-in. glass only: clear, 0.93; gray, 0.67; bronze, 0.65; green, 0.67.

Overall heat transfer coefficient of window area (air to air) is usually assumed to be 1.0 Btu/ft² hr, but it is lower if there is no wind.

[a]Transmittance of tinted glass depends on depth of tint..
[b]Two 1/2-in. panes with 1/2-in. air space, sealed.

---

*Tables compiled from several sources.

TABLE C.13 Properties and Uses of American Woods*

Species	Specific gravity		Characteristics	Uses	Weight		
	Green	Dry			lb/cu ft, green	lb/cu ft, air-dry 12%	lb/1000 board ft, air-dry 12%
Alder, red	0.37	0.41	Low shrinkage; moderate in strength, shock resistance, hardness, and weight†	Furniture; sash; doors; millwork	46	28	2330
Ash, black	0.45	0.49	Light in weight†	Cabinets; veneer; cooperage, containers	52	34	2830
Ash, Oregon	0.50	0.55	Similar to but lighter than white ash†	Similar to white ash	46	38	3160
Ash, white	0.54	0.58	Heavy; hard; stiff; strong; high shock resistance†	Handles; ladder rungs; baseball bats; farm implements; car parts	48	41	3420
Bald cypress (Southern cypress)			Moderate in strength, weight, hardness, and shrinkage**	Building construction; beams; posts; ties; tanks; ships; paneling	51	32	2670
Beech, American	0.56	0.64	Heavy; high strength, shock resistance, and shrinkage; uniform texture†	Flooring; furniture; handles; kitchenwear; ties (treated)	54	45	3750
Birch	0.57	0.63	Heavy; high strength, shock resistance, and shrinkage; uniform texture†	Interior finish; dowels; ties (treated); veneer; musical instruments	57	44	3670
Cottonwood	0.37	0.40	Uniform texture; does not split readily; moderate in weight, strength, hardness, and shrinkage	Crates; trunks; car parts; farm implements	49	28	2330
Douglas fir	0.41	0.44	Moderate in strength, weight, shock resistance, and shrinkage‡	Building and construction; poles; veneer; plywood; ships; furniture; boxes	38	34	2830
Elm	0.57	0.63	Moderate in strength, weight, and hardness; high shock resistance and shrinkage; good in bending†	Cooperage; baskets; crates; veneer; vehicle parts	54	34	2920
Hemlock, Eastern	0.38	0.40	Moderate in weight, strength, and hardness†	Building and construction; boxes	50	28	2330
Hemlock, Western	0.38	0.42	Moderate in weight, strength, and hardness†	Sash; doors; posts; piles; building and construction	41	29	2420
Hickory, true	0.65	0.73	High toughness, hardness, shock resistance, strength, and shrinkage†	Dowels; spokes; poles; shafts; gymnasium equipment	63	51	4250
Incense cedar	0.35		Uniform texture; easy to season; low shrinkage; shock resistance, weight, and stiffness**	Lumber; fence posts; ties; poles; shingles	45		
Larch, Western	0.48	0.52	Moderate in strength, weight, shock resistance, hardness, and shrinkage‡	Doors; sash; posts; pilings; building and construction	48	36	3000

TABLE C.13 (continued)   Properties and Uses of American Woods*

Species	Specific gravity		Characteristics	Uses	Weight		
	Green	Dry			lb/cu ft, green	lb/cu ft, air-dry 12%	lb/1000 board ft, air-dry 12%
Locust, black	0.66	0.69	High in shock resistance, weight, and hardness; very high strength; moderate shrinkage**	Mine timbers; posts; poles; ties	58	48	4000
Maple	0.44	0.48	High in hardness, weight, strength, shock resistance, and shrinkage; uniform texture†	Flooring; furniture; trim; spools; farm implements	54	40	3330
Oak, red and white	0.57	0.63	High in hardness, weight, strength, shock resistance, and shrinkage; red†, white‡	Trim; ships; flooring; ties; furniture; cooperage; piles	64	44	3670
Pine, jack			Coarse texture; low strength, stiffness, shock resistance, and shrinkage	Box lumber; fuel; mine timber; ties; poles; posts			
Pine, lodgepole	0.38	0.41	Moderate in weight, hardness, strength, shock resistance, and shrinkage; easy to work‡	Poles; mine timber; ties; construction	39	29	2420
Pine			High shrinkage; moderate strength, stiffness, hardness, and shock resistance	General construction; ties; poles; posts			
Pine, Ponderosa	0.38	0.40	Moderate in weight, shock resistance, shrinkage, and hardness; easy to work†	Building; paneling; sash; frames	45	28	2330
Pine, S. yellow	0.47	0.51	Moderate in shock resistance, shrinkage, and hardness; high in strength‡	Building and construction; poles; pilings; boxes	55	41	3420
Pine, sugar	0.35	0.36	Low shock resistance; easy to work; moderate strength†	Sash; counters; blinds; patterns	52	25	2080
Pine, Western white	0.36	0.38	Moderate in strength, shock resistance, shrinkage, and hardness; easy to work†	Building and construction; patterns; boxes	35	27	2250
Red cedar, Eastern and Western	0.44	0.47	High shock resistance; low stiffness and shrinkage; moderate in strength and hardness**	Fence posts; closet liners; chests; flooring	37	37	2750
Redwood	0.38	0.40	Low shrinkage; medium in weight, strength, hardness, and shock resistance**	Posts; doors; interiors; cooling towers	50	28	2330
Spruce, Eastern	0.38	0.40	Moderate in hardness, shock resistance, weight, shrinkage, and strength†	Building; millwork; boxes; ladders	34	28	2330

TABLE C.13 (continued) Properties and Uses of American Woods*

Species	Specific gravity		Characteristics	Uses	Weight		
	Green	Dry			lb/cu ft, green	lb/cu ft, air-dry 12%	lb/1000 board ft, air-dry 12%
Spruce, Engelmann	0.31	0.33	Generally straight grained; light in weight; low strength as a beam or post; low shock resistance; moderate shrinkage	Mine timber; ties; poles; flooring; studding; paper	39	23	1920
Spruce, Sitka	0.37	0.40	Moderate in weight, hardness, strength, shock resistance, and shrinkage†	Important in boat and plane construction; sash; doors; boxes; siding	33	28	2330
Sycamore	0.46	0.49	High shrinkage; moderate in weight, strength, hardness, and shock resistance†	Boxes; ties; posts; veneer; flooring; butcher blocks	52	34	2830
Tamarack	0.49	0.53	Coarse texture; moderate in strength, hardness, shrinkage, and shock resistance	Ties; mine timber; posts; poles; tanks; scaffolding	47	37	3080
Tupelo			Uniform texture; moderate in strength, hardness, shock resistance; high shrinkage; interlocked grain makes splitting difficult†	Flooring; planking; crates; furniture			
Walnut, black	0.51	0.55	Moderate shrinkage; high weight, strength, hardness, and shock resistance; easily worked and glued**	Gun stocks; cabinets; plywood; furniture; veneer	58	38	3170
White cedar	0.31	0.32	Low shrinkage, weight, shock resistance, and strength; soft; easily worked**	Poles; posts; ties; tanks; ships	24	23	1920
Willow, black			High strength and shock resistance; low beam strength and weight; interlocked grain	Lumber; veneer; charcoal; furniture; sub-flooring; studding			

†Decay resistance low.

‡Decay resistance medium.

**Decay resistance high.

*From: *Materials Data Book*, E.R. Parker, McGraw-Hill Book Company, 1967, pp. 252-255.

Note: For weight-density in kg/m³, multiply value in lb/ft³ by 16.02.

TABLE C.14 Properties of Natural Fibers*

Because there are great variations within a given fiber class, average properties may be misleading. The following typical values are only a rough comparative guide.

Name	Specific gravity	Tenacity, g/denier	Tensile strength, 10³ psi	Elongation at break (dry), %	Standard regain, % of dry[b]	Fiber diameter, microns	Fiber length, in.	Fiber shape and kind	Resistant to
ANIMAL ORIGIN									
Wool	1.32	1.0-1.7	17-29	23-35	15 18	17-40	1.5-5	Oval, crimped, scales	Age, weak acids, solvents
Silk	1.25	3.5 5	90	20-25	10	10-13		Flexible, soft, smooth	Heat, solvents, weak acids, wear
Cashmere						15-16	1-4	Round, scales, soft	
Mohair	1.32	1.2-1.5		30	13	24-50	6-12	Round, silky	Wear, age, solvents, weak acids
Camel hair	1.32	1.8		40	13	10-40	1-6	Oval, striated	Age, solvents
VEGETABLE ORIGIN									
Cotton	1.54	2-5	30-120	5-11	7.5-8.5	10-20	0.5-2	Flat, convoluted, ribbon	Age, heat, washing, wear, solvents, alkalies, insects
Jute (bast)	1.5		50	, -1.5	14	15-20		Woody, rough, polygon	
Sisal (leaf)	1.49	2.2	75	2-: 5	13	10-30	Strand 30-40	Stiff, straight	
Flax (bast)	1.52	4-7		2-3	12	15-18	Strand 40-50	Soft, fine	Age, solvents, washing, insects, weak acids, and alkalies
Kenaf (bast)			45			15-30		Polygon or oval	
Hemp (bast)	1.48			2		18-25	Strand 30-70	Polygon or oval, irregular	
Henequen (leaf)			60				Strand 30-60	Finer than sisal	
Abaca (leaf) (Manila)	1.48	2.3-2.9	100	2-3	13		Strand 30-120		
MINERAL ORIGIN									
Asbestos	2.5		40-200			Various	0.5-10	Smooth, straight	Heat to 400 deg C, acids, chemicals, organisms
Glass[a]	2.5	7-12	200-500	3-4.5	0	Various		Circular, smooth	Chemicals, insects
Silicate[a] (Ca, Al, Mg)	2.85				0				Heat to 900 deg C, most chemicals, insects, rot

Note: Wide variations may be expected, especially for different grades of cotton. Wet strength is lower (for rayon, very much lower), but it depends on the duration of soaking. The strength of yarn is only a fraction of the cumulative strength of all individual fibers.

Most fibers exhibit relaxation of stress at constant strain and also increase in elongation at constant load (creep). The stress-strain curve is greatly affected by the rate of extension. When the stress is removed, there is a quick elastic recovery, a delayed recovery, and a permanent set. Hence the elastic behavior of any fiber depends on its stress-strain history. The elastic recoveries of nylon and wool are high; those of cotton, flax, and rayon are much lower.

The heat capacity (specific heat) of most fibers is about one-third that of water.

Other fibers: Fur hair is slightly coarser than silk fibers. Camel and llama hairs are almost as coarse as wool but only about one-third the size of human hair. Horse hair is over 100 microns; hog bristles, over 200 microns. Jute, sisal, and hemp are intermediate between cotton and wool. These are rough average sizes, and many natural fibers range 50% above or below such averages.

[a]Here classified as natural fibers for convenience, although they are man-made by processing.
[b]Expected equilibrium moisture regain of dry fiber, in percent of dry weight, when exposed in air at 70 deg F, 65% relative humidity.

*Compiled from several sources.

TABLE C.15 Properties of Manufactured Fibers*

Chemical class; common name (sources)	Specific gravity	Tenacity, g/denier	Tensile strength, 10^3 psi	Elongation at break, %	Regain (standard)	Softening point, deg C	Melting point, deg C	Flammability	Brittleness temp, deg C
CELLULOSE FIBERS (NATURAL)									
Acetate	1.30	1.-1.3	18-25	20-30	6.5	140	230	Melts and burns	
Triacetate	1.32	1.2-1.4	20-28	25-30	3-4.5	225	300	Melts and burns	
Viscose rayon	1.51	2-2.6	30-46	17-25	13.		200ᵃ	Burns readily	
High-tenacity viscose	1.53	3 5	60-80	10-12	10		200ᵃ	Burns readily	< -114
Polynosic viscose	1.53	3-5	60-80	8-20	7		200ᵃ	Burns readily	
Cuprammonium rayon (cupro)	11.52	1.7-2.3	30-45	10-17	12.5		250ᵃ	Burns readily	
PROTEIN FIBERS (NATURAL)									
Animal: casein (milk)	1.3	1.0	15	60-70	14	100	150	Slow	
Vegetable—seed: soybeans, peanuts, corn	1.3	0.7-0.9	11-14	40-60	11-15	150	250	Slow	
Vegetable—latex: rubber (vulcanized)	1.0	0.4-0.6	4-7	700-900	0		300	Burns	-60
SYNTHETIC FIBERS									
Polyacrylonitrile (acrylic)	1.17	2-5	50-75	25-40	2	190	260	Burns	
Polyamide (nylon)	1.14	4-9	70-120	20-40	4	200	215-250	Slow	< -100
Polyester (PET dacron)	1.38	4-8	70-120	10-50	0.4	225	250-290	Low	
Polyethylene (olefin, low density)	0.92	3-6	40-70	25-40	0.15	90-120	120	Slow	-114
Polyethylene (olefin, high density)	0.95	5-7	60-80	10-20	0.01	120-130	140	Slow	-114
Polypropylene (olefin)	0.91	4.5-8	45-80	15-30	0-0.5	145	160-170	Self-ext. low	-70
Polyurethane (spandex)	1.1	0.5-1.0	7-16	500-700	1.0	190	250	Burns	
Polyvinyl chloride (PVC)	1.38	0.7-2	12-17	100-125	0.1	70	140ᵃ	No; chars	< -100
Polyvinyl alcohol (PVA)	1.3	3-7	60-90	15-28	5	230	240	Slow	
Polyvinylidene chloride (saran)	1.7	2	40	20-30	0.1	115-135	170	No	
Polytetrafluoroethylene (PTFE)	2.1	1.2-1.4	33	15-30	0	225	300ᵃ	No	

Note: Mechanical properties are for room temperature and humidity and based on unstressed cross section.

ᵃDecomposition; does not melt.

*Compiled from several sources.

TABLE C.16 Properties of Rubbers and Elastomers*

Elastomers cannot be classified in any brief and simple manner, nor are they well characterized by the usual mechanical tests. The terms *rubber* and *synthetic rubber* are loosely appiled to a great variety of elastic materials, from pure gum natural rubber and pure synthetics to cured, compounded, filled, and even reinforced products.

ASTM designations (D1418) by chemical polymer description are used in the following table; yet within each class the properties can very widely, depending on the exact composition, heat treatment service temperature, and application. Typical uses, such as rubber springs and cushioning, permit an almost unlimited number of combinations of design variables.

Mechanically, rubbers may be expected to lose strength rapidly with increase in temperature, to show a large hysteresis in stress-strain behavior, to exhibit marked creep and set, and to be greatly affected by rates of load application or frequency of repeated stress. "Heat build-up," i.e., increase in temperature in service, as well as deterioration from environment (sunlight, oils, ozone, etc.) will reduce the valuable properties of many rubbers, both natural and synthetic.

The following data apply to typical samples of commercial elastomers for common uses.

KEY:

A—Acetone	J—Alkalies	S—Salts
B—Benzene	K—Ketones	T—Heat or high temperature
C—Carbon tetrachloride	L—Alcohols	U—Ultraviolet
D—Carbon disulfide	M—Ammonia	V—Vegetable oils
E—Phenol	N—Turpentine	W—Weathering
F—Sulfur compounds	O—Coal derivatives; bitumens	X—Oxidation
G—Glycerol or glycol	P—Petroleum products	Y—Aging
H—Hexane	R—Aromatics	Z—Ozone
I—Acids		

Chemical name	Polyisoprene	Butadiene	Styrene-butadiene	Acrylonitrile butadiene
Other names	Natural (or synthetic) rubber NR (IR)	BR Cis 4	Buna S Styrene SBR, GR-S	Nitrile, Buna N Hycar NBR, GR-A
CHEMICAL AND PHYSICAL				
Specific gravity	0.93	1.0	1.0	1.0
Specific heat	0.40	0.45	0.40	0.47
Thermal conductivity				
W/cm·K	0.001 7	0.002 5	0.002 6	0.002 5
Btu/hr·ft·deg F	0.10	0.14	0.15	0.14
Service temperature, deg C				
min	−25	−40	−20	−20
max	90	90	75	110
Solvents, softeners	D,K,P,V	D,H,N,P	K,P,R,V	C,K,O,R
Resistant to	A,I,J,L	G,I,J,W,Y	G,I,L,S,X	G,I,K,L,P,S, T,V,W
Swelled by	D,P,V	A,P,V	P,V	A,E,N
MECHANICAL AND ELECTRICAL				
Tensile strength				
kg/cm² (max)	300.	210.	210.	295.
kpsi (max)	4.3	3.0	3.0	4.2
Elongation at break, %	600.	700.	600.	600.
Vol. resistivity, ohm-cm	10^{15}	10^{15}	10^{14}	10^{10}
Dielectric strength				
kV/cm	235		235	185
V/mil	600.		600.	475.
Dielectric constant	3.0	2.3	2.8	3.0
Power factor (50–100 Hz)	0.003	0.005	0.005	0.007
Rebound	Good	Good	Fair	Good
COMPARATIVE RATINGS—RESISTANCE TO				
Abrasion	Good	Excellent	Good	Excellent
Cold flow (set)	Excellent		Good	Good
Tearing	Good		Poor	Fair
Air permeability	Fair	Good	Fair	Excellent
Oxidation	Fair	Fair	Fair	Fair
Flame	Poor		Poor	Poor

*Compiled from several sources.

TABLE C.16 (continued)   **Properties of Rubbers and Elastomers**[*]

	Polychloro-prene	Isobutylene-isoprene	Polysulfide	Polymethane
Chemical name				
Other names	Neoprene[a] CR, GR-M	Butyl IIR, GR-I	Thiokol[a] PS, GR-P	Adiprene[a] PU
CHEMICAL AND PHYSICAL				
Specific gravity	1.25	0.95	1.4	1.2
Specific heat	0.5	0.45	0.31	0.45
Thermal conductivity				
W/cm·K	0.002 1	0.001 3	0.003	0.001 3
Btu/hr·ft·deg F	0.12	0.075	0.17	0.075
Service temperature, deg C				
min	-20	-40	-15	-35
max	100	120	90	120
Solvents, softeners	A,B,C,D,I,N,R	D,P	C	
Resistant to	G,L,P,S,T,U,V. W,Y,Z	E,G,J,S,U,V, W,X,Y,Z	L,P,U,Z	P,V,X,Z
Swelled by	C,D,N,R	D,H,P	C,R	B,C,K,R
MECHANICAL AND ELECTRICAL				
Tensile strength				
kg/cm² (max)	240.	175.	90.	350.
kpsi (max)	3.5	2.5	1.3	5.0
Elongation at break, %	800.	700.	500.	550.
Vol. resistivity, ohm-cm	10^{11}	10^{17}	10^8	10^{11}
Dielectric strength				
kV/cm	195	295	125	195
V/mil	500	750	325	500
Dielectric constant	7.	2.4	8.	7.
Power factor (50–100 Hz)	.04	0.004	0.02	0.04
Rebound	Good	Poor	Poor	
COMPARATIVE RATINGS—RESISTANCE TO				
Abrasion	Excellent	Fair	Poor	Excellent
Cold flow (set)	Excellent	Fair	Poor	Poor
Tearing	Good	Good	Poor	Excellent
Air permeability	Good	Excellent	Good	Excellent
Oxidation	Good	Good	Good	Good
Flame	Excellent	Poor	Poor	Poor

[*]Proprietary.

Appendix D. Gases and Vapors

TABLE D.1 SI Units — Definitions, Abbreviations and Prefixes

BASIC UNITS—MKS

Length	meter	m	Electric current	ampere	A
Mass	kilogram	kg	Thermodynamic temperature	kelvin	K
Time	second	s	Luminous intensity	candela	cd

DERIVED UNITS

Property	Units†	Abbreviations and dimensions	
Acceleration	meter per second squared	m/s^2	
Activity (of radioactive source)	1 per second	s^{-1}	
Angular acceleration	radian per second squared	rad/s^{-1}	
Angular velocity	radian per second	rad/s	
Area	square meter	m^2	
Density	kilogram per cubic meter	kg/m^3	
Dynamic viscosity	newton-second per sq meter	$N \cdot s/m^2$	
Electric capacitance	farad	F	$(A \cdot s/V)$
Electric charge	coulomb	C	$(A \cdot s)$
Electric field strength	volt per meter	V/m	
Electric resistance	ohm		(V/A)
Entropy	joule per kelvin	J/K	
Force	newton	N	$(kg \cdot m/s^2)$
Frequency	hertz	hz	(s^{-1})
Illumination	lux	lx	(lm/m^2)
Inductance	henry	H	$(V \cdot s/A)$
Kinematic viscosity	sq meter per second	m^2/s	
Luminance	candela per sq meter	cd/m^2	
Luminous flux	lumen	lm	$(cd \cdot sr)$
Magnetomotive force	ampere	A	
Magnetic field strength	ampere per meter	A/m	
Magnetic flux	weber	Wb	$(V \cdot s)$
Magnetic flux density	tesla	T	(Wb/m^2)
Power	watt	W	(J/s)
Pressure	newton per square meter	N/m^2	
Radiant intensity	watt per steradian	W/sr	
Specific heat	joule per kilogram kelvin	J/kg K	
Thermal conductivity	watt per meter kelvin	W/m K	
Velocity	meter per second	m/s	
Volume	cubic meter	m^3	
Voltage, potential difference, electromotive force	volt	V	(W/A)
Wave number	1 per meter	m^{-1}	
Work, energy, quantity of heat	joule	J	$(N \cdot m)$

PREFIX NAMES OF MULTIPLES AND SUBMULTIPLES OF UNITS

Decimal equivalent	Prefix	Pronun-ciation	Symbol	Exponential expression
1,000,000,000,000	tera	tĕr′á	T	10^{+12}
1,000,000,000	giga	jĭ′gá	G	10^{+9}
1,000,000	mega	mĕg′á	M	10^{+6}
1,000	kilo	kĭl′ō	k	10^{+3}
100	hecto	hĕk′tō	h	10^{+2}
10	deka	dĕk′á	da	10
0.1	deci	dĕs′ĭ	d	10^{-1}
0.01	centi	sĕn′tĭ	c	10^{-2}
0.001	milli	mĭl′ĭ	m	10^{-3}
0.000 001	micro	mī′krō	μ	10^{-6}
0.000 000 001	nano	năn′ō	n	10^{-9}
0.000 000 000 001	pico	pē′kō	p	10^{-12}
0.000 000 000 000 001	femto	fĕm′tō	f	10^{-15}
0.000 000 000 000 000 001	atto	ăt′tō	a	10^{-18}

Appendix E. Miscellaneous

TABLE E.1 Sizes and Allowable Unit Stresses for Softwood Lumber

American Softwood Lumber Standard. A voluntary standard for softwood lumber has been developing since 1922. Five editions of Simplified Practice Recommendation R16 were issued from 1924–53 by the Department of Commerce; the present NBS voluntary Product Standard PS 20-70, "American Softwood Lumber Standard," was issued in 1970. It was supported by the American Lumber Standards Committee, which functions through a widely representative National Grading Rule Committee.

Part a. Nominal and Minimum-Dressed Sizes of Lumber*

Item	Thicknesses			Face widths		
		Minimum-dressed			Minimum-dressed	
	Nominal	Dry,[a] inches	Green, inches	Nominal	Dry,[a] inches	Green, inches
Boards[b]				2	$1\frac{1}{2}$	$1\frac{9}{16}$
				3	$2\frac{1}{2}$	$2\frac{9}{16}$
				4	$3\frac{1}{2}$	$3\frac{9}{16}$
				5	$4\frac{1}{2}$	$4\frac{5}{8}$
	1	$\frac{3}{4}$	$\frac{25}{32}$	6	$5\frac{1}{2}$	$5\frac{5}{8}$
				7	$6\frac{1}{2}$	$6\frac{5}{8}$
	$1\frac{1}{4}$	1	$1\frac{1}{32}$	8	$7\frac{1}{4}$	$7\frac{1}{2}$
				9	$8\frac{1}{4}$	$8\frac{1}{2}$
	$1\frac{1}{2}$	$1\frac{1}{4}$	$1\frac{9}{32}$	10	$9\frac{1}{4}$	$9\frac{1}{2}$
				11	$10\frac{1}{4}$	$10\frac{1}{2}$
				12	$11\frac{1}{4}$	$11\frac{1}{2}$
				14	$13\frac{1}{4}$	$13\frac{1}{2}$
				16	$15\frac{1}{4}$	$15\frac{1}{2}$
Dimension				2	$1\frac{1}{2}$	$1\frac{9}{16}$
				3	$2\frac{1}{2}$	$2\frac{9}{16}$
				4	$3\frac{1}{2}$	$3\frac{9}{16}$
	2	$1\frac{1}{2}$	$1\frac{9}{16}$	5	$4\frac{1}{2}$	$4\frac{5}{8}$
	$2\frac{1}{2}$	2	$2\frac{1}{16}$	6	$5\frac{1}{2}$	$5\frac{5}{8}$
	3	$2\frac{1}{2}$	$2\frac{9}{16}$	8	$7\frac{1}{4}$	$7\frac{1}{2}$
	$3\frac{1}{2}$	3	$3\frac{1}{16}$	10	$9\frac{1}{4}$	$9\frac{1}{2}$
				12	$11\frac{1}{4}$	$11\frac{1}{2}$
				14	$13\frac{1}{4}$	$13\frac{1}{2}$
				16	$15\frac{1}{4}$	$15\frac{1}{2}$
Dimension				2	$1\frac{1}{2}$	$1\frac{9}{16}$
				3	$2\frac{1}{2}$	$2\frac{9}{16}$
				4	$3\frac{1}{2}$	$3\frac{9}{16}$
				5	$4\frac{1}{2}$	$4\frac{5}{8}$
	4	$3\frac{1}{2}$	$3\frac{9}{16}$	6	$5\frac{1}{2}$	$5\frac{5}{8}$
	$4\frac{1}{2}$	4	$4\frac{1}{16}$	8	$7\frac{1}{4}$	$7\frac{1}{2}$
				10	$9\frac{1}{4}$	$9\frac{1}{2}$
				12	$11\frac{1}{4}$	$11\frac{1}{2}$
				14		$13\frac{1}{2}$
				16		$15\frac{1}{2}$
Timbers	5 and thicker		$\frac{1}{2}$ off	5 and wider		$\frac{1}{2}$ off

[a] Maximum moisture content of 19% or less.

[b] Boards less than the minimum thickness for 1 in. nominal but 5/8-in. or greater thickness dry (11/16-in. green) may be regarded as American Standard Lumber, but such boards shall be marked to show the size and condition of seasoning at the time of dressing. They shall also be distinguished from 1-in. boards on invoices and certificates.

*Reprinted from: "American Softwood Lumber Standard," NBS PS 20-70, National Bureau of Standards, 1970; available from Superintendent of documents.

Note: This table applies to boards, dimensional lumber, and timbers. The thicknesses apply to all widths and all widths to all thicknesses.

TABLE E.1 (continued) Sizes and Allowable Unit Stresses for Softwood Lumber

The "American Softwood Lumber Standard", PS 20-70, gives the size and grade provisions for American Standard lumber and describes the organization and procedures for compliance enforcement and review. It lists commercial name classifications and complete definitions of terms and abbreviations.

Eleven softwood species are listed in PS 20-70, viz., cedar, cypress, fir, hemlock, juniper, larch, pine, redwood, spruce, tamarack, and yew. Five dimensional tables show the standard dressed (surface planed) sizes for almost all types of lumber, including matched tongue-and-grooved and shiplapped flooring, decking, siding, etc. Dry or seasoned lumber must have 19% or less moisture content, with an allowance for shrinkage of 0.7–1.0% for each four points of moisture content below the maximum. Green lumber has more than 19% moisture. Table A illustrates the relation between nominal size and dressed or green sizes.

National Design Specification. Part b is condensed from the 1971 edition of "National Design Specification for Stress-Grade Lumber and Its Fastenings," as recommended and published by the National Forest Products Association, Washington, D.C. This specification was first issued by the National Lumber Manufacturers Association in 1944; subsequent editions have been issued as recommended by the Technical Advisory Committee. The 1971 edition is a 65-page bulletin with a 20-page supplement giving "Allowable Unit Stresses, Structural Lumber," from which Part b has been condensed. The data on working stresses in this Supplement have been determined in accordance with the corresponding ASTM Standards, D245-70 and D2555-70.

Part b. Species, Sizes, Allowable Stresses, and Modulus of Elasticity of Lumber

Normal loading conditions: Moisture content not over 19%, No. 1 grade, visual grading. To convert psi to N/m², multiply by 6 895.

Species[a]	Sizes, nominal	Typical grading agency, 1971[b]	Allowable unit stresses, psi[d]				Modulus of elasticity, psi
			Extreme fiber in bending[c]	Tension parallel to grain	Compression perpendicular	Compression parallel	
CEDAR							
Northern white	2 × 4	NL, NH	1 100	600	205	675	800 000
	2 or 4 × 6+	NL, NH	1 000	575	205	675	800 000
Western	2 × 4	NC	1 450	725	285	975	1 100 000
	2 or 4 × 6+	NC, WW	1 250	725	285	975	1 100 000
FIR							
Balsam	2 × 4	NL, NH	1 300	675	170	825	1 200 000
	2 or 4 × 6+	NL, NH	1 150	650	170	825	1 200 000
Douglas (larch)	2 × 4	WC, NC	2 400	1 200	385	1 250	1 800 000
	2 or 4 × 6+	WC, NC	1 750	1 000	385	1 250	1 800 000
HEMLOCK							
Eastern (tamarack)	2 × 4	NL, NH	1 750	900	365	1 050	1 300 000
	2 or 4 × 6+	NL, NH	1 500	875	365	1 050	1 300 000
Hem-fir	2 × 4	WC, NC	1 600	825	245	1 000	1 500 000
	2 or 4 × 6+	WC, NC	1 400	800	245	1 000	1 500 000
Mountain	2 × 4	WC, WW	1 700	850	370	1 000	1 300 000
	2 or 4 × 6+	WC, WW	1 450	850	370	1 000	1 300 000
PINE							
Idaho white	2 × 4	WW	1 400	725	240	925	1 400 000
	2 or 4 × 6+	WW	1 200	700	240	925	1 400 000
Lodgepole	2 × 4	WW	1 500	750	250	900	1 300 000
	2 or 4 × 6+	WW	1 300	750	250	900	1 300 000

TABLE E.1 (continued) Sizes and Allowable Unit Stresses for Softwood Lumber

Species[a]	Sizes, nominal	Typical grading agency, 1971[b]	Allowable unit stresses, psi[d]				Modulus of elasticity, psi
			Extreme fiber in bending[c]	Tension parallel to grain	Compression perpendicular	Compression parallel	
PINE *(continued)*							
Northern	2 × 4	NL, NH	1 600	825	280	975	1 400 000
	2 or 4 × 6+	NL, NH	1 400	800	280	975	1 400 000
Ponderosa (sugar)	2 × 4	WW, NC	1 400	700	250	850	1 200 000
	2 or 4 × 6+	WW, NC	1 200	700	250	850	1 200 000
Red	2 × 4	NC	1 350	700	280	825	1 300 000
	2 or 4 × 6+	NC	1 150	675	280	825	1 300 000
Southern	2 × 4	SP	2 000	1 000	405	1 250	1 800 000
	2 or 4 × 6+	SP	1 750	1 000	405	1 250	1 800 000
REDWOOD							
California	2 or 4 × 2 or 4	RI	1 950	1 000	425	1 250	1 400 000
	2 or 4 × 6 to 12	RI	1 700	1 000	425	1 250	1 400 000
SPRUCE							
Eastern	2 × 4	NL, NH	1 500	750	255	900	1 400 000
	2 or 4 × 6+	NL, NH	1 250	750	255	900	1 400 000
Engelmann	2 × 4	WW	1 300	675	195	725	1 200 000
	2 or 4 × 6+	WW	1 150	650	195	725	1 200 000
Sitka	2 × 4	WC	1 550	775	280	925	1 500 000
	2 or 4 × 6+	WC	1 300	775	280	925	1 500 000

Note: Allowable unit stresses in horizontal shear are in the range of 60–100 psi for No. 1 grade.

[a] Grade designations are not entirely uniform. Values in the table apply approximately to "No. 1." There is seldom more than one better grade than No. 1, and this may be designated as select, select structural, dense, or heavy. In addition to lower grades 2 and 3, there may be other lower grades, designated as construction, standard, stud, and utility. In bending and tension the allowable unit stresses in the lowest recognized grade (utility) are of the order of $\frac{1}{3}$ to $\frac{1}{2}$ of the allowable stresses for grade No. 1. The tabular values for allowable bending stress are for the extreme fiber in "repetitive member uses," and edgewise use. The original tables give correction factors, which are less than unity for moist locations and for short-time loading; they are greater than unity if the moisture content of the wood in service is 15% or less. In general, all data apply to uses within covered structures. From the extensive tables, only the No. 1 grade in nominal 2 × 4 size and 2-in. or 4-in. planks, 6 in., and wider have been selected for illustration.

In a few cases the allowable stresses specified for the Canadian products will vary slightly from those given here for the same species by the U.S. agencies.

[b] Grading agencies represented by letters in this column are as follows:
   NC = National Lumber Grades Authority (a Canadian agency)
   NH = Northern Hardwood and Pine Manufacturers Association
   NL = Northern Lumber Manufacturers Association
   RI = Redwood Inspection Service
   SP = Southern Pine Inspection Bureau
   WC = West Coast Lumber Inspection Bureau
   WW = Western Wood Products Association

[c] It is assumed that all members are so framed, anchored, tied, and braced that they have the necessary rigidity.

[d] For short term loads, these values may be increased: add 15% for 2-month snow load; add 33% for wind or earthquake; add 100% for impact load.

REFERENCES

"Wood Handbook," Handbook No. 72, U.S. Department of Agriculture, 1955.

"Timber Construction Manual," American Institute of Timber Construction, John Wiley & Sons, Inc., 1966.

"National Design Specification for Stress-Grade Lumber and its Fastenings," National Forest Products Association, Washington D.C., 1971.

TABLE E.2 Standard Grades of Bolts

Part a: SAE Grades for Steel Bolts

SAE grade no.	Size range incl.	Proof strength,[†] kpsi	Tensile strength,[†] kpsi	Material	Head marking
1	$\frac{1}{4}$–$1\frac{1}{2}$			Low- or medium-carbon steel	
2	$\frac{1}{4}$–$\frac{3}{4}$	55	74		
	$\frac{7}{8}$–$1\frac{1}{2}$	33	60		
5	$\frac{1}{4}$–1	85	120	Medium-carbon steel, Q & T	
	$1\frac{1}{8}$–$1\frac{1}{2}$	74	105		
5.2	$\frac{1}{4}$–1	85	120	Low-carbon martensite steel, Q & T	
7	$\frac{1}{4}$–$1\frac{1}{2}$	105	133	Medium-carbon alloy steel, Q & T[‡]	
8	$\frac{1}{4}$–$1\frac{1}{2}$	120	150	Medium-carbon alloy steel, Q & T	
8.2	$\frac{1}{4}$–1	120	150	Low-carbon martensite steel, Q & T	

[†]Minimum values.
[‡]Roll threaded after heat treatment.
SOURCES: See "Helpful Hints," by Russell, Burdsall & Ward Corp., Mentor, Ohio 44060; and Chap. 23.

TABLE E.2 (continued) **Standard Grades of Bolts**

Part b: ASTM Grades for Steel Bolts

ASTM designation	Size range incl.	Proof strength,† kpsi	Tensile strength,† kpsi	Material	Head marking
A307	¼ to 4			Low-carbon steel	
A325 type 1	½ to 1	85	120	Medium-carbon steel, Q & T	 A325
	1⅛ to 1½	74	105		
A325 type 2	½ to 1	85	120	Low-carbon martensite steel, Q & T	 A325
	1⅛ to 1½	74	105		
A325 type 3	½ to 1	85	120	Weathering steel, Q & T	 A325
	1⅛ to 1½	74	105		
A354 grade BC				Alloy steel, Q & T	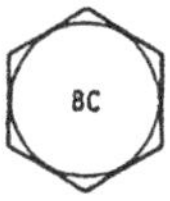 BC
A354 grade BD	¼ to 4	120	150	Alloy steel, Q & T	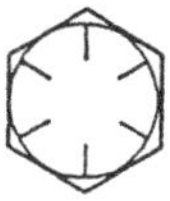
A449	¼ to 1	85	120	Medium-carbon steel, Q & T	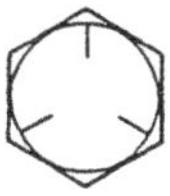
	1⅛ to 1½	74	105		
	1¾ to 3	55	90		
A490 type	½ to 1½	120	150	Alloy steel, Q & T	 A490
A490 type 3				Weathering steel, Q & T	 A490

†Minimum value.

Sources: See "Helpful Hints," by Russell, Burdsall & Ward Corp., Mentor, Ohio 44060; and Chapter 23.

TABLE E.2 (continued) **Standard Grades of Bolts**

Part c: Metric Mechanical Property Classes for Steel Bolts, Screws, and Studs

Property class	Size range incl.	Proof strength, MPa	Tensile strength, MPa	Material	Head marking
4.6	M5–M36	225	400	Low- or medium-carbon steel	4.6
4.8	M1.6–M16	310	420	Low- or medium-carbon steel	4.8
5.8	M5–M24	380	520	Low- or medium-carbon steel	5.8
8.8	M16–M36	600	830	Medium-carbon steel, Q & T	8.8
9.8	M1.6–M16	650	900	Medium-carbon steel, Q & T	9.8
10.9	M5–M36	830	1040	Low-carbon martensite steel, Q & T	10.9
12.9	M1.6–M36	970	1220	Alloy steel, Q & T	12.9

Sources: "Helpful Hints," by Russell, Burdsall & Ward Corp., Mentor, Ohio 44060; see also Chapter 23 and SAE standard J1199, and ASTM standard F569.

TABLE E.3 Steel Pipe Sizes

Nominal Pipe Size, in.	Outside Diameter, in.	Schedule Number or Weight	Wall Thickness, in.	Inside Diameter, in.	Surface Area		Areas and Weights Cross-sectional		Weight
					Outside, ft²/ft	Inside, ft²/ft	Metal Area, in.²	Flow Area, in.²	Pipe lb/ft
¾	1.05	40	0.113	0.824	0.275	0.216	0.333	0.533	1.131
		80	0.154	0.742	0.275	0.194	0.434	0.432	1.474
1	1.315	40	0.133	1.049	0.344	0.275	0.494	0.864	1.679
		80	0.179	0.957	0.344	0.250	0.639	0.719	2.172
1¼	1.660	40	0.140	1.38	0.434	0.361	0.668	1.496	2.273
		80	0.191	1.278	0.434	0.334	0.881	1.283	2.997
1½	1.900	40	0.145	1.61	0.497	0.421	0.799	2.036	2.718
		80	0.200	1.50	0.497	0.393	1.068	1.767	3.632
2	2.375	40	0.154	2.067	0.622	0.541	1.074	3.356	3.653
		80	0.218	1.939	0.622	0.508	1.477	2.953	5.022
2½	2.875	40	0.203	2.469	0.753	0.646	1.704	4.79	5.794
		80	0.276	2.323	0.753	0.608	2.254	4.24	7.662
3	3.5	40	0.216	3.068	0.916	0.803	2.228	7.30	7.58
		80	0.300	2.900	0.916	0.759	3.016	6.60	10.25
3½	4.0	40	0.226	3.548	1.047	0.929	2.680	9.89	9.11
		80	0.318	3.364	1.047	0.881	3.678	8.89	12.51
4	4.5	40	0.237	4.026	1.178	1.054	3.17	12.73	10.79
		80	0.337	3.826	1.178	1.002	4.41	11.50	14.99
5	5.563	10 S	0.134	5.295	1.456	1.386	2.29	22.02	7.77
		40	0.258	5.047	1.456	1.321	4.30	20.01	14.62
		80	0.375	4.813	1.456	1.260	6.11	18.19	20.78
6	6.625	10 S	0.134	6.357	1.734	1.664	2.73	31.7	9.29
		40	0.280	6.065	1.734	1.588	5.58	28.9	18.98
		80	0.432	5.761	1.734	1.508	8.40	26.1	28.58
8	8.625	10 S	0.148	8.329	2.258	2.180	3.94	54.5	13.40
		30	0.277	8.071	2.258	2.113	7.26	51.2	24.7
		80	0.500	7.625	2.258	1.996	12.76	45.7	43.4
10	10.75	10 S	0.165	10.420	2.81	2.73	5.49	85.3	18.7
		30	0.279	10.192	2.81	2.67	9.18	81.6	31.2
		Extra heavy	0.500	9.750	2.81	2.55	16.10	74.7	54.7
12	12.75	10 S	0.180	12.390	3.34	3.24	7.11	120.6	24.2
		30	0.330	12.09	3.34	3.17	12.88	114.8	43.8
		Extra heavy	0.500	11.75	3.34	3.08	19.24	108.4	65.4
14	14.0	10	0.250	13.5	3.67	3.53	10.80	143.1	36.7
		Standard	0.375	13.25	3.67	3.47	16.05	137.9	54.6
		extra heavy	0.500	13.00	3.67	3.40	21.21	132.7	72.1
16	16.0	10	0.250	15.50	4.19	4.06	12.37	188.7	42.1
		Standard	0.375	15.25	4.19	3.99	18.41	182.7	62.6
		extra heavy	0.500	15.00	4.19	3.93	24.35	176.7	82.8
18	18.0	10 S	0.188	17.624	4.71	4.61	10.52	243.9	35.8
		Standard	0.375	17.25	4.71	4.52	20.76	233.7	70.6
		extra heavy	0.500	17.00	4.71	4.45	27.49	227.0	93.5
20	20.0	10 S	0.218	19.564	5.24	5.12	13.55	300.6	46.1
		Standard	0.375	19.25	5.24	5.04	23.12	291	78.6
		extra heavy	0.500	19.00	5.24	4.97	30.6	283.5	104.1
22	22.0	10	0.250	21.50	5.76	5.63	17.1	363	58.1
		Standard	0.375	21.25	5.76	5.56	25.5	355	86.6
		extra heavy	0.500	21.00	5.76	5.50	33.8	346	114.8
24	24.0	10	0.250	23.50	6.28	6.15	18.7	434	63.4
		Standard	0.375	23.25	6.28	6.09	27.8	425	94.6
		extra heavy	0.500	23.00	6.28	6.02	36.9	415	125.5
26	26.0	Standard	0.375	25.25	6.81	6.61	30.2	501	102.6
		extra heavy	0.500	25.00	6.81	6.54	40.1	491	136.2
30	30.0	10	0.312	29.376	7.85	7.69	29.1	678	98.9
		Standard	0.375	29.250	7.85	7.66	34.9	672	118.7
		extra heavy	0.500	29.00	7.85	7.59	46.3	661	157.6
34	34.0	Standard	0.375	33.250	8.90	8.70	39.6	868	134.7
		extra heavy	0.500	33.00	8.90	8.64	52.6	855	178.9
36	36.0	Standard	0.375	35.25	9.42	9.23	42.0	976	142.7
		extra heavy	0.500	35.00	9.42	9.16	55.8	962	189.6
42	42.0	Standard	0.375	41.25	11.0	10.8	49.0	1336	166.7
		extra heavy	0.500	41.00	11.0	10.73	65.2	1320	221.6

*Reprinted with permission, from: "Design Properties of Pipe," ©1958, Chemetron Corporation.

TABLE E.4 Commercial Copper Tubing*

The following table gives dimensional data and weights of copper tubing used for automotive, plumbing, refrigeration, and heat exchanger services. For additional data see the standards handbooks of the Copper Development Association, Inc., the ASTM standards, and the "SAE Handbook."

Dimensions in this table are actual specified measurements, subject to accepted tolerances. Trade size designations are usually by actual OD, except for water and drainage tube (plumbing), which measures 1/8-in. larger OD. A 1/2-in. plumbing tube, for example, measures 5/8-in. OD, and 2-in. plumbing tube measures 2 1/8-in. OD.

KEY TO GAGE SIZES

Standard-gage wall thicknesses are listed by numerical designation (14 to 21), BWG or Stubs gage. These gage sizes are standard for tubular heat exchangers. The letter A designates SAE tubing sizes for automotive service. Letter designations K and L are the common sizes for plumbing services, soft or hard temper.

OTHER MATERIALS

These same dimensional sizes are also common for much of the commercial tubing available in aluminum, mild steel, brass, bronze, and other alloys. Tube weights in this table are based on copper at 0.323 lb/in³. For other materials the weights should be multiplied by the following approximate factors:

aluminum	0.30		monel	0.96
mild steel	0.87		stainless steel	0.89
brass	0.95			

Size, OD		Wall Thickness			Flow Area		Metal Area,	Surface Area		Weight,
								Inside,	Outside,	
in.	mm	in.	mm	gage	in.²	mm²	in.²	ft²/ft	ft²/ft	lb/ft
1/8	3.2	.030	0.76	A	0.003	1.9	0.012	0.017	0.033	0.035
3/16	4.76	.030	0.76	A	0.013	8.4	0.017	0.034	0.049	0.058
1/4	6.4	.030	0.76	A	0.028	18.1	0.021	0.050	0.066	0.080
1/4	6.4	.049	1.24	18	0.018	11.6	0.031	0.038	0.066	0.120
5/16	7.94	.032	0.81	21A	0.048	31.0	0.028	0.065	0.082	0.109
3/8	9.53	.032	0.81	21A	0.076	49.0	0.033	0.081	0.098	0.134
3/8	9.53	.049	1.24	18	0.060	38.7	0.050	0.072	0.098	0.195
1/2	12.7	.032	0.81	21A	0.149	96.1	0.047	0.114	0.131	0.182
1/2	12.7	.035	0.89	20L	0.145	93.6	0.051	0.113	0.131	0.198
1/2	12.7	.049	1.24	18K	0.127	81.9	0.069	0.105	0.131	0.269
1/2	12.7	.065	1.65	16	0.108	69.7	0.089	0.97	0.131	0.344
5/8	15.9	.035	0.89	20A	0.242	156	0.065	0.145	0.164	0.251
5/8	15.9	.040	1.02	L	0.233	150	0.074	0.143	0.164	0.285
5/8	15.9	.049	1.24	18K	0.215	139	0.089	0.138	0.164	0.344
3/4	19.1	.035	0.89	20A	0.363	234	0.079	0.178	0.196	0.305
3/4	19.1	.042	1.07	L	0.348	224	0.103	0.174	0.196	0.362
3/4	19.1	.049	1.24	18K	0.334	215	0.108	0.171	0.196	0.418
3/4	19.1	.065	1.65	16	0.302	195	0.140	0.162	0.196	0.542
3/4	19.1	.083	2.11	14	0.268	173	0.174	0.151	0.196	0.674
7/8	22.2	.045	1.14	L	0.484	312	0.117	0.206	0.229	0.455
7/8	22.2	.065	1.65	16K	0.436	281	0.165	0.195	0.229	0.641
7/8	22.2	.083	2.11	14	0.395	255	0.206	0.186	0.229	0.800
1	25.4	.065	1.65	16	0.594	383	0.181	0.228	0.262	0.740
1	25.4	.083	2.11	14	0.546	352	0.239	0.218	0.262	0.927
1 1/8	28.6	.050	1.27	L	0.825	532	0.176	0.268	0.294	0.655

*Compiled and computed.

TABLE E.4 (continued) **Commercial Copper Tubing***

Size, OD		Wall Thickness			Flow Area		Metal Area,	Surface Area		Weight,
in.	mm	in.	mm	gage	in.²	mm²	in.²	Inside, ft²/ft	Outside, ft²/ft	lb/ft
1 1/8	28.6	.065	1.65	16K	0.778	502	0.216	0.261	0.294	0.839
1 1/4	31.8	.065	1.65	16	0.985	636	0.242	0.293	0.327	0.938
1 1/4	31.8	.083	2.11	14	0.923	596	0.304	0.284	0.327	1.18
1 3/8	34.9	.055	1.40	L	1.257	811	0.228	0.331	0.360	0.884
1 3/8	34.9	.065	1.65	16K	1.217	785	0.267	0.326	0.360	1.04
1 1/2	38.1	.065	1.65	16	1.474	951	0.294	0.359	0.393	1.14
1 1/2	38.1	.083	2.11	14	1.398	902	0.370	0.349	0.393	1.43
1 5/8	41.3	.060	1.52	L	L779	1148	0.295	0.394	0.425	1.14
1 5/8	41.3	.072	1.83	K	1.722	1111	0.351	0.388	0.425	1.36
2	50.8	.083	2.11	14	2.642	1705	0.500	0.480	0.628	1.94
2	50.8	.109	2.76	12	2.494	1609	0.620	0.466	0.628	2.51
2 1/8	54.0	.070	1.78	L	3.095	1997	0.449	0.520	0.556	1.75
2 1/8	54.0	.083	2.11	14K	3.016	1946	0.529	0.513	0.556	2.06
2 5/8	66.7	.080	2.03	L	4.77	3078	0.645	0.645	0.687	2.48
2 5/8	66.7	.095	2.41	13K	4.66	3007	0.760	0.637	0.687	2.93
3 1/8	79.4	.090	2.29	L	6.81	4394	0.950	0.771	0.818	3.33
3 1/8	79.4	.109	2.77	12K	6.64	4284	1.034	0.761	0.818	4.00
3 5/8	92.1	.100	2.54	L	9.21	5942	1.154	0.897	0.949	4.29
3 5/8	92.1	.120	3.05	11K	9.00	5807	1.341	0.886	0.949	5.12
4 1/8	104.8	.110	2.79	L	11.92	7691	1.387	1.022	1.080	5.38
4 1/8	104.8	.134	3.40	10K	11.61	7491	1.682	1.009	1.080	6.51

TABLE E.5 Standard Gages for Wire, Sheet, and Twist Drills

Gage	(1) Mfrs. steel sheet	(2) USS steel sheet (old)	(3) Birmingham or Stub	(4) W & M or Roebling steel wire	(5) AWG or B & S non-ferrous wire or sheet	Numbered twist drills	Copper wire (AWG) Circular mils	Ohms/ 1000 ft, 77°F	Lb/1000 ft	Sheet steel Lb/sq ft
0000000		0.500		0.4900						20.00
000000		0.469		0.4615	0.580					18.75
00000		0.438		0.4305	0.516					17.50
0000		0.406	.454	0.3938	0.460		212,000	0.0500	641.0	16.25
000		0.375	.425	0.3625	0.410		168,000	0.0630	508.0	15
00		0.344	.380	0.3310	0.365		133,000	0.0795	403.0	13.75
0		0.313	.340	0.3065	0.325		106,000	0.100	319.0	12.50
1		0.281	.300	0.2830	0.289	0.2280	83,700	0.126	253.0	11.25
2		0.266	.284	0.2625	0.258	0.2210	66,400	0.159	201.0	10.625
3	.2391	0.250	.259	0.2437	0.229	0.2130	52,600	0.201	159.0	10
4	.2242	0.234	.238	0.2253	0.204	0.2090	41,700	0.253	126.0	9.375
5	.2092	0.219	.220	0.2070	0.182	0.2055	33,100	0.319	100.0	8.75
6	.1943	0.203	.203	0.1920	0.162	0.2040	26,300	0.403	79.5	8.125
7	.1793	0.188	.180	0.1770	0.144	0.2010	20,800	0.508	63.0	7.5
8	.1644	0.172	.165	0.1620	0.128	0.1990	16,500	0.641	50.0	6.875
9	.1495	0.156	.148	0.1483	0.114	0.1960	13,100	0.808	39.6	6.25
10	.1345	0.141	.134	0.1350	0.102	0.1935	10,400	1.02	31.4	5.625
11	.1196	0.125	.120	0.1205	0.0907	0.1910	8,230	1.28	24.9	5
12	.1046	0.109	.109	0.1055	0.0808	0.1890	6,530	1.62	19.8	4.375
13	.0897	0.0937	.095	0.0915	0.0720	0.1850	5,180	2.04	15.7	3.75
14	.0747	0.0781	.083	0.0800	0.0641	0.1820	4,110	2.58	12.4	3.125
15	.0673	0.0703	.072	0.0720	0.0571	0.1800	3,260	3.25	9.86	2.813
16	.0598	0.0625	.065	0.0625	0.0508	0.1770	2,580	4.09	7.82	2.5
17	.0538	0.0562	.058	0.0540	0.0453	0.1730	2,050	5.16	6.20	2.25
18	.0478	0.0500	.049	0.0475	0.0403	0.1695	1,620	6.51	4.92	2
19	.0418	0.0437	.042	0.0410	0.0359	0.1660	1,290	8.21	3.90	1.75
20	.0359	0.0375	.035	0.0348	0.0320	0.1610	1,020	10.4	3.09	1.50
21	.0329	0.0344	.032	0.0318	0.0285	0.1590	810	13.1	2.45	1.375
22	.0299	0.0312	.028	0.0286	0.0253	0.1570	642	16.5	1.94	1.25
23	.0269	0.0281	.025	0.0258	0.0226	0.1540	509	20.8	1.54	1.125
24	.0239	0.0250	.022	0.0230	0.0201	0.1520	404	26.2	1.22	1
25	.0209	0.0219	.020	0.0204	0.0179	0.1495	320	33.0	0.970	0.875
26	.0179	0.0187	.018	0.0181	0.0159	0.1470	254	41.6	0.769	0.75
27	.0164	0.0172	.016	0.0173	0.0142	0.1440	202	52.5	0.610	0.6875
28	.0149	0.0156	.014	0.0162	0.0126	0.1405	160	66.2	0.484	0.625
29	.0135	0.0141	.013	0.0150	0.0113	0.1360	127	83.4	0.384	0.5625
30	.0120	0.0125	.012	0.0140	0.0100	0.1285	101	105	0.304	0.5
31	.0105	0.0109	.010	0.0132	0.0089	0.1200	79.7	133	0.241	0.4375
32	.0097	0.0102	.009	0.0128	0.0080	0.1160	63.2	167	0.191	0.4063
33	.0090	0.0094	.008	0.0118	0.0071	0.1130	50.1	211	0.152	0.375
34	.0082	0.0086	.007	0.0104	0.0063	0.1110	39.8	266	0.120	0.3438
35	.0075	0.0078	.005	0.0095	0.0056	0.1100	31.5	335	0.0954	0.3125
36	.0067	0.0070	.004	0.0090	0.0050	0.1065	25.0	423	0.0757	0.2813
37	.0064	0.0066		0.0085	0.0045	0.1040	19.8	533	0.0600	0.2656
38	.0060	0.0062		0.0080	0.0040	0.1015	15.7	673	0.0476	0.25
39				0.0075	0.0035	0.0995	12.5	848	0.0377	
40				0.0070	0.0031	0.0980	9.9	1070	0.0200	
41				0.0066	0.0028	0.0960				
42				0.0062	0.0025	0.0935				
43				0.0060	0.0022	0.0890				
44				0.0058	.0020	0.0860				
45				0.0055	.0018	0.0820				
46				0.0052	.0016	0.0810				
47				0.0050	.0014	0.0785				
48				0.0048	.0012	0.0760				
49				0.0046	.0011	0.0730				
50				0.0044	.0010	0.0700				

Note: The present trend, especially for sheet and strip, is to quote thickness as decimal or fraction of an inch rather than gage number. ANSI Standard preferred thicknesses have been adopted. These preferred sizes for thickness of uncoated sheet, strip, and plate under 0.25 in. are as follows: .224, .220, .180, .160, .140, .125, .112, .100, .090, .080, .071, .063, .056, .050, .045, .040, .036, .032, .028, .025, .022, .020, .018, .016, .014, .012, .011, .010, .009, .008, .007, .006, .005, .004.

KEY: (1) Manufacturer's standard for hot- and cold-rolled uncoated carbon steel sheet and most alloy steel sheet.
(2) U.S. Standard for cold-rolled steel strip and stainless and nickel alloy sheet.
(3) Birmingham or Stub for hot-rolled carbon and alloy steel strip and tubing.
(4) Washburn and Moen, Roebling, or U.S. Steel for steel wire.
(5) American wire gage or Brown and Sharpe for non-ferrous wire, sheet, and strip.

Dimensions in approximate decimals of an inch.

TABLE E.6 Properties of Typical Gaseous and Liquid Commercial Fuels*

Gaseous fuels	Composition, percent by volume								Mol wt of fuel	Theor. air/fuel ratio by wt	Higher heating value, Btu/lb_m	Density, lb_m/ft^3
	H_2	N_2	O_2	CH_4	CO	CO_2	C_2H_4	C_6H_6				
Blast furnace gas	1.0	60.0	—	—	27.5	11.5	—	—	29.6	0.667	1,170	.075 5 [a]
Blue water gas	47.3	8.3	0.7	1.3	37.0	5.4	—	—	16.4	3.759	6,550	.042 2 [a]
Carb. water gas	40.5	2.9	0.5	10.2	34.0	3.0	6.1	2.8	18.3	7.299	11,350	.046 6 [a]
Coal gas	54.5	4.4	0.2	24.2	10.9	3.0	1.5	1.3	12.1	10.87	16,500	.031 1 [a]
Coke-oven gas	46.5	8.1	0.8	32.1	6.3	2.2	3.5	0.5	13.7	17.24	17,000	.032 6 [a]
Natural gas (15.8% C_2H_6)	—	0.8	—	83.4	—	—	—	—	18.3	17.24	24,100	.045 1 [a]
Producer gas	14.0	50.9	0.6	3.0	27.0	4.5	—	—	24.7	14.29	2,470	.063 6 [a]

Liquid commercial fuels	Vapor		Gravity, API, 60°F	Distillation			Flash point, °F	Viscosity, centistokes, 100°F	Mol wt of fuel	Theor. air/fuel ratio by wt	Higher heating value, Btu/lb_m	Density, lb_m/ft^3
	c_p, 60°F	c_p/c_v, 60°F		10%, °F	90%, °F	End point, °F						
	(approximately)											
Gasoline	0.4	1.05	63	121	320	397	0	—	113	14.93	20,460	43.8 [b]
Gasoline	0.4	1.05	63	118	330	410	0	—	126 [c]	14.97	20,260	46.1 [b]
Kerosene	0.4	1.05	41.9	370	510	546	130	—	154 [c]	14.99	19,750	51.5 [b]
Diesel oil (1-D)	0.4	1.05	42	—	550	—	100	1.4–2.5	170	15.02	19,240	54.6 [b]
Diesel oil (2-D)	0.4	1.05	36	—	540–576	—	125	2.0–5.8	184	15.06	19,110	57.4 [b]
Diesel oil (4-D)	0.4	1.05	—	—	—	—	130	5.8–26.4	198	14.93	18,830	59.9 [b]

[a] Based on dry air at 25°C and 760 mm Hg.
[b] Based on H_2O at 60°F, 1 atm ($\rho = 62.367$ lb_m/ft^3).
[c] Estimated.
* Abridged from: "Engineering Experimentation," G.L. Tuve and L.C. Domholdt, Mc/Graw-Hill Book Company, 1966; and "The Internal Combustion Engine," 2nd ed. C.F. Taylor and E.S. Taylor, International Textbook Co., 1961.
Note: For heating value in J/kg, multiply the value in Btu/lb_m by 2324. For density in kg/m³, multiply the value in lb/ft^3 by 16.02.

TABLE E.7 Combustion Data for Hydrocarbons*

Hydrocarbon	Formula	Higher heating value (vapor), Btu/lb_m	Theor. air/fuel ratio, by mass	Max flame speed, ft/sec	Adiabatic flame temp (in air), $°F$	Ignition temp (in air), $°F$	Flash point, $°F$	Flammability limits (in air), % by volume	
PARAFFINS OR ALKANES									
Methane	CH_4	23875	17.195	1.1	3484	1301	gas	5.0	15.0
Ethane	C_2H_6	22323	15.899	1.3	3540	968–1166	gas	3.0	12.5
Propane	C_3H_8	21669	15.246	1.3	3573	871	gas	2.1	10.1
n-Butane	C_4H_{10}	21321	14.984	1.2	3583	761	−76	1.86	8.41
iso-Butane	C_4H_{10}	21271	14.984	1.2	3583	864	−117	1.80	8.44
n-Pentane	C_5H_{12}	21095	15.323	1.3	4050	588	< −40	1.40	7.80
iso-Pentane	C_5H_{12}	21047	15.323	1.2	4055	788	< −60	1.32	9.16
Neopentane	C_5H_{12}	20978	15.323	1.1	4060	842	gas	1.38	7.22
n-Hexane	C_6H_{14}	20966	15.238	1.3	4030	478	−7	1.25	7.0
Neohexane	C_6H_{14}	20931	15.238	1.2	4055	797	−54	1.19	7.58
n-Heptane	C_7H_{16}	20854	15.141	1.3	3985	433	25	1.00	6.00
Triptane	C_7H_{16}	20824	15.141	1.2	4035	849	—	1.08	6.69
n-Octane	C_8H_{18}	20796	15.093	—	—	428	56	0.95	3.20
iso-Octane	C_8H_{18}	20770	15.093	1.1	—	837	10	0.79	5.94
OLEFINS OR ALKENES									
Ethylene	C_2H_4	21636	14.807	2.2	4250	914	gas	2.75	28.6
Propylene	C_3H_6	21048	14.807	1.4	4090	856	gas	2.00	11.1
Butylene	C_4H_8	20854	14.807	1.4	4030	829	gas	1.98	9.65
iso-Butene	C_4H_8	20737	14.807	1.2	—	869	gas	1.8	9.0
n-Pentene	C_5H_{10}	20720	14.807	1.4	4165	569	—	1.65	7.70
AROMATICS									
Benzene	C_6H_6	18184	13.297	1.3	4110	1044	12	1.35	6.65
Toluene	C_7H_8	18501	13.503	1.2	4050	997	40	1.27	6.75
p-Xylene	C_8H_{10}	18663	13.663	—	4010	867	63	1.00	6.00
OTHER HYDROCARBONS									
Acetylene	C_2H_2	21502	13.297	4.6	4770	763–824	gas	2.50	81
Naphthalene	$C_{10}H_8$	17303	12.932	—	4100	959	174	0.90	5.9

*Based largely on: "Gas Engineers' Handbook," American Gas Association, Inc., Industrial Press, 1967.

REFERENCES

"American Institute of Physics Handbook," 2nd ed., D.E. Gray, Ed., McGraw-Hill Book Company, 1963

"Chemical Engineers' Handbook," 4th ed., R.H. Perry, C.H. Chilton, and S.D. Kirkpatrick, Eds., McGraw-Hill Book Company, 1963.

"Handbook of Chemistry and Physics," 54rd ed., R.C. Weast, Ed., The Chemical Rubber Company, 1972; gives the heat of combustion of 500 organic compounds.

"Handbook of Laboratory Safety," 2nd ed., N.V. Steere Ed., The Chemical Rubber Company, 1971.

"Physical Measurements in Gas Dynamics and Combustion," Princeton University Press, 1954.

Note: For heating value in J/kg, multiply the value in Btu/lb_m by 2324. For flame speed in m/s, multiply the value in ft/s by 0.3048.

Index

A

Absorbers, 142
Absorption, 165
Absorption chiller/heater, 145
Absorption chillers, 141
    direct-fired, 141
Absorption cycle, 142
Absorption heaters, 141
Absorption heat pumps, 141
Absorption heat transformers, 141
Absorption systems, 141
Acceptable indoor air quality, 5
Activated alumina, 166
Activated carbon, 166
Activated carbon filters, 80
Adiabatic mixing process, 26, 31
    two-stream, 23
Adiabatic saturation process ideal, 14
Adjustable frequency AC drives, 132
Adsorption, 80
Adsorption path, 172
Air
    primary, 130
    recirculating, 26
    return, 26
    secondary, 130
    transfer, 130
Air cleaning, 77
Air-conditioners
    packaged, 2
    packaged terminal (PTAQ) systems, 148
    self-contained, 2
Air-conditioner systems, room, 147
Air-conditioning, 1
Air-conditioning cycle, 26
    basic, 26
Air-conditioning design, 5, 6
Air-conditioning or HVAC&R system, 1
Air-conditioning process, 19
Air-conditioning system selection, 158
Air-conditioning systems
    comfort, 2
    off-peak, 123
    processing, 2
Air conditions, standard, 29
Air-cooled condenser, 88
    moist, 13
Air economizer, 136
    enthalpy-based, 136
    temperature-based, 136
Air economizer mode, 34
Air filters, 77 to 80
Air filtration, 77

Air handling units (AHU), 2, 71
Air purge, 34
Air quality
    acceptable indoor, 6
    improving indoor, 54
Air skin central systems, 158
Air-source heat pumps, 115
Air system, 2, 160
Air temperature (mean daily outdoor), 52
Air units, 161
Air washer, 81
Ambient air quality standards, 55
Annual fuel utilization efficiency, 105
Atmosphere, 11
Atmospheric burner, 104
Atmospheric dusts, 78
Atmospheric vent combustion system, 103
A-weighted sound level, 56
Azeotropic blends, 36
    near, 37

B

Baseboard heaters, 108
Basic air-conditioning cycle, 26
Beam radiant heaters, 109
Biocides, 98
Biologicals, 54
Blade pitch modulation, 132
Block load, 60
Blow-through unit, 71
Boilers, 106 to 108
Branch pipes, 98
Brine, 125
Brine-coil, 126
Brine-coil ice-storage system, case study of, 126
Building loop, 98, 99
Built-in volume ratio, 120
Built-up systems, 2
Bundles, 120
Bypass mixing process, 25

C

Campus-type chilled water system, 101, 102
Capillary tube, 91
Cascade system, 48
Case study of a brine-coil ice-storage system, 126
Cast iron sectional boilers, 106
Central hydronic systems, 2
Central systems, 2

Centrifugal chiller, 118
    incorporating heat recovery, 120
Centrifugal compressor, 117
    performance map, 119
Centrifugal vapor compression refrigeration system, 116
Charging, 125
Chilled water storage, 128
Chilled water system, 95
Chillers, 141
    priority, 125
Chimney, 107
Chlorine, 98
Chlorofluorocarbons, 36, 37
Clean Air Act Amendments, 37
Clean rooms, 55
Closed cycle, 43
Closed system, 95
Codes, 8
Coefficient
    conduction transfer function, 65
    heat transfer
        inside, 64
        outdoor, 64
    room transfer function, 68
    shading, 65
    space air transfer function, 69
    transfer function, 63
Coefficient of performance, 45
Coil accessories, 77
Coil cleanliness, 77
Coil freeze-up protection, 77
Coil load, 60, 114
    cooling, 60
    heating, 60
    direct expansion (DX), 75
    dry, 75
    dry-wet, 75
    water cooling, 75
    water heating, 75
    wet, 75
Cold air distribution, 125
Cold air supply with space humidity control, 34
Cold air supply without humidity control, 31
Combustion products, 54
Combustion system
    atmospheric vent, 103
    natural vent, 103
Comfort air-conditioning systems, 2
Comparison of various systems, 160
Compound system, 46
Compression efficiency, 84
Compression ratio, 46
Compressor displacement, 42
Compressor efficiency, 84
Compressors, 83 to 87
Compressor short cycling, 115
Computer-aided design (CAD), 8
Computer-aided drafting, 8
Concentration, 141
    solution, 141
Condensate, 23
Condensate drain tap, 77
Condenser, 142
Condenser temperature difference, 88

Condenser water system, 95
Condensing heat exchangers, 104
Condensing unit, 73
Conduction, transfer function coefficients, 65
Conductivity, thermal, 43
Constant, dielectric, 43
Constant flow, 98
Constant volume systems, 132
Constant volume zone-reheat central systems, 155
Constant-volume zone reheat packaged systems, 155
Construction, 63
Control
    device, 3
    systems, 3, 161
Controls, 113, 114
Convective heat, 60
Coolants, 74
Cool-down mode, 34
Cool-down period, 26
Cooling, 166
    direct, 125
Cooling coil load, 60
Cooling load, 60
    latent, 59
    sensible, 59
    space, 62
Cooling medium, 35
Cooling mode operation, 116
Cooling process, 26
Corrosion, 98, 142
Crankcase heater, 115
Crystallization controls, 145
Crystallization line, 144
Custom-built units, 72
Cycle
    absorption, 142
    air-conditioning, 26
    closed, 43
    ideal, 44
    open, 43
    refrigeration, 42
Cylinder unloader, 113

D

Dalton's law, 12
Damper modulation, 132
Defrosting, 115
Degree day, 52
Degree of saturation, 13
Dehumidification
    desiccant, 166
    vapor compression systems, 166
Dehumidified air, 166
Dehumidifying process, 26
Desiccant air conditioning, 165
Desiccant dehumidification, 166
Desiccant material, 171
Desiccants, 165
Desiccant wheel, 171
Design
    computer-aided, 8
    criteria, indoor, 52
    load, 26

supply volume flow rate, 27
Design-bid, 6
Design-build, 6
Desirable characteristics, 166
Desorbers, 142
Desorption, 166
Desorption path, 172
Dew point temperature, 14
Dielectric constant, 43
Diluted solution, 141
Direct cooling, 125
Direct evaporative cooler, 92
Direct expansion (DX) coils, 91
Direct expansion ice makers, 91
Direct expansion refrigeration system, 111
Direct expansion systems, 111
Direct fired, 141
Direct return, 98
Direct-fired absorption chiller, 141
Direct-fired generator, 142
Direct-indirect evaporative cooler, 93
Discharge line, 113
Displacement
    nonpositive, 83
    positive, 83
Disposable filters, 78
DOP penetration and efficiency test, 78
Double effect, 141
Double-strength sheet glass, 66
Double-tube condenser, 88
Downflow, 103
Drafting, computer-aided, 8
Draperies, 65
Draw-through unit, 71
Drawings and specifications, 7
Dry bulb, 14
Dry bulb temperature
    summer design, 51
    winter outdoor design, 52
Dry coil, 75
Dry filters, 78
Dry-wet coil, 75
Dual-duct constant volume central system, 157
Dual-duct VAV central systems, 157
Dual-temperature water system, 95
Dust holding capacity, 77
Dust removal, 78
Dusts, atmospheric, 78

E

Economizer, 136
EER, 74
Effectiveness of refrigeration cycle, 42
Efficiency, 78, 84
Electric boilers, 106
Electric infrared heater, 110
Energy conservation recommendations, 161
Energy efficiency ratio, 74
Enthalpy-based air economizer, 136
Enthalpy of moist air, 13
Environmental tobacco, 54
Equilibrium, sorption, 166
Evaporative condenser, 90

direct-indirect, 93
    indirect, 92
Evaporative-cooled system, 95
Evaporative cooling, 22
Evaporative cooling system, 92
Evaporative heat loss, 52
Evaporator, 142
External shading devices, 65

F

Fan-coil system
    four-pipe, 148
    two-pipe, 149
    water cooling electric heating, 149
Fan-duct system, 129
Fan energy use, 136
Fan-powered VAV boxes, 157
Fan-powered VAV central systems, 160
Fan speed modulation, 132
Field built-up units, 72
Figure of merit, 128
Filter dryer, 113
Filters
    activated carbon, 80
    air, 77
    disposable, 78
    dry, 78
    HEPA, 79
    low-efficiency, 78
    medium-efficiency, 78
    reusable, 78
    ULPA, 79
Finite difference method, 64
Fins, 74
    density, 74
    spacing, 74
Fin-tube heaters, 108
Flammability, 37, 42
Flash cooler, 118
Float valve, 91, 118
Flooded liquid coolers, 91
Flooded shell-and-tube liquid coolers, 91
Flow
    constant, 98
    horizontal, 103
    parallel, 142
    rate, 29
    resistance, 129
    reverse-parallel, 142
    series, 142
    variable, 98
Formaldehyde gases, 54
Fouling factor, 87
Four-pipe fan-coil systems, 148
Four-pipe system, 98
Fuel utilization efficiency, 105
Full-load, 26
Full-storage, 125

G

Gaseous and liquid commercial fuels, properties of typical, E-85
Gases
    formaldehyde, 54
    organic, 54
    properties of A-18
Gases and vapors, D-74
Gas infrared heaters, 110
Gas properties, equations for, A-34
Generators, 142
Glide, 37

H

Halogenated hydrocarbons, 36
Heat exchange, sensible, 52
Heat exchangers
    condensing, 104
    high-temperature, 142
    low-temperature, 142
    primary, 104
    secondary, 104
Heat extraction, sensible, 68
Heat gain, 66
Heat loss, evaporative, 52
Heat pumps, 115
Heat pump systems, air-source, 152
Heat recovery, centrifugal chillers incorporating, 120
Heat rejection
    factor, 87
    total, 87
Heat transfer, radiative, 60
Heat transfer coefficient
    indoor, 64
    outdoor, 64
Heat transformers, absorption, 141
Heaters, 108
Heating bundles, 120
Heating coil load, 60
Heating load, 69
Heating load, space, 60
Heating mode operation, 116
Heating seasonal performance factor, 74
Heating system, low-pressure warm air, 108
Heating systems, 3, 161
Heating systems, types of, 103
Heavy construction, 63
HEPA filters, 79
Hermetic compressors, 83
High velocity induction space conditioning systems, 158
High-pressure boilers, 106
High-temperature heat exchangers, 141
Horizontal air handling units, 71
Horizontal flow, 103
Hot water boiler, 106
Hot water heating system, 95
Human bioeffluents, 54
Humidifier, 80
    steam, 81
Humidifying process, 21
Humidity
    indoor, 53

    relative, 12
Humidity control, 34
Humidity ratio, 12
HVAC&R, 1
Hybrid cycles, 171
Hydrofluorocarbons (HFCs), 36

I

Ice burning, 125
Ice-harvester, 126
Ice makers, 91
Ice making, 125
Ice-on-coil, 126
Ice-storage system, 125
Ideal adiabatic saturation process, 14
Ideal cycle, 44
Ideal single-stage vapor compression cycle, 44
Impurities, 98
Indirect evaporative cooler, 92
Individual room, 2
Individual systems, 2, 147
Indoor air contaminants, 54
Indoor air quality procedure, 54
Indoor coil, 116
Indoor design criteria, 52
Indoor humidity, 53
Indoor packaged unit, 73
Indoor shading devices, 65
Infiltration, 67
Infrared heating, 109
Initial pressure drop, 77
Inlet cone modulation, 132
Inlet vanes modulation, 132
Inorganic compounds, 37
Inside heat transfer coefficient, 64
Intermittently operated mode, 26
Interstage, pressure, 46
Isentropic efficiency, 84
Isotherm, 166

L

Latent cooling load, 59
Latent heat of moist air, 13
Law, Dalton's, 12
Light construction, 63
Liquid absorbent, 36
Liquid coolers, 91
Liquid desiccants, 166
Liquid line, 113
Liquid overfeed ice makers, 91
Liquid slugging, 114
Liquid spray tower, 168
Load
    block, 60
    coil, 60
    cooling, 59
    design, 26
    heating, 69
    heating coil, 60
    latent cooling, 59
    peak, 60

profile, 60
ratio, 119
refrigeration, 60
sensible cooling, 59
shift, 125
space cooling, 59
space heating, 59
Load-leveling, 125
Loops, 98
Loop systems
    plant-distribution-building, 101
    two-pipe individual, 109
Low-efficiency filters, 78
Low NO_x emissions, 106
Low-pressure boilers, 106
Low-temperature control, 114
Low-temperature cut-out control, 145
Low-temperature generator, 142
Low-temperature heat exchangers, 142
Lumber, E-75

M

Main pipe, 98
Make-up air handling units, 72
Make-up air unit, 155
Mean daily outdoor air temperature, 52
Mean daily range, 52
Mean radiant temperature, 49
Mechanical efficiency, 84
Mechanical engineer's responsibilities, 7
Mechanical refrigeration, 35
Mechanical services, 6
Medium-efficiency filters, 78
Medium-pressure boilers, 106
Medium construction, 63
Metabolic rate, 52
Methods
    finite differences, 64
    total equivalent temperature differential/time averaging, 64
Minimum outdoor ventilation air control, 139
Mixing process
    adiabatic, 26
    bypass, 25
    two-stream adiabatic, 23
Mixing VAV box, 160
Mode
    air economizer, 26, 34
    cool-down, 34
    intermittently operated, 26
    summer, 26
    warm-up, 34
    winter, 26
Modulation, 132 to 133
    blade pitch, 132
    damper, 132
    fan speed, 132
    inlet cone, 132
    inlet vanes, 132
Moist air
    density, 13
    latent heat of, 13
    sensible heat of, 13

specific heat of, 14
Moisture content, 60
Moisture transfer, 60
Moist volume, 13
Montreal protocol, 37
Motor overload control, 114
Multiple orifices, 91
Multistage system, 46
Multizone central systems, 158

N

Natural vent, 103
Near azeotropic blends, 37
Negative pressure, 56
Nighttime shutdown period, 26
Noise, 56
    control, 56
    criteria (NC curves), 56
Nonazeotropic, 37
Normalized loading, 166
Numbering system for refrigerants, 36

O

Octave, 56
Off-peak hours, 123
Oil lubrication, 113
Oil miscibility, 42
Oil pressure failure control, 114
Oil separator, 122
Once-through system, 96
On-off control, 113
Open compressors, 83
Open cycle, 43
Open system, 96
Operating point, 119
Operating temperature, 52
Or-equal specifications, 8
Organic gases, 54
Orifice plates, 118
Outdoor coil, 116
Outdoor heat transfer coefficient, 64
Outdoor shading devices, 65
Overcompression, 120
Oversaturation, 22
Ozone depletion potential (ODP), 36

P

Packaged air-conditioner, 2
Packaged systems, 2, 154
Packaged terminal air conditioner (PTAC) systems, 148
Packaged units, 2
Partial storage, 125
Particulate concentration, total, 54
Part load, 26
Peak load, 60
Performance
    coefficient of, 45
    factor, 92
    specification, 8
Period

cool-down, 26
    nighttime shutdown, 26
    warm-up, 26
Pipes, 98
    steel, E-81
Plant-building loop, 101
Plastics, trade names and composition, C-60
Porous matrix gas infrared heater, 110
Positive displacement, 83
Positive pressure, 55
Power burners, 104
Power consumption, 84
Power input to the compressor, 84
Power vent combustion systems, 104
Pressure
    interstage, 46
    negative, 56
    positive, 55
Pressure drop, 77
Pressure drop of water in pipes, 96
Primary-secondary loop, 101
Primary air, 130
Primary heat exchangers, 104
Primary surfaces, 74
Processes
    cooling, 26
    dehumidifying, 26
    refrigeration, 43
    sensible heating, 26
Processing air-conditioning systems, 2
Products of combustion, 54
Properties of commercial nylon resins, C-62
Properties of dry air at atmospheric pressure, A-2
Properties of gases, A-18
Properties of liquid, B-35
Properties of manufactured fibers, C-70
Properties of moist air, A-10
Properties of natural fibers, C-70
Properties of rubbers and elastomers, C-72
Properties of saturated water and steam, A-12
Properties of silicate glasses, C-63
Properties of solids, C-38
Properties of superheated steam, A-15
Properties of typical gaseous and liquid commercial fuels, E-85
Properties and uses of American woods, C-67
Properties of window glass, C-66
Protocols, Montreal, 37
Psychrometer, sling-type, 14
Psychrometric chart, 16
Psychrometrics, 11
Pump-down control, 114
Pure metals, thermal properties of, C-41
Purge unit, 118

R

Radiative heat transfer, 60
Radon, 54
Rated volume, 29
Ratio
    built-in volume, 120
    compression, 46
    humidity, 12
    load, 119

    sensible heat (SHR), 19
    turn-down, 107
    variable volume, 120
Reactivation, 166
Recirculating air, 26
Recirculating air unit, 155
Recirculating mode, 172
Recirculating units, 72
Refrigerant charge, 115
Refrigerant leakage, 43
Refrigerant piping, 113
Refrigerants, numbering system for, 36
Refrigeration, 35
Refrigeration compressor, 83
Refrigeration condenser, 87
Refrigeration cycles, 43
Refrigeration feed, 113
Refrigeration load, 60
Refrigeration process, 43
Refrigeration system, 35, 161
    centrifugal vapor compression, 116
    direct expansion, 111
Regeneration, 166
Relative humidity, 12
Relief fan, 135
Return air, 135
Reusable filters, 78
Reverse-parallel flow, 142
Reverse return, 98
Rooftop air handling units, 72
Rooftop package unit, 73
Room air-conditioner systems, 147
Room transfer function coefficients, 68
Rubbers and elastomers, properties of, C-72
R value, 64

S

Safety, 42
Safety controls, 145
Saturation, degree of, 13
Saturation effectiveness, 92
Saturation process, ideal adiabatic, 14
Scale, 98
Scotch marine boiler, 106
Screw chiller, 120
Secondary air, 130
Secondary heat exchangers, 104
Secondary surfaces, 74
Self-contained air-conditioner, 2
Semihermetic compressors, 83
Sensible cooling process, 59
Sensible heat exchange, 53
Sensible heat extraction, 68
Sensible heat gain, 59
Sensible heating process, 20, 26
Sensible heat of moist air, 13
Sensible heat ratio (SHR), 19
Service life, 77
Shading devices, 65
Shell-and-tube, flooded liquid coolers, 91
Shell-and-tube, water-cooled condenser, 90
Sight glass, 113
Silica gel, 166

Single loop, 101
Single zone, 26
Single zone constant volume central system, 155
Sling-type psychrometer, 14
Sol-air temperatures, 64
Solar cooling load factor, 63
Solar heat gain factors (SHGF), 66
Solid desiccant dehumidifier, 169
Solid desiccants, 166
Solid packed tower, 169
Solids
    density of, C-39
    properties of, C-38
Solution and refrigerant flow, 144
Solutions, 141
Sorbents, 165
Sorption, 166
Sorption equilibrium, 166
Sound power level, 56
Space air transfer function coefficients, 69
Space conditioning process, 20
Space-conditioning systems, 148
Space cooling load, 59
Space heat extraction, 59
Space heat gain, 59
Space heating load, 59
Space pressure differential, 56
Space sensible cooling load, 20
Space systems, 2
Spark ignition, 104
Specific heat of moist air, 14
Specific stiffness of metals and alloys, C-40
Specifications, 7
    or-equal, 8
    performance, 8
Speed modulation, 114
Stack, 107
Standard air conditions, 29
Standing pilot, 104
Steady-state efficiency, 105
Steam humidifier, 81
Storage priority, 125
Strainer, 114
Stratified tanks, 128
Subcooling, 46
Summer design dry bulb temperature, 51
Summer mode, 26
Summer outdoor wet bulb temperature, 52
Superheated steam, properties of, A-15
Superheating, 46
Supply temperature differential, 30
Surfaces
    primary, 74
    secondary, 74
Surge, 119
System
    closed, 95
    open, 96
System curve, 130
System effect, 130
System head curve, 119
System lead lifts, 119
System-operating point, 130
System selection, air-conditioning, 158

Systems
    air, 2
    central hydronic, 2
    control, 3
    heating, 3
    individual, 2
    packaged, 2
    space-conditioning, 2
    unitary packaged, 2
    water, 2

T

Temperature
    air, mean daily outdoor, 52
    dew point, 14
    dry bulb, 52, 53
    mean radiant, 52
    operating, 52
    sol-air, 64
    thermodynamic wet bulb, 14
    wet bulb, 15, 48
Temperature-based air economizer, 136
Temperature difference, 63
    condenser, 88
Terminal unit, 71
Thermal conductivity, 43
Thermal efficiency, 105
Thermal storage system, 125
Thermocline, 128
Thermodynamic wet bulb temperature, 14
Thermostatic expansion valve, 91
Throttling devices, 142
Time-of-day structure, 125
Ton-hour, 125
Total equivalent temperature, differential/time averaging
    method, 64
Total heat rejection, 87
Total particulate concentration, 54
Tower bundles, 120
Toxicity, 42
Transfer air, 130
Transfer function, 63
    coefficients, 63
Tube feeds, 74
Tubes, capillary, 91
Turn-down ratio, 107
Two-pipe fan-coil system, 149
Two-pipe individual loop systems, 109
Two-pipe system, 98
Two-stream adiabatic mixing process, 23
Types of heating systems, 103

U

ULPA filters, 79
Undercompression, 120
Unitary packaged systems, 2
Upflow, 103

V

Vapor compression cycle, ideal single-stage, 44
Vapors and gases, D-74
Variable-air-volume (VAV) systems, 132
Variable flow, 98
Variable volume ratio, 120
VAV box, 155
VAV central system, 155 to 157
Venetian blinds, 65
Vent
    natural, 103
    power, 104
Ventilating mode, 172
Ventilation air control, minimum outdoor, 139
Volume, rated, 29
Volumetric efficiency, 84

W

Warm air furnace, 103
Warm air supply, 34
Warm-up mode, 34
Warm-up period, 26
Water circuits, 74
Water-cooled condenser, 88
Water cooling coil, 75

Water-cooling electric heating fan-coil system, 149
Water economizer, 136
Water heating coil, 75, 76
Water in pipes, pressure drop of, 96
Water systems, 95
Water tube boilers, 106
Water vapor, at low pressures, A-11
Weight arrestance test, 78
Weighting factors, 63
Wet bulb, 52
Wet coil, 75
Wet suction, 122
WG, 56
Wheels, desiccant, 171
Winter mode, 26
Winter outdoor design dry bulb temperature, 52

Y

Year-round operation, 136

Z

Zeotropic, 37
Zone, 59

Printed and bound by CPI Group (UK) Ltd, Croydon, CR0 4YY

23/10/2024

01778259-0001